Carlos Cotta and Jano van Hemert (Eds.)

Recent Advances in Evolutionary Computation for Combinatorial Optimization

Studies in Computational Intelligence, Volume 153

Editor-in-Chief

Prof. Janusz Kacprzyk
Systems Research Institute
Polish Academy of Sciences
ul. Newelska 6
01-447 Warsaw
Poland
E-mail: kacprzyk@ibspan.waw.pl

Carlos Cotta
Jano van Hemert
(Eds.)

Recent Advances in Evolutionary Computation for Combinatorial Optimization

 Springer

Carlos Cotta
ETSI Informatica (3.2.49)
Campus de Teatinos
Universidad de Malaga
29071, Malaga
Spain
Email: ccottap@lcc.uma.es

Jano van Hemert
National e-Science Centre
University of Edinburgh
15 South College Street
Edinburgh EH8 9AA
United Kingdom
Email: jano@vanhemert.co.uk

ISBN 978-3-540-70806-3   e-ISBN 978-3-540-70807-0

DOI 10.1007/978-3-540-70807-0

Studies in Computational Intelligence   ISSN 1860949X

Library of Congress Control Number: 2008931010

Typeset & Cover Design: Scientific Publishing Services Pvt. Ltd., Chennai, India.

Printed in acid-free paper

9 8 7 6 5 4 3 2 1

springer.com

"Invention consists in avoiding the constructing of useless contraptions and in constructing the useful combinations which are in infinite minority."

Henri Poincaré (1854–1912)

Preface

Combinatorial optimisation is a ubiquitous discipline. Its applications range from telecommunications to logistics, from manufacturing to personnel scheduling, from bioinformatics to management, and a long et cetera. Combinatorial optimisation problems (COPs) are characterised by the need of finding an optimal or quasi-optimal assignment of values to a number of discrete variables, with respect to a certain objective function or to a collection of objective functions. The economical, technological and societal impact of this class of problems is out of question, and has driven the on-going quest for effective solving strategies.

Initial approaches for tackling COPs were based on exact methods, but the intrinsic complexity of most problems in the area make such methods unaffordable for realistic problem instances. Approximate methods have been defined as well as, but in general these are far from practical too, and do not provide a systematic line of attack to deal with COPs. Parameterised complexity algorithms allow efficiently solving certain COPs for which the intrinsic hardness is isolated within an internal structural parameter, whose value can be kept small. For the remaining problems (most COPs actually), practical solving requires the use of metaheuristic approaches such as, evolutionary algorithms, swarm intelligence and local search techniques. Dating back to the last decades of the twentieth century, these methods trade completeness for pragmatic effectiveness, thereby providing probably optimal or quasi-optimal solutions to a plethora of hard COPs.

The application of metaheuristics to COPs is an active field in which new theoretical developments, new algorithmic models, and new application areas are continuously emerging. In this sense, this volume presents recent advances in the area of metaheuristic combinatorial optimisation. The most popular metaheuristic family is evolutionary computation (EC), and as such an important part of the volume deals with EC approaches. However, contributions in this volume are not restricted to EC, and comprise metaheuristics as a whole. Indeed, among the essential lessons learned in the last years, the removal of dogmatic artificial barriers between metaheuristic families has been a key factor for the success of these techniques, and this is also reflected in this book. Most articles in this collection are extended versions of selected papers from the 7th Conference on Evolutionary Computation and Metaheuristics in Combinatorial Optimization (EvoCOP'2007 – Valencia, Spain). First organised in 2001 and held annually since then, EvoCOP has grown to be a reference event in metaheuristic combinatorial optimisation, with a very strong selection process and high quality contributions each year. This quality is reflected in all contributions comprised here.

The volume is organised in five blocks. The first one is devoted to theoretical developments and methodological issues. In the first paper, Craven addresses a problem in the area of combinatorial group theory, and analyzes the behavior of evolutionary algorithms on it. Ridge and Kudenko deal with an important issue in metaheuristic combinatorial optimization, namely determining whether a problem characteristic affects heuristic performance. They use a design-of-experiments approach for this purpose. Aguirre and Tanaka analyze the behavior of evolutionary algorithms within the well-known framework of NK-landscapes, paying special attention to scalability. Finally, Balaprakash, Birattari and Stützle propose a principled approach to the design of stochastic local search algorithm, based on a structured engineering-like methodology.

The second block is centered on hybrid metaheuristics. Pirkwieser, Raidl and Puchinger propose a combination of evolutionary algorithms and Lagrangean decomposition, and show its effectiveness on a constrained variant of the spanning tree problem. Nepomuceno, Pinheiro, and Coelho tackle a constrained cutting problem using a master-slave combination of metaheuristics and exact methods based on mathematical programming. Alba and Luque address a problem from sequence analysis using a genetic algorithm that incorporates a specific local search method as evolutionary operator. Finally, Neri, Kotilainen and Vapa use memetic algorithms to train neural networks that are used in turn to locate resources in P2P networks.

The third block focuses specifically on constrained problems. Musliu presents an iterated local search algorithm for finding small-width tree decompositions of constraint graphs. This algorithm is successfully tested on vertex coloring instances. Luna, Alba, Nebro and Pedraza consider a frequency assignment problem in GSM networks. They approach this problem via mutation-based evolutionary algorithms and simulated annealing. Finally, Juhos and van Hemert present some reduction heuristics for the graph coloring problem, and show that these can be used to improve the performance of other solvers for this problem.

The fourth block comprises contributions on the travelling salesman problem (TSP) and routing problems. Firstly, Fischer and Merz describe a strategy for simplifying TSP instances based on fixing some edges, and show that substantial time gains are possible with little or none performance degradation. Subsequently, Manniezzo and Roffilli address a routing problem in the area of waste collection, and compare different metaheuristics on large-scale real-world instances. Labadi, Prins and Reghioui consider a related routing problem too, and propose a memetic algorithm with population management that is shown to be very competitive. Finally, Julstrom consider a facility location problem, and use an evolutionary algorithm with a permutational decoder for its resolution. The algorithm features different heuristics for reordering elements in the solutions.

Lastly, the fifth block is devoted to scheduling problems. Firstly, Ballestín consider a resource reting problem with time lags, and compare the performance of several metaheuristics. It is shown that some evolutionary metaheuristics can outperform truncated branch-and-bound of this problem. Fernandes and Lourenço tackle job-shop scheduling problems using an algorithm that combines GRASP and branch and bound. The hybrid algorithm compares favorably with other heuristics for this problem. Finally, Xhafa and Durán consider a job scheduling problem in computational grids, and propose the use of parallel memetic algorithms. The results indicate the usefulness of this approach.

We would like to thank all the people who made this volume possible, starting by the authors who contributed the technical content of the book. We also thank Prof. Janusz Kacprzyk for his support to the development of this project. Last, but not least, we thank Dr. Thomas Ditzinger and the editorial staff of Springer for their kind attention and help.

Málaga (Spain), Edinburgh (Scotland, UK)   Carlos Cotta
March 2008   Jano van Hemert

Contents

List of Contributors

Hernán Aguirre
Shinshu University
4-17-1 Wakasato
Nagano, 380-8553
Japan
ahernan@shinshu-u.ac.jp

Enrique Alba
Universidad de Málaga
Campus Teatinos
29071 Málaga
Spain
eat@lcc.uma.es

Prasanna Balaprakash
Université Libre de Bruxelles
Av. F. Roosevelt 50
B-1050 Brussels
Belgium
pbalapra@ulb.ac.be

Francisco Ballestín
Public University of Navarra
Campus de Arrosada
31006 Pamplona
Spain
francisco.ballestin@unavarra.es

Mauro Birattari
Université Libre de Bruxelles
Av. F. Roosevelt 50
B-1050 Brussels
Belgium
mbiro@ulb.ac.be

André L.V. Coelho
Universidade de Fortaleza
Av. Washington Soares 1321
Sala J-30, Fortaleza, CE
Brazil
acoelho@unifor.br

Matthew J. Craven
University of Exeter
North Park Road
Exeter EX4 4QF
United Kingdom
m.j.craven@ex.ac.uk

Bernat Duran
Polytechnic University of Catalonia
C/Jordi Girona 1-3
08034 Barcelona
Spain
bduran@lsi.upc.edu

Susana Fernandes
Universidade do Algarve
8000-117 Faro
Portugal
sfer@ualg.pt

Thomas Fischer
University of Kaiserslautern
P.O. Box 3049
D-67653 Kaiserslautern
Germany
fischer@informatik.uni-kl.de

Jano van Hemert
University of Edinburgh
15 South College Street
EH8 9AA Edinburgh
United Kingdom
j.vanhemert@ed.ac.uk

István Juhos
University of Szeged
P.O. Box 652
6701 Szeged
Hungary
paper@juhos.info

Bryant A. Julstrom
St. Cloud State University
St. Cloud, MN 56301
United States of America
julstrom@stcloudstate.edu

Niko Kotilainen
University of Jyväskylä
P.O. Box 35 (Agora), FI-40014
Finland
niko.kotilainen@jyu.fi

Daniel Kudenko
University of York
York YO10 5DD
United Kingdom
Kudenko@cs.york.ac.uk

Nacima Labadi
University of Technology of Troyes
BP 2060
10010 Troyes Cedex
France
nacima.labadi@utt.fr

Helena R. Lourenço
Univertitat Pompeu Fabra
Ramon Trias Fargas 25-27
08005 Barcelona
Spain
helena.ramalhinho@upf.edu

Francisco Luna
Universidad de Málaga
Campus Teatinos
29071 Málaga
Spain
flv@lcc.uma.es

Gabriel Lugue
Universidad de Málaga
Campus Teatinos
29071 Málaga
Spain
gabriel@lcc.uma.es

Vittorio Maniezzo
University of Bologna
Mura Anteo Zamboni, 7
40127 Bologna
Italy
vittorio.maniezzo@unibo.it

Peter Merz
University of Kaiserslautern
P.O. Box 3049
D-67653 Kaiserslautern
Germany
pmerz@informatik.uni-kl.de

Nysret Musliu
Vienna University of Technology
Favoritenstrasse 9–11/1861
A-1040 Vienna
Austria
musliu@dbai.tuwien.ac.at

Antonio J. Nebro
Universidad de Málaga
Campus Teatinos
29071 Málaga
Spain
antonio@lcc.uma.es

Napoleão Nepomuceno
Universidade de Fortaleza
Av. Washington Soares 1321
Sala J-30, Fortaleza, CE
Brazil
napoleao@edu.unifor.br

Ferrante Neri
University of Jyväskylä
P.O. Box 35 (Agora), FI-40014
Finland
neferran@cc.jyu.fi

Salvador Pedraza
Optimi Corp.
Parque Tecnologico de Andalucia
29590 Campanillas Málaga
Spain
Salvador.Pedraza@optimi.com

Plácido Pinheiro
Universidade de Fortaleza
Av. Washington Soares 1321
Sala J-30, Fortaleza, CE
Brazil
placido@unifor.br

Sandro Pirkwieser
Vienna University of Technology
Favoritenstrae 9–11
A-1040 Vienna
Austria
pirkwieser@ads.tuwien.ac.at

Christian Prins
University of Technology of Troyes
BP 2060
10010 Troyes Cedex
France
christian.prins@utt.fr

Jakob Puchinger
University of Melbourne
Vic 3010
Australia
jakobp@csse.unimelb.edu.au

Günther Raidl
Vienna University of Technology
Favoritenstrae 9–11
A-1040 Vienna
Austria
raidl@ads.tuwien.ac.at

Mohamed Reghioui
University of Technology of Troyes
BP 2060
10010 Troyes Cedex
France
mohamed.reghioui_hamzaoui@utt.fr

Enda Ridge
University of York
York YO10 5DD
United Kingdom
ERidge@cs.york.ac.uk

Matteo Roffilli
University of Bologna
Mura Anteo Zamboni, 7
40127 Bologna
Italy
roffilli@unibo.it

Thomas Stützle
Université Libre de Bruxelles
Av. F. Roosevelt 50
B-1050 Brussels
Belgium
stuetzle@ulb.ac.be

Kiyoshi Tanaka
Shinshu University
4-17-1 Wakasato
Nagano, 380-8553
Japan
ktanaka@shinshu-u.ac.jp

Mikko Vapa
University of Jyväskylä
P.O. Box 35 (Agora), FI-40014
Finland
mikko.vapa@jyu.fi

Fatos Xhafa
Polytechnic University of Catalonia
C/Jordi Girona 1-3
08034 Barcelona
Spain
fatos@lsi.upc.edu

Part I

Theory and Methodology

1

An Evolutionary Algorithm for the Solution of Two-Variable Word Equations in Partially Commutative Groups

Matthew J. Craven

Mathematical Sciences, University of Exeter, North Park Road, Exeter EX4 4QF, UK
m.j.craven@ex.ac.uk

Summary. We describe an implementation of an evolutionary algorithm on partially commutative groups and apply it to solve certain two-variable word equations on a subclass of these groups, transforming a problem in combinatorial group theory into one of combinatorial optimisation. We give results which indicate efficient and successful behaviour of the evolutionary algorithm, hinting at the presence of a new degenerate deterministic solution and a framework for further results in similar group-theoretic problems. This paper is an expanded and updated version of [4], presented at EvoCOP 2007.

Keywords: Evolutionary Algorithm, Infinite Group, Partially Commutative, Semi-deterministic.

1.1 Introduction

1.1.1 History and Background

Genetic algorithms were introduced by Holland [7] and have enjoyed a recent renaissance in many applications including engineering, scheduling and attacking problems such as the travelling salesman and graph colouring problems. However, the use of these algorithms in group theory [1, 10, 11] has been in operation for a comparatively short time.

This paper discusses an adaptation of genetic algorithms to solve certain two-variable word equations from combinatorial group theory. We shall augment the genetic algorithm with a semi-deterministic predictive method of choosing specific kinds of reproductive algorithms to execute, so making an evolutionary algorithm (hereafter referred to as an EA).

We work over a subclass of partially commutative groups (which are also known as graph groups [13]), omitting a survey of the theory of the groups here and focusing on certain applications.

There exists an explicit solution for many problems in this setting. The biautomaticity of the partially commutative groups was established in [13], so as a corollary the conjugacy problem is solvable. Wrathall [16] gave a fast algorithm for the word problem based upon restricting the problem to a monoid generated

C. Cotta and J. van Hemert (Eds.): Recent Advances in Evol. Comp., SCI 153, pp. 3–19, 2008.
springerlink.com

by the group generators and their formal inverses. In [15], an algorithm is given for the conjugacy problem; it is linear time by a stack-based computation model.

Our work is an original experimental investigation of EAs in this setting to determine why they seem to be effective in certain areas of combinatorial group theory and to determine bounds for what happens for given problems. This is done by translating problems involving two-variable word equations to combinatorial optimisation problems. To our knowledge, our work and the distinct algebraic approach of [2] provide the only attacks on the above problem in partially commutative groups.

Firstly we outline the pertinent theory of partially commutative groups which we use directly.

1.1.2   Partially Commutative Groups

Let $X = \{x_1, x_2, \ldots, x_n\}$ be a finite set and define the operation of multiplication of $x_i, x_j \in X$ to be the juxtaposition $x_i x_j$. As in [15], we specify a *partially commutative group* $G(X)$ by X and the collection of all elements from X that *commute*; that is, the set of all pairs (x_i, x_j) such that $x_i, x_j \in X$ and $x_i x_j = x_j x_i$. For example, take $X = \{x_1, x_2, x_3, x_4\}$ and suppose that $x_1 x_4 = x_4 x_1$ and $x_2 x_3 = x_3 x_2$. Then we denote this group $G(X) = \langle X : [x_1, x_4], [x_2, x_3] \rangle$.

The elements of X are called *generators* for $G(X)$. Note that for general $G(X)$ some generators commute and some do not, and there are no other non-trivial relations between the generators. We concentrate on a particular subclass of the above groups. For a set X with n elements as above, define the group

$$V_n = \langle X : [x_i, x_j] \text{ if } |i - j| \geq 2 \rangle .$$

For example, in the group V_4 the pairs of elements that commute with each other are $(x_1, x_3), (x_1, x_4)$ and (x_2, x_4). We may also write this as $V(X)$ assuming an arbitrary set X. The elements of V_n are represented by group *words* written as products of generators. We use x_i^μ to denote μ successive multiplications of the generator x_i; for example, $x_2^4 = x_2 x_2 x_2 x_2$. Denote the empty word $\varepsilon \in V_n$. The *length*, $l(u)$, of a word $u \in V_n$ is the minimal number of single generators from which u can be written. For example $u = x_1^2 x_2 x_1^{-1} x_4 \in V_4$ is a word of length five.

For a subset, Y, of the set X we say the group $V(Y)$ is a *parabolic subgroup* of $V(X)$. It is easily observed that any partially commutative group G may be realised as a subgroup of V_n given a sufficiently large rank n.

Vershik [14] solved the word problem (that is, deciding whether two given words are equal) in V_n by means of reducing words to their *normal form*. The Knuth-Bendix normal form of a word $u \in V_n$ of length $l(u)$ may be thought of as the "shortest form" of u and is given by the unique expression

$$\overline{u} = x_{i_1}^{\mu_1} x_{i_2}^{\mu_2} \ldots x_{i_k}^{\mu_k}$$

such that all $\mu_i \neq 0, l(\overline{u}) = \sum |\mu_i|$ and

 i) if $i_j = 1$ then $i_{j+1} > 1$;
 ii) if $i_j = m < n$ then $i_{j+1} = m - 1$ or $i_{j+1} > m$;
 iii) if $i_j = n$ then $i_{j+1} = n - 1$.

The name of the above form follows from the Knuth-Bendix algorithm with ordering $x_1 < x_1^{-1} < x_2 < x_2^{-1} < \ldots < x_n < x_n^{-1}$. We omit further discussion of this here; the interested reader is referred to [8] for a description of the algorithm.

The algorithm to produce the above normal form is essentially a restriction of the stack-based (or heap-based) algorithm of [16], and we thus conjecture that the normal form of a word $u \in V_n$ may be computed efficiently in time $O\left(l(u) \log l(u)\right)$ for the "average case". By computational experiment, this seems to be the case, with a "worst case" complexity of $O\left(l(u)^2\right)$ in exceptional circumstances. From now on we write $\overline{u}$ to mean the normal form of the word u, and use the expression $u \doteq v$ (equivalence) to mean that $\overline{u} = \overline{v}$. For a word $u \in V_n$, we say that

$$RF(u) = \{x_i^\alpha : l(\overline{ux_i^{-\alpha}}) = l(\overline{u}) - 1, \alpha = \pm 1\}$$

is the *roof of* u and

$$FL(u) = \{x_i^\alpha : l(\overline{x_i^{-\alpha}u}) = l(\overline{u}) - 1, \alpha = \pm 1\}$$

is the *floor of* u. The roof (respectively the floor) of u correspond to the generators which may be cancelled after their inverses are juxtaposed to the right (respectively the left) end of u to create the word u', which is then reduced to its normal form $\overline{u'}$. For instance, if $u = x_1^{-1}x_2x_6x_5^{-1}x_4x_1$ then $RF(u) = \{x_1, x_4\}$ and $FL(u) = \{x_1^{-1}, x_6\}$. To verify x_4 is in the roof of u we juxtapose x_4^{-1} to the right of u and reduce to normal form, as follows:

$$ux_4^{-1} = x_1^{-1}x_2x_6x_5^{-1}x_4x_1x_4^{-1}$$
$$\rightarrow x_1^{-1}x_2x_1x_6x_5^{-1} = \overline{ux_4^{-1}}$$

Thus $l(\overline{u}) = 6$ and $l(\overline{ux_4^{-1}}) = 5$, confirming the above.

1.2   Statement of Problem

Given the group V_n and two words a, b in the group, we wish to determine whether a and b lie in the same double coset with respect to given subgroups. In other words, consider the following problem:

The Double Coset Search Problem (DCSP)

Given two parabolic subgroups $V(Y)$ and $V(Z)$ of the group V_n and two words $a, b \in V_n$ such that $b \in V(Y)\,a\,V(Z)$, find words $x \in V(Y)$ and $y \in V(Z)$ such that $b \doteq xay$.

As an application of our work, note that our defined groups, V_n, are inherently related to the braid groups, a rich source of primitives for algebraic cryptography. In particular, the DCSP over the group V_n is an analogue of an established braid group primitive. The problem is also not a decision problem; we already know that $b \doteq xay$ for some x and y. The reader is invited to consult [9] for further details of the braid group primitive.

We attack this two-variable group word problem by transforming it into one of combinatorial optimisation. In the following exposition, an *instance* of the DCSP is specified by a pair (a, b) of given words, each in V_n, and the notation $\mathcal{M}((a, b))$ denotes the set of all *feasible solutions* to the given instance. We will use a GA to iteratively produce "approximations" to solutions to the DCSP, and denote an "approximation" for a solution $(x, y) \in \mathcal{M}((a, b))$ by $(\chi, \zeta) \in V(Y) \times V(Z)$.

Combinatorial Optimisation DCSP

> Input: Two words $a, b \in V_n$.
> Constraints: $\mathcal{M}((a, b)) = \{(\chi, \zeta) \in V(Y) \times V(Z) : \chi a \zeta \doteq b\}$.
> Costs: The function $C((\chi, \zeta)) = l(\overline{\chi a \zeta b^{-1}}) \geq 0$.
> Goal: Minimise C.
> Output: The pair (χ, ζ) for which C is minimal.

As above, the cost of the pair (χ, ζ) is a non-negative integer imposed by the above function C. The length function defined on V_n takes non-negative values; hence an *optimal solution* for the instance is a pair (χ, ζ) such that $C((\chi, \zeta)) = 0$. Therefore our goal is to minimise the cost function C.

Regarding the complexity of the above problem, it is clear that the number of possible words in the group V_n is infinite. Moreover, [14] gives a formula for the reduced number of "non-equivalent" group words which form the search space, and suggests this number is still very large. For example, there are around 3.005×10^{17} non-equivalent words of length at most twenty in V_{10}; this number, and by implication the size of the search space, grows rapidly as the bound on word length is increased. Hence without considering search space structure, a purely brute force attack would be unproductive for "longer" words. To the best of our knowledge, there were no other proposed attacks before our experimental work.

In the next section we expand these notions and detail the method we use to solve this optimisation problem.

1.3   Genetic Algorithms on the Group V_n

1.3.1   An Introduction to the Approach

For brevity we do not discuss the elementary concepts of genetic algorithms here, but refer the reader to [7,12] for a discussion of their many interesting properties and remark that we use standard terms such as *cost-proportionate selection* and *reproductive method* in the usual way.

We give a brief introduction to our approach. We begin with an initial population of "randomly generated" pairs of words, each pair of which is treated as an approximation to a solution $(x, y) \in \mathcal{M}((a, b))$ of an instance (a, b) of the DCSP. We explicitly note that the GA does not know either of the words x or y. Each pair of words in the population is ranked according to some cost function which measures how "closely" the given pair of words approximates (x, y). After that we systematically imitate natural selection and breeding methods to produce a new population, consisting of modified pairs of words from our initial population. Each pair of words in this new population is then ranked as before. We continue to iterate populations in this way to gather steadily closer approximations to a solution (x, y) until we arrive at a solution (or otherwise) to our equation.

1.3.2 The Representation and Computation of Words

We work over the group V_n and two given parabolic subgroups $V(Y)$ and $V(Z)$, and wish the genetic algorithm to find an exact solution to a posed instance (a, b). This is where we differ a little from the established genetic algorithm paradigm. We naturally represent a group word $u = x_{i_1}^{\mu_1} x_{i_2}^{\mu_2} \ldots x_{i_k}^{\mu_k}$ of arbitrary length by a string of integers, where we consecutively map each generator of the word u according to the map

$$x_i^{\epsilon_i} \rightarrow \begin{cases} +i & \text{if } \epsilon_i = +1 \\ -i & \text{if } \epsilon_i = -1 \end{cases}.$$

Thus a generator power $x_i^{\mu_i}$ is represented by a string of $|\mu_i|$ repetitions of the integer $\text{sign}(\mu_i)\, i$. For example, if $u = x_1^{-1} x_4 x_2 x_3^{-2} x_7 \in V_7$ then u is represented by the string $-1\ 4\ 2\ -3\ -3\ 7$. In this context the length of u is equal to the number of integers in its natural string representation.

We define a *chromosome* to be the genetic algorithm representation of a pair, (χ, ζ), of words, and note, again, that each word is naturally of variable length. A *population* is a multiset consisting of a fixed number, p, of chromosomes. The genetic algorithm has two populations in memory, the *current population* and the *next generation*. As is traditional, the current population contains the chromosomes under consideration at the current iteration of the algorithm, and the next generation has chromosomes deposited into it by the algorithm, which form the current population on the next iteration. A *subpopulation* is a submultiset of a given population.

We use the natural representation for ease of algebraic operation, acknowledging that faster or more sophisticated data structures exist (for example the stack-based data structure of [15]). However we believe the simplicity of our representation yields relatively uncomplicated reproductive algorithms. In contrast, we believe a stack-based data structure yields reproductive methods of considerable complexity. We give our reproductive methods in the next subsection.

Besides normal form reduction of a word u we use *pseudo-reduction* of u. Let $\{x_{i_{j_1}}, x_{i_{j_1}}^{-1}, \ldots, x_{i_{j_m}}, x_{i_{j_m}}^{-1}\}$ be the generators which would be removed from u if we were to reduce u to normal form. Pseudo-reduction of u is defined as simply removing the above generators from u, with no reordering of the resulting word (as there would be for normal form). For example, if $u = x_6x_8x_1^{-1}x_2x_8^{-1}x_2^{-1}x_6x_4x_5$ then its *pseudo-normal form* is $\tilde{u} = x_6x_1^{-1}x_6x_4x_5$ and the normal form of u is $\overline{u} = x_1^{-1}x_4x_6^2x_5$. Clearly, we have $l(\tilde{u}) = l(\overline{u})$. This form is also efficiently computable, with complexity at most that of the algorithm used to compute the normal form $\overline{u}$. By saying we find the pseudo-normal form of a chromosome (χ, ζ), we refer to the calculation of the pseudo-normal form of each of the components χ and ζ. Note that a word is not assumed to be in any given form unless we state otherwise.

1.3.3   Reproduction

The following reproduction methods are adaptations of standard genetic algorithm reproduction methods. The methods act on a subpopulation to give a child chromosome, which we insert into the next population (more details are given in Section 1.5).

1. Sexual (*crossover*): by some selection function, input two parent chromosomes c_1 and c_2 from the current population. Choose one random segment from c_1, one from c_2 and output the concatenation of the segments.
2. Asexual: input a parent chromosome c, given by a selection function, from the current population. Output one child chromosome by one of the following:
   a) *Insertion* of a random generator into a random position of c.
   b) *Deletion* of a generator at a random position of c.
   c) *Substitution* of a generator located at a random position in c with a random generator.
3. Continuance: return several chromosomes $c_1, c_2, \ldots, c_m$ chosen by some selection algorithm, such that the first one returned is the "fittest" chromosome (see the next subsection). This method is known as *partially elitist*.
4. Non-Local Admission: return a random chromosome by some algorithm. Each component of the chromosome is generated by choosing generators $g_1, \ldots, g_k \in X$ uniformly at random and taking the component to be the word $g_1 \ldots g_k$.

It is clear that before the cost (and so the normal form) of a chromosome c is calculated, insertion increases the length of c and deletion decreases the length of c. This gives a ready analogy of "growing" solutions of DCSP instances from the chromosomes, and, by calculating the roof and floor of c, we may deduce that over several generations the lengths of respective chromosomes change at varying rates.

With the exception of continuance, the methods are repeated for each child chromosome required.

1.3.4   The Cost Function

In a sense, a cost function induces a partial metric over the search space to give a measure of the "distance" of a chromosome from a solution. Denote the solution of an instance of the DCSP by (x, y) and a chromosome by (χ, ζ). Let $E(\chi, \zeta) = \chi a \zeta b^{-1}$; for simplicity we denote this expression by E. The normal form of the above expression is denoted $\overline{E}$. When (χ, ζ) is a solution to an instance, we have $\overline{E} = \varepsilon$ (the empty word) with defined length $l(\overline{E}) = 0$.

The cost function we use is as follows: given a chromosome (χ, ζ) its cost is given by the formula $C((\chi, \zeta)) = l(\overline{E})$. This value is computed for every chromosome in the current population at each iteration of the algorithm. This means we seek to minimise the value of $C((\chi, \zeta))$ as we iterate the genetic algorithm.

1.3.5 Selection Algorithms

We realise continuance by roulette wheel selection. This is cost proportionate. As we will see in Algorithm 1.2, we implicitly require the population to be ordered best cost first. To this end, write the population as a list $\{(\chi_1, \zeta_1), \ldots (\chi_p, \zeta_p)\}$ where $C((\chi_1, \zeta_1)) \leq C((\chi_2, \zeta_2)) \leq \ldots \leq C((\chi_p, \zeta_p))$. The selection algorithm is given by Algorithm 1.1.

Algorithm 1.1. Roulette Wheel Selection

INPUT: The population size p; the population chromosomes (χ_i, ζ_i); their costs $C((\chi_i, \zeta_i))$; and n_s, the number of chromosomes to select

OUTPUT: n_s chromosomes from the population

1. Let $W \leftarrow \sum_{i=1}^{p} C((\chi_i, \zeta_i))$;
2. Compute the sequence $\{p_s\}$ such that $p_s((\chi_i, \zeta_i)) \leftarrow \frac{C((\chi_i, \zeta_i))}{W}$;
3. Reverse the sequence $\{p_s\}$;
4. For $j = 1, \ldots, p$, compute $q_j \leftarrow \sum_{i=1}^{j} p_s((\chi_i, \zeta_i))$;
5. For $t = 1, \ldots, n_s$, do
   a) If $t = 1$ output (χ_1, ζ_1), the chromosome with least cost. End.
   b) Else
      i. Choose a random $r \in [0, 1]$;
      ii. Output (χ_k, ζ_k) such that $q_{k-1} < r < q_k$. End.

Observe that this algorithm respects the requirement that chromosomes with least cost are selected more often. For crossover we use *tournament selection*, where we input three randomly chosen chromosomes in the current population and select the two with least cost. If all three have identical cost, then select the first two chosen. Selection of chromosomes for asexual reproduction is at random from the current population.

1.4 Traceback

In a variety of ways, cost functions are a large part of a GA. The reproduction methods often specify that a random generator is chosen, and depending upon certain conditions, some generators may be more "advantageous" than others. Under a given cost function, a generator that is advantageous is one which when inserted or substituted into a given chromosome produces a decrease in cost of that chromosome (rather than an increase). Consider the following simple example.

Let $w = f_1 f_2 f_3^{-1} f_7 f_5^{-1}$ be a "chromosome", with a simple cost function $C(w) = \overline{w}$ and suppose we wish to insert a generator between positions three and four in w. Note that w has length $l(w) = 5$. If we insert f_2 into w at that point then we produce $w' = f_1 f_2 f_3^{-1} f_2 f_7 f_5^{-1}$ and $l(\overline{w'}) = 6$ (an increase in chromosome length). On the other hand, inserting f_1^{-1} at the same point produces $w' = f_1 f_2 f_3^{-1} f_1^{-1} f_7 f_5^{-1}$ and $l(\overline{w'}) = 4$ (a decrease in chromosome length). In fact, the insertion of any generator g from the set $\{f_3, f_5, f_7^{-1}\}$ in the above position of w causes a reduction in the length of "chromosome". That is, an advantageous generator g is the inverse of any generator from the set $RF(f_1 f_2 f_3^{-1}) \cup FL(f_7 f_5^{-1})$; this is just union of the roof of subword $f_1 f_2 f_3^{-1}$ to the left of the insertion point and the floor of the subword $f_7 f_5^{-1}$ to the right. The number of advantageous generators for the above word w is three, which is certainly less than $2n$ (the total number of available generators for insertion). Hence reducing the number of possible choices of generator may serve to guide a GA based on the above cost function and increase the likelihood of reducing cost.

Unfortunately, the situation for our cost function is not so simple because what we must consider are four words a, b, χ, ζ and also all the positions in those four words. To compute the cost of (χ, ζ) for every possible generator (all $2n$ of them) and position in the components χ and ζ is clearly inefficient on a practical scale. We give a possible semi-deterministic (and more efficient) approach to reducing the number of choices of generator for insertion, and term the approach *traceback*.

In brief, we take the instance given by (a, b) and use the component words a and b to determine properties of a feasible solution $(x, y) \in \mathcal{M}((a, b))$ to the instance. This approach exploits the "geometry" of the search space by tracking the process of reduction of the expression $E = \chi a \zeta b^{-1}$ to its normal form in V_n and proceeds as follows.

1.4.1   General Description of Traceback Procedure

Recall Y and Z respectively denote the set of generators of the parabolic subgroups $G(Y)$ and $G(Z)$. Suppose we have a chromosome (χ, ζ) at some stage of the GA computation. Form the expression $E = \chi a \zeta b^{-1}$ associated to the given instance of the DCSP and label each generator from χ and ζ with its position in the product $\chi \zeta$. Then reduce E to its normal form $\overline{E}$; during reduction the labels travel with their associated generators. As a result some generators from χ or ζ may be cancelled and some may not, and at the end of the reduction procedure the set of labels of the non-cancelled generators of χ and ζ give the original positions.

The generators in V_n which commute mean that the chromosome may be split into *blocks* $\{\beta_i\}$. Each block is formed from at least one consecutive generator of χ and ζ which move together under reduction of the expression E. Let B be the set of all blocks from the above process. Now a block $\beta_m \in B$ and a position q (which we call the *recommended position*) at either the left or right end of that block are randomly chosen. Depending upon the position chosen take the subword δ between either the current and next block β_{m+1} or the current and prior block β_{m-1} (if available). If there is just a single block, then take the subword δ to be between (the single block) β_1 and the end or beginning of $\overline{E}$.

Now identify the word χ or ζ from which the position q originated and its associated generating set $S = Y$ or $S = Z$. The position q is at either the left or right end of the chosen block. So depending on the end of the block chosen, randomly select the inverse of a generator from $RF(\delta) \cap S$ or $FL(\delta) \cap S$. Call this the *recommended generator* g. Note if both χ and ζ are entirely cancelled (and so B is empty), we return a random recommended generator and position.

With these, the insertion algorithm inserts the inverse of the generator on the appropriate side of the recommended position in χ or ζ. In the cases of substitution and deletion, we substitute the recommended generator or delete the generator at the recommended position. We now give an example for the DCSP over the group V_{10} with the two parabolic subgroups of $V(Y) = V_7$ and $V(Z) = V_{10}$.

1.4.2   Example of Traceback on a Given Instance

Take the short DCSP instance

$$(a, b) = (x_2^2 x_3 x_4 x_5 x_4^{-1} x_7 x_6^{-1} x_9 x_{10}, \ x_2^2 x_4 x_5 x_4^{-1} x_3 x_7 x_6^{-1} x_{10} x_9)$$

and let the current chromosome be $(\chi, \zeta) = (x_3 x_2^{-1} x_3^{-1} x_5 x_7, \ x_5 x_2 x_3 x_7^{-1} x_{10})$. Represent the labels of the positions of the generators in χ and ζ by the following numbers immediately above each generator:

$$\underbrace{\overset{0}{x_3} \ \overset{1}{x_2^{-1}} \ \overset{2}{x_3^{-1}} \ \overset{3}{x_5} \ \overset{4}{x_7}}_{\chi} \quad \underbrace{\overset{5}{x_5} \ \overset{6}{x_2} \ \overset{7}{x_3} \ \overset{8}{x_7^{-1}} \ \overset{9}{x_{10}}}_{\zeta}$$

Forming the expression E and reducing it to its Knuth-Bendix normal form gives

$$\overline{E} = \overset{0}{x_3} \ \overset{1}{x_2^{-1}} \ \overset{2}{x_3^{-1}} \ x_2 \ x_2 \ x_3 \ x_2^{-1} \ \overset{3}{x_5} \ x_4 \ x_5 \ x_4^{-1} \ x_7 \ \overset{4}{x_7}$$
$$\overset{5}{x_6^{-1}} \ x_5 \ x_4 \ \overset{8}{x_7^{-1}} \ x_6 \ x_5^{-1} \ x_4^{-1} \ x_7^{-1} \ x_9 \ x_{10} \ x_{10} \ \overset{9}{x_9^{-1}} \ x_{10}^{-1}$$

which contains eight remaining generators from (χ, ζ). Take the cost to be $C((\chi, \zeta)) = l(\overline{E}) = 26$, the number of generators in $\overline{E}$ above. There are three blocks for χ:

$$\beta_1 = \overset{0}{x_3} \ \overset{1}{x_2^{-1}} \ \overset{2}{x_3}, \ \beta_2 = \overset{3}{x_5}, \ \beta_3 = \overset{4}{x_7}$$

and three for ζ:

$$\beta_4 = \frac{5}{x_5}, \ \beta_5 = \frac{8}{x_7^{-1}}, \ \beta_6 = \frac{9}{x_{10}}.$$

Suppose we choose the position labeled '8', which is in the component ζ and gives the block β_5. This is a block of length one; by the stated rule, we may take the word to the left or the right of the block as our choice for δ.

Suppose we choose the word to the right, so $\delta = x_6 x_5^{-1} x_4^{-1} x_7^{-1} x_9 x_{10}$ and in this case, $S = \{x_1, \ldots, x_{10}\}$. So we choose a random generator from $FL(\delta) \cap S = \{x_6, x_9\}$. Choosing $g = x_6$ and inserting its inverse on the right of the generator labeled by '8' in the component ζ gives

$$\zeta' = x_5 x_2 x_3 x_7^{-1} \mathbf{x_6^{-1}} x_{10},$$

and we take the other component of the chromosome to be $\chi' = \chi$. The cost becomes $C((\chi', \zeta')) = l(\overline{\chi' a \zeta' b^{-1}}) = 25$. Note that we could have taken any block and the permitted directions to create the subword δ. In this case, there are eleven choices of δ, clearly considerably fewer than the total number of subwords of $\overline{E}$. Traceback provides a significant increase in performance over merely random selection (this is easily calculated in the above example to be by a factor of 38). We shall use traceback in conjunction with our genetic algorithm to form the evolutionary algorithm.

In the next section we give the pseudocode of the EA and detail the methods we use to test its performance.

1.5   Setup of the Evolutionary Algorithm

1.5.1   Specification of Output Alphabet

Let $n = 2m$ for some integer $m > 2$. Define the subsets of generators $Y = \{x_1, \ldots, x_{m-1}\}$, $Z = \{x_{m+2}, \ldots, x_n\}$ and two corresponding parabolic subgroups $G(Y) = \langle Y \rangle, G(Z) = \langle Z \rangle$. Clearly $G(Y)$ and $G(Z)$ commute as groups: if we take any $m > 2$ and any words $x_y \in G(Y)$, $x_z \in G(Z)$ then $x_y x_z = x_z x_y$. We direct the interested reader to [9] for information on the importance of the preceding statement. Given an instance (a, b) of the DCSP with parabolic subgroups as above, we will seek a representative for each of the two words $x \in G(Y)$ and $y \in G(Z)$ that are a solution to the DCSP. Let us label this problem (P).

1.5.2   The Algorithm and Its Parameters

Given a chromosome (χ, ζ) we choose crossover to act on either χ or ζ at random, and fix the other component of the chromosome. Insertion is performed according to the position in χ or ζ given by traceback and substitution is with a random generator, both such that if the generator chosen cancels with an immediately neighbouring generator from the word then another random generator is chosen. We choose to use pseudo-normal form for all chromosomes to remove all redundant generators while preserving the internal ordering of (χ, ζ).

By experiment, EA behaviour and performance is mostly controlled by the *parameter set* chosen. A parameter set is specified by the population size p and numbers of children begat by each reproduction algorithm. The collection of numbers of children is given by a multiset of non-negative integers $P = \{p_i\}$, where $\sum p_i = p$ and each p_i is given, in order, by the number of crossovers, selections, substitutions, deletions, insertions and random chromosomes. The parameter set is held constant throughout a GA run; clearly self-adaptation of the parameter set may be a fruitful area of future research in this area. The EA is summarised in Algorithm 1.2.

The positive integer σ is an example of a *termination criterion*, where the EA terminates if more than σ populations have been generated. We also refer to this as *suicide*. In all cases here, σ is chosen by experimentation; EA runs that continued beyond σ populations were found to be unlikely to produce a successful conclusion. Note that step 4(e) of the above algorithm covers all reproductive algorithms as detailed in Section 1.3.3; for instance, we require consideration of two chromosomes from the entire population for crossover and consideration of just one chromosome for an asexual operation.

By deterministic search we found a population size of $p = 200$ and parameter set $P = \{5, 33, 4, 128, 30, 0\}$ for which the EA performs well when $n = 10$. This

---

Algorithm 1.2. EA for DCSP

INPUT: The parameter set, words a, b and their lengths $l(a), l(b)$, termination control σ, initial length L_I
OUTPUT: A solution (χ, ζ) or timeout; i, the number of populations computed

1. Generate the initial population P_0, consisting of p random (unreduced) chromosomes (χ, ζ) of initial length L_I;
2. $i \leftarrow 0$;
3. Reduce every chromosome in the population to its pseudo-normal form.
4. While $i < \sigma$ do
    a) For $j = 1, \ldots, p$ do
        i. Reduce each pair $(\chi_j, \zeta_j) \in P_i$ to its pseudo-normal form $(\tilde{\chi}_j, \tilde{\zeta}_j)$;
        ii. Form the expression $E = \tilde{\chi}_j \, a \, \tilde{\zeta}_j \, b^{-1}$;
        iii. Perform the traceback algorithm to give $C((\chi_j, \zeta_j))$, recommended generator g and recommended position q;
    b) Sort current population P_i into least-cost-first order and label the chromosomes $(\tilde{\chi}_1, \tilde{\zeta}_1), \ldots, (\tilde{\chi}_p, \tilde{\zeta}_p)$;
    c) If the cost of $(\tilde{\chi}_1, \tilde{\zeta}_1)$ is zero then return solution (χ_1, ζ_1) and the generation count i. END.
    d) $P_{i+1} \leftarrow \emptyset$;
    e) For $j = 1, \ldots, p$ do
        i. Using the data obtained in step $4(a)(iii)$, perform the appropriate reproductive algorithm on either the population P_i or the chromosome(s) within and denote the resulting chromosome (χ'_j, ζ'_j);
        ii. $P_{i+1} \leftarrow P_{i+1} \cup \{(\chi'_j, \zeta'_j)\}$;
    f) $i \leftarrow i + 1$.
5. Return failure. END.

means that the three most beneficial reproduction algorithms were insertion, substitution and selection.

We observed that the EA exhibits the well-known common characteristic of sensitivity to changes in parameter set; we consider this in future work. Once again by experimentation, we found an optimal length of one for each word in our initial population, P_0. We now devote the remainder of the paper to our results of testing the EA and analysis of the data collected.

1.5.3   Method of Testing

We wished to test the performance of the EA on "randomly generated" instances of problem (P). Define the length of an instance of the problem (P) to be the set of lengths $\{l(\overline{a}), l(\overline{x}), l(\overline{y})\}$ of words $a, x, y \in V_n$ used to create that instance. Each of the words a, x and y are generated by the following simple random walk on V_n. To generate a word $\overline{u}$ of given length $k = l(\overline{u})$, first randomly generate an unreduced word u_1 of unreduced length $l(u_1) = k$. Then if $l(\overline{u_1}) < k$, randomly generate a word u_2 of unreduced length $k - l(\overline{u_1})$, take $u = u_1 u_2$ and repeat this procedure until we produce a word $u = u_1 u_2 \ldots u_r$ with $l(\overline{u})$ equal to the required length k.

We identified two key input data for the EA: the length of an instance of (P) and the group rank, n. Two types of tests were performed, varying these data:

1. Test of the EA with long instances while keeping the rank small;
2. Test of the EA with instances of moderate length while increasing the rank.

The algorithms and tests were developed and conducted in GNU C++ on a Pentium IV 2.53 GHz computer with 1GB of RAM running Debian Linux 3.0.

1.5.4   Results

Define the *generation count* to be the number of populations (and so iterations) required to solve a given instance; see the counter i in Algorithm 1.2. We present the results of the tests and follow this in Section 1.5.5 with discussion of the results.

Increasing Length

We tested the EA on eight randomly generated instances (I1)–(I8) with the rank of the group V_n fixed at $n = 10$. The instances (I1)–(I8) were generated beginning with $l(\overline{a}) = 128$ and $l(\overline{x}) = l(\overline{y}) = 16$ for instance (I1) and progressing to the following instance by doubling the length $l(\overline{a})$ or both of the lengths $l(\overline{x})$ and $l(\overline{y})$. The EA was run ten times on each instance and the mean runtime $\overline{t}$ in seconds and mean generation count $\overline{g}$ across all runs of that instance was taken. For each collection of runs of an instance we took the standard deviation σ_g of the generation counts and the mean time in seconds taken to compute each population. A summary of results is given by Table 1.1.

Table 1.1. Results of increasing instance lengths for fixed rank $n = 10$

Instance	$l(\bar{a})$	$l(\bar{x})$	$l(\bar{y})$	$\bar{g}$	$\bar{t}$	σ_g	sec/gen
I1	128	16	16	183	59	68.3	0.323
I2	128	32	32	313	105	198.5	0.339
I3	256	64	64	780	380	325.5	0.515
I4	512	64	64	623	376	205.8	0.607
I5	512	128	128	731	562	84.4	0.769
I6	1024	128	128	1342	801	307.1	0.598
I7	1024	256	256	5947	5921	1525.3	1.004
I8	2048	512	512	14805	58444	3576.4	3.849

Increasing Rank. These tests were designed to keep the lengths of computed words relatively small while allowing the rank n to increase. We no longer impose the condition of $l(\bar{x}) = l(\bar{y})$. Take s to be the arithmetic mean of the lengths of $\bar{x}$ and $\bar{y}$. Instances were constructed by taking $n = 10, 20$ or 40 and generating a random word a of maximal length 750, random words x and y of maximal length 150 and then reducing the new $b = xay$ to its normal form $\bar{b}$.

We then ran the EA once on each of 505 randomly generated instances for $n = 10$, with 145 instances for $n = 20$ and 52 instances for $n = 40$. We took the time t in seconds to produce a solution and the respective generation count g. The data collected is summarised on Table 1.2 by grouping the lengths, s, of instances into intervals of length fifteen. For example, the range 75–90 means all instances where $s \in [75, 90)$. Across each interval we computed the means $\bar{g}$ and $\bar{t}$ along with the standard deviation σ_g. We now give a discussion of the results and some conjectures.

1.5.5 Discussion

Firstly, the mean times (in seconds) given by Tables 1.1 and 1.2 depend upon the time complexity of the underlying algebraic operations. We conjecture for

Table 1.2. Results of increasing rank from $n = 10$ (upper rows) to $n = 20$ (centre rows) and $n = 40$ (lower rows)

s	15–30	30–45	45–60	60–75	75–90	90–105	105–120	120–135	135–150
$\bar{g}$	227	467	619	965	1120	1740	1673	2057	2412
$\bar{t}$	44	94	123	207	244	384	399	525	652
$\bar{g}$	646	2391	2593	4349	4351	8585	8178	8103	10351
$\bar{t}$	251	897	876	1943	1737	3339	3265	4104	4337
$\bar{g}$	1341	1496	2252	1721	6832	14333	14363	-	-
$\bar{t}$	949	1053	836	1142	5727	10037	11031	-	-

$n = 10$ that these have time complexity no greater than $O(k \log k)$ where k is the mean length of all words across the entire run of the EA that we wish to reduce.

Table 1.1 shows we have a good method for solving large scale problem instances when the rank is $n = 10$. By Table 1.2 we observe that the EA operates very well in most cases across problem instances where the mean length of x and y is less than 150 and rank at most forty. Fixing s in a given range, the mean generation count increases at an approximately linearithmic rate as n increases. This seems to hold for all n up to forty, so we conjecture that for a mean instance of problem (P) with given rank n and instance length s the generation count for an average run of the EA lies between $O(sn)$ and $O(sn \log n)$. This conjecture means the EA generation count depends linearly on s.

To verify this conjecture we present Figure 1.1, a scatter diagram of the data used to produce the statistics on the $n = 10$ rows of Table 1.2. Each point on the diagram records an instance length s and its corresponding generation count g, and for comparison has the three lines $g = sn, g = sn^2$ and $g = sn \log n$ superimposed.

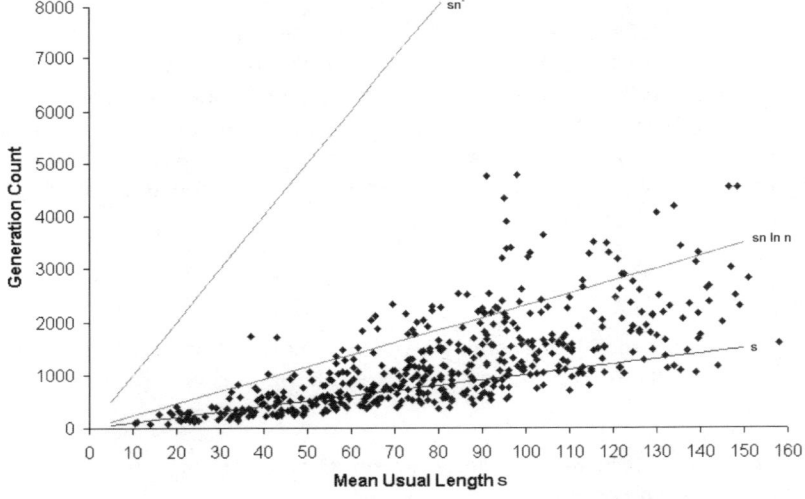

Fig. 1.1. Scatter diagram depicting the instance lengths and corresponding generation counts of all experiments for $n = 10$

From the figure we notice that the conjecture seems to be a reasonable one, with the vast majority of points lying beneath the line $g = sn \log n$ and in all cases beneath the line $g = sn^2$ (which would correspond to quadratic dependency). The corresponding scatter diagrams for $n = 20$ and $n = 40$ show identical behaviour; for brevity, we do not give the corresponding scatter diagrams here.

As n increases across the full range of instances of problem (P), increasing numbers of suicides tend to occur as the EA encounters increasing numbers of local minima. These may be partially explained by observing traceback. For n large, we are likely to have many more blocks than for n small (as the likelihood of two arbitrary generators commuting is larger). While traceback is much more efficient than a purely random method, this means that there are a greater number of chances to read the subword δ between blocks. Indeed, there may be so many possible subwords δ that it may take many EA iterations to reduce cost. This is one situation in which the EA is said to encounter a *likely* local minimum, and is manifest as the cost value of the top chromosome in the current population not changing over a (given) number of generations. We give an example of this situation.

Consider the following typical GA output, where c denotes the cost of the chromosome and l denotes the length of the components shown. Suppose the best chromosomes from populations 44 and 64 (before and after a likely local minimum) are:

```
Gen 44 (c = 302) : x = 9 6 5 6 7 4 5 -6 7 5 -3 -3 (l = 12)

y = -20 14 12 14 -20 -20 (l = 6)

Gen 64 (c = 300) : x = 9 8 1 7 6 5 6 7 4 5 -6 7 9 5 -3 -3
(l = 16)

y = 14 12 12 -20 14 15 -14 -14 -16 17 15 14 -20 15 -19 -20
-20 -19 -20 18 -17 -16 (l = 22)
```

In this case, the population stagnates for twenty generations. Notice that cost reduction (from a cost value of 302 to 300) is not made by a small change in chromosome length, but by a large one. Hence the likely local minimum has been overcome: the component χ grows by four generators and ζ by sixteen. This overcoming of a likely local minimum may be accomplished by several different methods, of which we suggest two below.

It happened that the cost reduction in the above example was made when a chromosome of greater cost in the ordered population was selected and then mutated, as the new chromosome at population 64 is far longer. We call this *promotion*, where a chromosome is "promoted" by an asexual reproduction algorithm from a lower position in the ordered population and becomes of lower cost than the current "best" chromosome. In this situation it seems traceback acts as a topological sorting method on the generators of the expression E, giving complex systems of cancellations in E which result in a cost deduction greater than one. This suggests that finetuning the parameter set to focus more on reproduction lower in the population and reproduction which causes larger changes in word length may improve performance.

The growth of components may also be due to crossover, where two random subwords from components of two chromosomes are juxtaposed. This means that there is potential for much larger growth in chromosome length than that obtained by simple mutation. Indeed, [3] conjectures that

> "It seems plausible to conjecture that sexual mating has the purpose to overcome situations where asexual evolution is stagnant."

Bremermann [3, p. 102]

In our experience this tends to be more effective than promotion. Consequently, this implies the EA performs well in comparison to asexual hillclimbing methods. Indeed, this is the case in practice: by making appropriate parameter choices we may simulate such a hillclimb, which experimentally we find encounters many more local minima. These local minima seem to require substantial changes in the forms of chromosome components χ and ζ (as above); this clearly cannot be done by mere asexual reproduction.

1.6   Conclusions

We have shown that we have a successful EA for solution of the DCSP over partially commutative groups. The evidence for this is given by Tables 1.1 and 1.2, which indicate that the EA successfully solves the DCSP for a large number of different ranks, n, and instance lengths ranging from the very short to "long" (where the hidden words x and y are of length 512).

We have at least for reasonable values of n an indication of a good underlying deterministic algorithm based on traceback (which is itself semi-deterministic). In fact, such deterministic algorithms were developed in [2] as the result of the analysis of experimental data arising from our work. This hints that the search space has a "good" structure, in some sense, and may be exploited by appropriately sensitive EAs and other artificial intelligence technologies in our framework.

Indeed, the refinement of our EA into other classes of metaheuristic algorithms through the inclusion of certain sub-algorithms provides tantalising possibilities for future work. For example, a specialised local search may be included to create a memetic algorithm. This would be especially useful in the identification of local minima within appropriately defined neighbourhoods, and may encourage increased efficiency of algorithmic solution of the problem as well as shed light upon the underlying search space structure.

A second algorithm, which is degenerate and, at worst, semi-deterministic, may be produced from our EA by restricting the EA to a small population of chromosomes and performing (degenerate versions of) only those reproductive algorithms necessary to achieve diversity of the population and overcome local minima. We found the necessary reproductive algorithms to be those of insertion, selection and crossover. Notice that only five crossover operations are performed in the parameter set given in Section 1.5.2; interestingly, we found that only a small number of crossover operations were beneficial for our EA. A larger number of crossover operations encouraged very high population diversity and large generation counts, whereas performing no crossover operations at all transformed the EA into a hillclimb which falls victim to multiple local minima (as detailed above).

Our framework may be easily extended to other two-variable equations over the partially commutative groups, and with some work, to other group structures, creating a platform for algebraic computation. If we consider partially commutative structures by themselves, there have been several fruitful programmes producing theoretical methods for the decidability and the solvability of general equations [5,6]. The tantalising goal is to create an evolutionary platform for solution of general equations over various classes of group structure, thus creating an experimental study of solution methods and in turn motivating improvements on the classification of the types of equations that may be solved over specific types of group.

2

Determining Whether a Problem Characteristic Affects Heuristic Performance
A Rigorous Design of Experiments Approach

Enda Ridge and Daniel Kudenko

The Department of Computer Science, The University of York, UK
ERidge@cs.york.ac.uk, Kudenko@cs.york.ac.uk

Summary. This chapter presents a rigorous Design of Experiments (DOE) approach for determining whether a problem characteristic affects the performance of a heuristic. Specifically, it reports a study on the effect of the cost matrix standard deviation of symmetric Travelling Salesman Problem (TSP) instances on the performance of Ant Colony Optimisation (ACO) heuristics. Results demonstrate that for a given instance size, an increase in the standard deviation of the cost matrix of instances results in an increase in the difficulty of the instances. This implies that for ACO, it is insufficient to report results on problems classified only by problem size, as has been commonly done in most ACO research to date. Some description of the cost matrix distribution is also required when attempting to explain and predict the performance of these heuristics on the TSP. The study should serve as a template for similar investigations with other problems and other heuristics.

Keywords: Design of Experiments, Nest Hierarchical Design, Problem Difficulty, Ant Colony Optimisation.

2.1 Introduction and Motivation

Ant Colony Optimisation (ACO) algorithms [17] are a relatively new class of stochastic metaheuristic for typical Operations Research (OR) problems of discrete combinatorial optimisation. To date, research has yielded important insights into ACO behaviour and its relation to other heuristics. However, there has been no rigorous study of the relationship between ACO algorithms and the difficulty of problem instances. Specifically, in researching ACO algorithms for the Travelling Salesperson Problem (TSP), it has generally been assumed that problem instance size is the main indicator of problem difficulty. Cheeseman *et al* [18] have shown that there is a relationship between the standard deviation of the edge lengths of a TSP instance and the difficulty of the problem for an *exact* algorithm. This leads us to wonder whether the standard deviation of edges lengths may also have a significant effect on problem difficulty for the ACO heuristics. Intuitively, it would seem so. An integral component of the construct solutions phase of ACO algorithms depends on the relative lengths of edges in

C. Cotta and J. van Hemert (Eds.): Recent Advances in Evol. Comp., SCI 153, pp. 21–35, 2008.

the TSP. These edge lengths are often stored in a TSP *cost matrix*. The probability with which an artificial ant chooses the next node in its solution depends, among other things, on the relative length of edges connecting to the nodes being considered. This study hypothesises that a high variance in the distribution of edge lengths results in a problem with a different difficulty to a problem with a low variance in the distribution of edge lengths.

This research question is important for several reasons. Current research on ACO algorithms for the TSP does not report the problem characteristic of standard deviation of edge lengths. Assuming that such a problem characteristic affects performance, this means that for instances of the same or similar sizes, differences in performance are confounded with possible differences in standard deviation of edge lengths. Consequently, too much variation in performance is attributed to problem size and none to problem edge length standard deviation. Furthermore, in attempts to model ACO performance, *all* important problem characteristics must be incorporated into the model so that the relationship between problems, tuning parameters and performance can be understood. With this understanding, performance on a new instance can be satisfactorily predicted given the salient characteristics of the instance.

This study focuses on Ant Colony System (ACS) [19] and Max-Min Ant System (MMAS) [20] since the field frequently cites these as its best performing algorithms. This study uses the TSP. The difficulty of solving the TSP to optimality, despite its conceptually simple description, has made it a very popular problem for the development and testing of combinatorial optimisation techniques. The TSP "has served as a testbed for almost every new algorithmic idea, and was one of the first optimization problems conjectured to be 'hard' in a specific technical sense" [21, p. 37]. This is particularly so for algorithms in the Ant Colony Optimisation (ACO) field where 'a good performance on the TSP is often taken as a proof of their usefulness' [17, p. 65].

The study emphasises the use of established Design of Experiment (DOE) [22] techniques and statistical tools to explore data and test hypotheses. It thus addresses many concerns raised in the literature over the field's lack of experimental rigour [23, 24, 25]. The designs and analyses from this paper can be applied to other stochastic heuristics for the TSP and other problem types that heuristics solve. In fact, there is an increasing awareness of the need for the experiment designs and statistical analysis techniques that this chapter illustrates [26].

The next Section gives a brief background on the study's ACO algorithms, ACS and MMAS. Section 2.3 describes the research methodology. Sections 2.4 and Section 2.5 describe the results from the experiments. Related work is covered in Section 2.6. The chapter ends with its conclusions and directions for future work.

2.2   Background

Given a number of cities and the costs of travelling from any city to any other city, the Travelling Salesperson Problem (TSP) is the problem of finding the cheapest round-trip route that visits each city exactly once. This has application in problems of traffic routing and manufacture among others [27]. The TSP is best represented by a graph of nodes (representing cities) and edges (representing the costs of visiting cites).

Ant Colony Optimisation algorithms are discrete combinatorial optimisation heuristics inspired by the foraging activities of natural ants. Broadly, the ACO algorithms work by placing a set of artificial ants on the TSP nodes. The ants build TSP solutions by moving between nodes along the graph edges. These movements are probabilistic and are influenced both by a heuristic function and the levels of a real-valued marker called a *pheromone*. Their movement decisions also favour nodes that are part of a candidate list, a list of the least costly cities from a given city. The iterated activities of artificial ants lead to some combinations of edges becoming more reinforced with pheromone than others. Eventually the ants converge on a solution.

It is common practice to hybridise ACO algorithms with local search [28] procedures. This study focuses on ACS and MMAS as constructive heuristics and so omits any such procedure. This does not detract from the proposed experiment design, methodology and analysis. The interested reader is referred to a recent text [17] for further information on ACO algorithms.

2.3 Method

This section describes the general experiment design issues relevant to this chapter. Others have discussed these in detail [29] for heuristics in general. Further information on Design of Experiments in general [22] and its adaptation for heuristic tuning [30] is available in the literature.

2.3.1 Response Variables

Response variables are those variables we measure because they represent the effects which interest us. A good reflection of problem difficulty is the solution quality that a heuristic produces and so in this study, solution quality is the response of interest. Birattari [31] briefly discusses measures of solution quality. He dismisses the use of relative error since it is not invariant under some transformations of the problem, as first noted by Zemel [32]. An example is given of how an affine transformation[1] of the distance between cities in the TSP, leaves a problem that is essentially the same but has a different relative error of solutions. Birattari instead uses a variant of Zemel's differential approximation measure [33] defined as:

$$c_{de}(c, i) = \frac{c - \bar{c}_i}{c_i^{rnd} - \bar{c}_i} \tag{2.1}$$

---

[1] An affine transformation is any transformation that preserves collinearity (i.e., all points lying on a line initially still lie on a line after transformation) and ratios of distances (e.g., the midpoint of a line segment remains the midpoint after transformation). Geometric contraction, expansion, dilation, reflection, rotation, and shear are all affine transformations.

where $c_{de}(c, i)$ is the differential error of a solution instance i with cost c, $\bar{c}_i$ is the optimal solution cost and c_i^{rnd} is the expected cost of a random solution to instance i. An additional feature of this Adjusted Differential Approximation (ADA) is that its value when applied to a randomly generated solution will be 1. The measure therefore indicates how good a method is relative to the most trivial random method. In this way it can be considered as incorporating a lower bracketing standard [34].

Nonetheless, percentage relative error from optimum is still the more popular solution quality measure and so this study records and analyses both measures. The measures will be highly correlated but it is worthwhile to analyse them separately and to see what effect the choice of response has on the study's conclusions.

Concorde [35] was used to calculate the optima of the instances. Expected values of random solutions were calculated as the average of 200 solutions generated by randomly permuting the order of cities to create tours.

2.3.2   Instances

Problem instances were created using a modification of the **portmgen** generator from the DIMACS TSP challenge [36]. The original **portmgen** created a cost matrix by choosing edge lengths uniformly randomly within a certain range. We adjusted the generator so that edge costs could be drawn from any distribution. In particular, we followed Cheeseman *et al*'s [18] approach and drew edge lengths from a Log-Normal distribution. Although Cheeseman *et al* did not state their motivation for using such a distribution, a plot of the normalised relative frequencies of the normalised edge costs of instances from a popular online benchmark library, TSPLIB [37], shows that the majority have a Log-Normal shape (Figure 2.1).

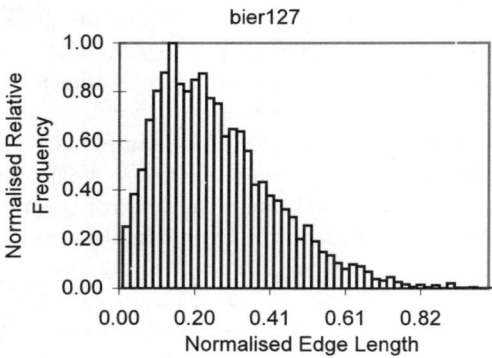

Fig. 2.1. Normalised relative frequency of normalised edge lengths of TSPLIB instance bier127. The normalised distribution of edge lengths demonstrates a characteristic Log-Normal shape.

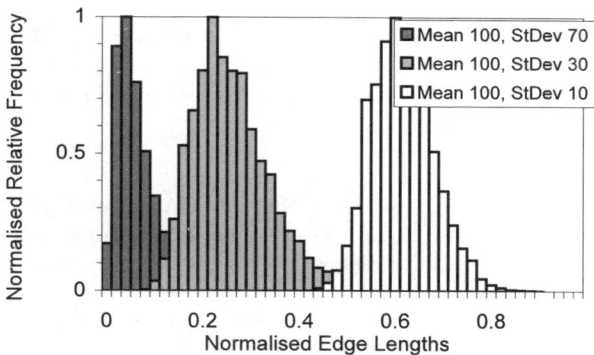

Fig. 2.2. Normalised relative frequencies of normalised edge lengths for 3 instances of the same size and same mean cost. Instances are distinguished by their standard deviation.

An appropriate choice of inputs to our modified **portmgen** results in edges with a Log-Normal distribution and a desired mean and standard deviation. Figure 2.2 shows normalised relative frequencies of the normalised edge costs of three generated instances.

Standard deviation of TSP edge lengths was varied across 5 levels: 10, 30, 50, 70 and 100. Three problem sizes; 300, 500 and 700 were used in the experiments. The same set of instances was used for the ACS and MMAS heuristics and the same instance was used for replicates of a design point.

2.3.3 Factors, Levels and Ranges

There are two types of factors or variables thought to influence the responses. Design factors are those that we vary in anticipation that they have an effect on the responses. Held-constant factors may have an effect on the response but are not of interest and so are fixed at the same value during all experiments.

Design Factors

There were two design factors. The first was the standard deviation of edge lengths in an instance. This was a fixed factor, since its levels were set by the experimenter. Five levels: 10, 30, 50, 70 and 100 were used. The second factor was the individual instances with a given level of standard deviation of edge lengths. This was a random factor since instance uniqueness was caused by the problem generator and so was not under the experimenter's direct control. Ten instances were created within each level of edge length standard deviation.

Held-constant Factors

Computations for pheromone update were limited to the candidate list length (Section 2.2). Both problem size and edge length mean were fixed for a given experiment. The held constant tuning parameter settings for the ACS and MMAS heuristics are listed in Table 2.1.

Table 2.1. Parameter settings for the ACS and MMAS algorithms. Values are taken from the original publications [17, 20]. See these for a description of the tuning parameters and the MMAS and ACS heuristics.

Parameter	Symbol	ACS	MMAS
Ants	m	10	25
Pheromone emphasis	α	1	1
Heuristic emphasis	β	2	2
Candidate List length		15	20
Exploration threshold	q_0	0.9	N/A
Pheromone decay	ρ_{global}	0.1	0.8
Pheromone decay	ρ_{local}	0.1	N/A
Solution construction		Sequential	Sequential

These tuning parameter values were used because they are listed in the field's main book [17] and are often adopted in the literature. It is important to stress that this research's use of parameter values from the literature by no means implies support for such a 'folk' approach to parameter selection in general. Selecting parameter values as done here strengthens the study's conclusions in two ways. It shows that results were not contrived by searching for a unique set of tuning parameter values that would demonstrate the hypothesised effect. Furthermore, it makes the research conclusions applicable to all other research that has used these tuning parameter settings without the justification of a methodical tuning procedure. Recall from the motivation (Section 2.1) that demonstrating an effect of edge length standard deviation on performance with even one set of tuning parameter values is sufficient to merit the factor's consideration in parameter tuning studies. The results from this research have been confirmed by such studies [38, 39].

2.3.4 Experiment Design, Power and Replicates

This study uses a *two-stage nested* (or *hierarchical*) design. Consider this analogy. A company receives stock from several suppliers. They test the quality of this stock by taking 10 samples from each supplier's batch. They wish to determine whether there is a significant overall difference in supplier quality and whether there is a significant quality difference in samples within a supplier's batch. Supplier and sample are factors. A full factorial design[2] of the supplier and sample factors is inappropriate because samples are unique to their supplier. The nested design accounts for this uniqueness by grouping samples *within* a given level of supplier. Supplier is termed the *parent* factor and batches are the *nested* factor.

A similar situation arises in this research. An algorithm encounters TSP instances with different levels of standard deviation of cost matrix. We want to determine whether there is a significant overall difference in algorithm solution quality for different levels of standard deviation. We also want to determine whether there is a significant difference in algorithm quality between instances that have the same standard deviation. Figure 2.3 illustrates the two-stage nested design schematically.

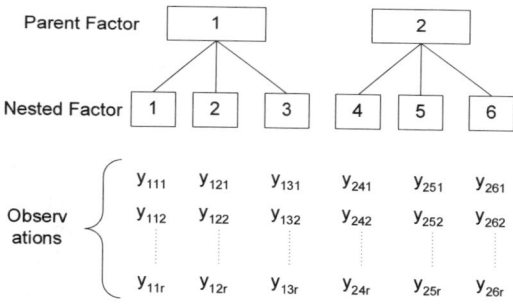

Fig. 2.3. Schematic for the Two-Stage Nested Design with r replicates. (adapted from [22]). Note the nested factor numbering to emphasise the uniqueness of the nested factor levels within a given level of the parent factor.

The standard deviation of the generated instance is the parent factor and the individual instance number is the nested factor. Therefore, an individual treatment consists of running the algorithm on a particular instance generated with a particular standard deviation. This design applies to an instance of a given size and therefore cannot capture possible interactions between instance size and instance standard deviation. Capturing such interactions would require a more complicated *crossed nested design*. This research uses the simpler design to demonstrate that standard deviation is important for a given size. Interactions are captured in the more complicated designs used in tuning ACS [38] and MMAS [39].

The heuristics are probabilistic and so repeated runs with identical inputs (instances, parameter settings etc.) will produce different results. All treatments are thus replicated in a work up procedure [64]. This involved adding replicates

---

[2] A full factorial design is one in which all levels of all factors are completely crossed with one another.

to the designs until sufficient power of 80% was reached given the study's significance level of 1% and the variability of the data collected. Power was calculated with Lenth's power calculator [65]. For all experiments, 10 replicates were sufficient to meet these requirements.

2.3.5   Performing the Experiment

Randomised run order

Available computational resources necessitated running experiments across a variety of similar machines. Runs were executed in a randomised order across these machines to counteract any uncontrollable nuisance factors[3]. While such randomising is strictly not necessary when measuring a machine-independent response, it is good practice nonetheless.

Stopping Criterion

The choice of stopping criterion for an experiment run is difficult when heuristics can continue to run and improve indefinitely. CPU time is not encouraged by some as a reproducible metric [29] and so some time independent metric such as a combinatorial count of an algorithm operation is often used.

A problem with this approach is that our choice of combinatorial count can bias our results. Should we stop after 1000 iterations or 1001? We mitigate this concern by taking evenly spaced measurements over 5000 iterations of the algorithms and separately analysing the data at all measurements. Note that a more formal detection of possible differences introduced by different stopping criteria would have required a different experiment design.

2.4   Analysis

The two-stage nested designs were analysed with the General Linear Model. Standard deviation was treated as a fixed factor since we explicitly chose its levels and instance was treated as a random factor. The technical reasons for this decision in the context of experiments with heuristics have recently been well explained in the heuristics literature [26].

2.4.1   Transformations

To make the data amenable to statistical analysis, a transformation of the responses was required for each analysis. The transformations were either a $\log_{10}$ transformation, inverse square root transformation or a square root transformation.

---

[3] A nuisance factor is one that we know affects the responses but we have no control over it. Two typical examples in experiments with heuristics are network traffic and background CPU processes.

2.4.2   Outliers

An outlier is a data value that is either unusually large or unusually small relative to the rest of the data. Outliers are important because their presence can distort data and render statistical analyses inaccurate. There are several approaches to dealing with outliers. In this research, outliers were deleted and the model building repeated until the models passed the usual ANOVA diagnostics for the ANOVA assumptions of model fit, normality, constant variance, time-dependent effects and leverage. Figure 2.4 lists the number of outliers deleted during the analysis of each experiment.

Algorithm	Problem size	ADA	Relative Error
ACS	300	3	3
	500	0	2
	700	5	4
MMAS	300	3	7
	500	1	2
	700	4	2

Fig. 2.4. Number of outliers deleted during the analysis of each experiment. Each experiment had a total of 500 data points.

Further details on the analyses and diagnostics are available in many textbooks [22].

2.5   Results

Figures 2.5 to 8 illustrate the results for both algorithms on problems of size 300 and size 700. Problems of size 500 showed the same trend and so are omitted. In all cases, the effect of Standard Deviation on solution quality was deemed statistically significant at the $p < 0.01$ level. The effect of instance was also deemed statistically significant. However, an examination of the plots that follow shows that only Standard Deviation has a practically significant effect. In each of these box-plots, the horizontal axis shows the standard deviation of the instances' cost matrices at five levels. This is repeated along the horizontal axis at the measurement points used. The vertical axis shows the solution quality response in its original scale. There is a separate plot for each algorithm and each problem size. Outliers have been included in the plots.

An increase in problem standard deviation gives an increase in the relative error of solutions produced. Instances with a higher standard deviation are more difficult to solve. Conclusions from the data were the same at all measurement points.

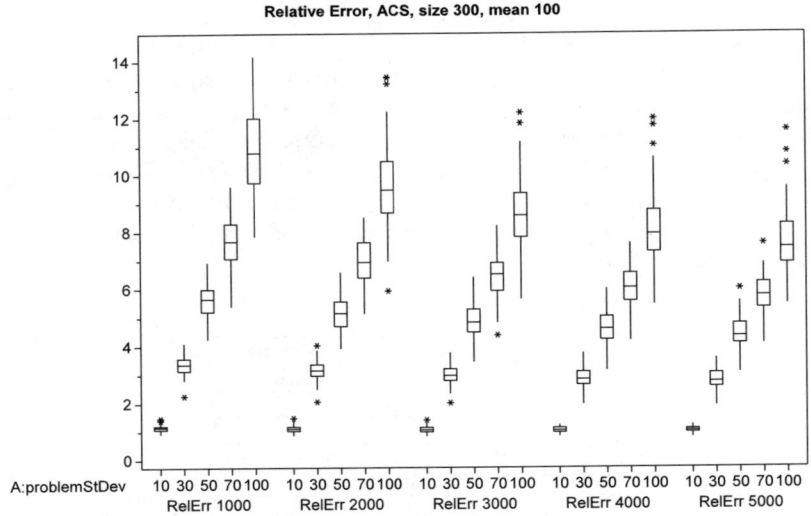

Fig. 2.5. Results of ACS applied to problems of size 300

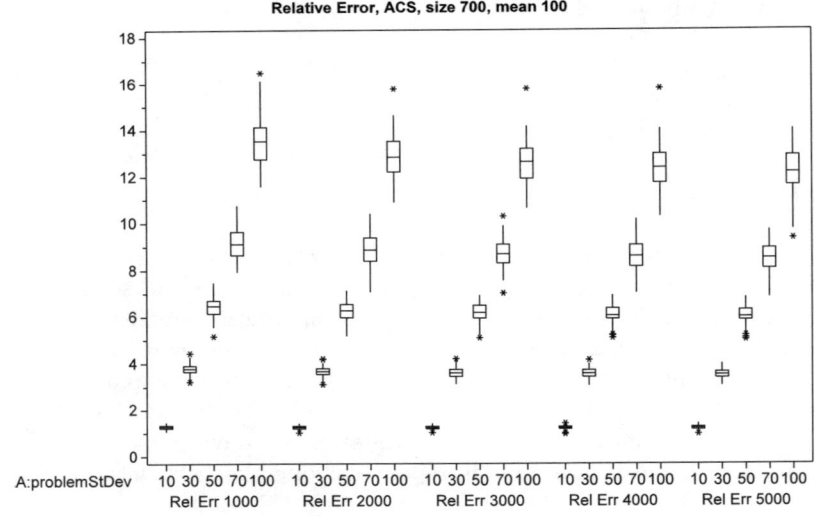

Fig. 2.6. Results of ACS applied to problems of size 700

The opposite trend was observed in the Adjusted Differential Approximation response. Figure 9 illustrates one such trend. The ADA value decreases as edge length standard deviation increases. This can be understood by examining a plot of the components of ADA (Equation (2.1)), the absolute solution value, optimal solution value and expected random solution value (Section 2.3.1). While

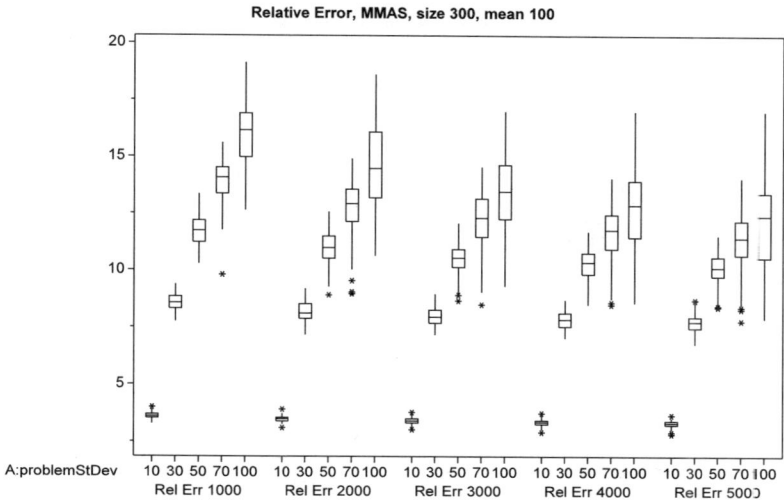

Fig. 2.7. Results of MMAS applied to problems of size 300

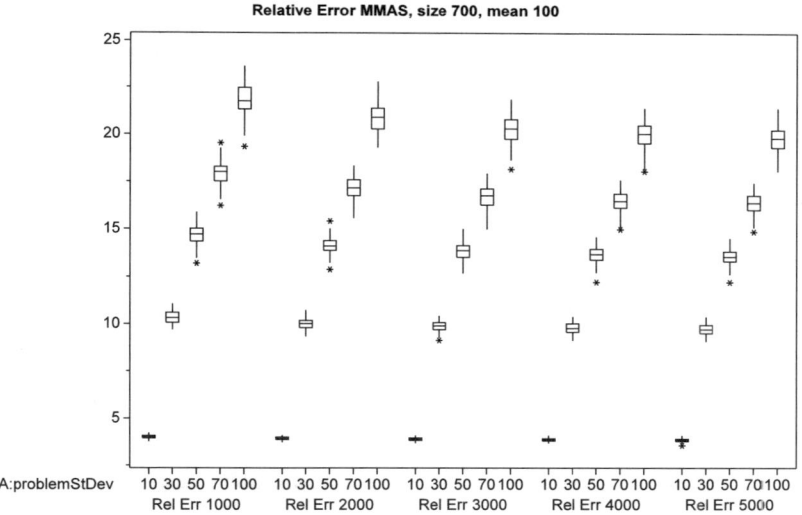

Fig. 2.8. Results of MMAS applied to problems of size 700

increasing standard deviation has no effect on the expected random solution, it has a strong effect on both the optimal and absolute solution values. As standard deviation increases, the ACO heuristics' performance deteriorates, coming closer to the performance of a random solution.

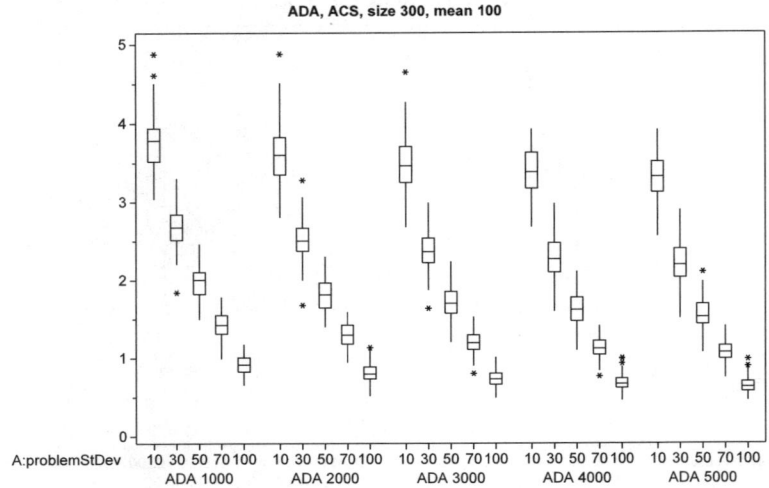

Fig. 2.9. Results of ACS applied to problems of size 300

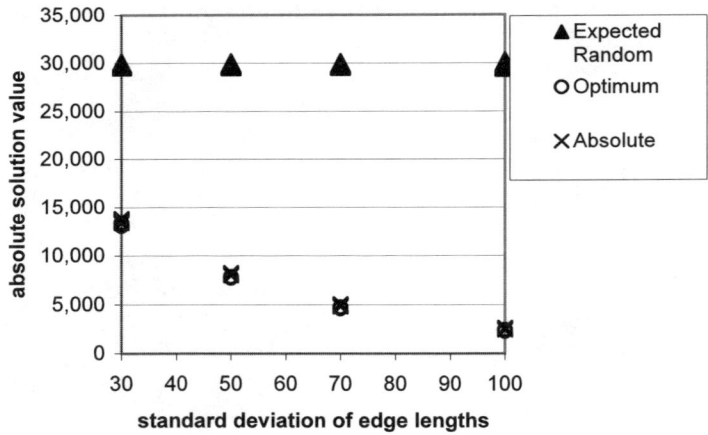

Fig. 2.10. Plots of the values affecting ADA calculation. The values of absolute solution and optimum solution decrease with increasing standard deviation of edge lengths. The expected random solution value remains unchanged.

2.6   Related Work

There has been some related work on problem difficulty for exact and heuristic algorithms. Cheeseman *et al* [18] investigated the effect of cost matrix standard deviation on the difficulty of Travelling Salesperson Problems for an exact algorithm. Three problem sizes of 16, 32 and 48 were investigated. For each problem size, many instances were generated such that each instance had the same mean cost but a varying standard deviation of cost. This varying standard deviation followed a Log-Normal distribution. The computational effort for an exact algorithm to solve each of these instances was measured and plotted against the standard deviation of cost matrix. This study differs from that of Cheeseman *et al* [18] in that it uses larger problem sizes and a heuristic algorithm rather than exact algorithm. Furthermore, its conclusions are reinforced with a rigorous DOE approach and statistical analyses.

Fischer *et al* [66] investigated the influence of Euclidean TSP structure on the performance of two algorithms, one exact and one heuristic. The former was branch-and-cut [35] and the latter was the iterated Lin-Kernighan algorithm [67]. In particular, the TSP structural characteristic investigated was the distribution of cities in Euclidean space. The authors varied this distribution by taking a structured problem instance and applying an increasing perturbation to the city distribution until the instance resembled a randomly distributed problem. There were two perturbation operators. A reduction operator removed between 1% to 75% of the cities in the original instance. A shake operator offset cities from their original location. Using 16 original instances, 100 perturbed instances were created for each of 8 levels of the perturbation factor. Performance on perturbed instances was compared to 100 instances created by uniformly randomly distributing cities in a square. Predictably, increased perturbation lead to increased solution times that were closer to the times for a completely random instance of the same size. It was therefore concluded that random Euclidean TSP instances are relatively hard to solve compared to structured instances. The reduction operator confounded changed problem structure with a reduction in problem size, a known factor in problem difficulty. This paper fixes problem size and cost matrix mean and controls cost matrix standard deviation, thus avoiding any such confounding.

Most recently, Van Hemert [68] evolved problem instances of a fixed size that were difficult to solve for two heuristics: Chained Lin-Kernighan and Lin Kernighan with Cluster Compensation. TSP instances of size 100 were created by uniform randomly selecting 100 coordinates from a 400x400 grid. This seems to be a similar generation approach to the **portgen** generator from the DIMACS TSP challenge. An initial population of such instances was evolved for each of the algorithms where higher fitness was assigned to instances that required a greater effort to solve. This effort was a combinatorial count of the algorithms' most time-consuming procedure.

Van Hemert then analysed the evolved instances using several interesting metrics. His aim was to determine whether the evolutionary procedure made the instances more difficult to solve and whether that difficulty was specific to the algorithm. To verify whether difficult properties were shared between algorithms, each algorithm was run on the other algorithm's evolved problem set. A set evolved for one algorithm was less difficult for the other algorithm. However, the alternative evolved set still required more effort than the random set indicating that some difficult instance properties were shared by both evolved problem sets. Our approach began with a specific hypothesis about a single problem characteristic and its effect on problem hardness. Van Hemert's, by contrast, evolved hard instances and then attempted to infer, *post-hoc*, which characteristics might be responsible for that hardness. If the researcher's aim is to stress test a heuristic, then we believe Van Hemert's approach is more appropriate. The approach presented here is appropriate when isolating a specific problem characteristic that may affect problem hardness.

To our knowledge, this chapter presents the first rigorous experiments on the hardness of problem instances for ACO heuristics. Recent results in screening factors affecting ACO performance [69] and on modelling the parameter, problem instance and performance relationship [38, 39] confirm the results of this study.

2.7   Conclusions

Our conclusions from the aforementioned results are as follows. For the Stützle implementations of ACS and MMAS, applied to symmetric TSP instances generated with log-normally distributed edge lengths such that all instances have a fixed cost matrix mean of 100 and a cost matrix standard deviation varying from 10 to 100:

1. a change in cost matrix standard deviation leads to a statistically and practically significant change in the difficulty of the problem instances for these algorithms.
2. there is no practically significant difference in difficulty between instances that have the same size, cost matrix mean and cost matrix standard deviation.
3. there is no practically significant difference between the difficulty measured after 1000 algorithm iterations and 5000 algorithm iterations.
4. conclusions were the same for the relative error and Adjusted Differential Approximation responses.

2.7.1   Implications

These results are important for the ACO community for the following reasons:

- They demonstrate in a rigorous, designed experiment fashion, that quality of solution of an ACO TSP algorithm is affected by the standard deviation of the cost matrix.
- They demonstrate that cost matrix standard deviation must be considered as a factor when building predictive models of ACO TSP algorithm performance.
- They clearly show that performance analysis papers using ACO TSP algorithms must report instance cost matrix standard deviation as well as instance size since two instances with the same size can differ significantly in difficulty.

- They motivate an improvement in benchmarks libraries so that they provide a wider crossing of both instance size and instance cost matrix standard deviation.

2.7.2 Assumptions and Restrictions

For completeness and for clarity, we state that this research does *not* examine the following issues.

- It did not examine clustered problem instances or grid problem instances. These are other common forms of TSP in which nodes appear in clusters and in a very structured grid pattern respectively. The conclusions should not be applied to other TSP types without a repetition of this case study.
- Algorithm performance was not being examined since no claim was made about the suitability of the parameter values for the instances encountered. Rather, the aim was to demonstrate an effect for standard deviation and so argue that it should be included as a factor in experiments that do examine algorithm performance.
- We cannot make a direct comparison between algorithms since algorithms were not tuned methodically. That is, we are not entitled to say that ACS did better than MMAS on, say, instance X with a standard deviation of Y.
- We cannot make a direct comparison of the response values for different sized instances. Clearly, 3000 iterations explores a bigger fraction of the search space for 300-city problems than for 500 city problems. Such a comparison *could* be made if it was clear how to scale iterations with problem size. Such scaling is an open question.

2.8 Future Work

There are several avenues of future work leading from this study.

The same analysis is worthwhile for other popular ACO algorithms. The code provided by Stützle also has implementations of Best-Worst Ant System, Ant System, Rank-Based Ant System and Elitist Ant System. The use of a well-established Design of Experiments approach with two-stage nested designs and analysis with the General Linear Model could also be applied to other heuristics for the TSP. It is important that we introduce such rigour into the field so that we can move from the competitive testing of highly engineered designs to the scientific evaluation of hypotheses about algorithms.

Recall that we have used fixed algorithm parameter settings from the literature. Screening experiments and Response Surface models have since established which of these parameters interact with cost matrix standard deviation to affect performance [38,39]. Further Design of Experiments studies are needed to confirm these results and to investigate similar research questions for other heuristics.

3

Performance and Scalability of Genetic Algorithms on NK-Landscapes

Hernán Aguirre and Kiyoshi Tanaka

Shinshu University, Faculty of Engineering, 4-17-1 Wakasato, Nagano, 380-8553 Japan
ahernan@shinshu-u.ac.jp, ktanaka@shinshu-u.ac.jp

Summary. This work studies the working principles, performance, and scalability of genetic algorithms on NK-landscapes varying the degree of epistasis interactions. Previous works that have focused mostly on recombination have shown that simple genetic algorithms, and some improved ones, perform worse than random bit climbers and not better than random search on landscapes of increased epistasis. In our work, in addition to recombination, we also study the effects on performance of selection, mutation, and drift. We show that an appropriate selection pressure and postponing drift make genetic algorithms quite robust on NK-landscapes, outperforming random bit climber on a broad range of classes of problems. We also show that the interaction of parallel varying-mutation with crossover improves further the reliability of the genetic algorithm.

Keywords: Genetic Algorithms, NK-Landscapes, Epistasis, Nonlinear Fitness Functions, Selection, Drift, Mutation, Recombination.

3.1 Introduction

Nonlinear correlations among variables and its impact on the structure of fitness landscapes and performance of optimizers is an important area of research in evolutionary combinatorial optimization. In the context of genetic algorithms (GAs), nonlinearities in fitness functions due to changes in the values of interacting bits are referred as epistasis. Epistasis has been recognized as an important factor that makes a problem difficult for combinatorial optimization algorithms and its influence on the performance of GAs is being increasingly investigated. Particularly, Kauffman's NK-landscapes model of epistatic interactions [40] has been used in several studies to test the performance of GAs on stochastically generated fitness functions on bit strings, parameterized with N bits and K epistatic interactions among bits (see Section 3.3 for closely related works).

This work studies the working principles, performance, and scalability of GAs on NK-landscapes varying the degree of epistasis interactions. Previous related works have shown that the performance of simple GAs and some improved ones are worse than random bit climbers for most values of epistatic interactions K and not better than random search for medium and high epistasis. These

C. Cotta and J. van Hemert (Eds.): Recent Advances in Evol. Comp., SCI 153, pp. 37–52, 2008.
springerlink.com

previous works have mostly focused on recombination. In our work, in addition to recombination, we also study the effects on performance of selection, mutation, and drift. Adaptive evolution is a search process driven by selection, drift, mutation, and recombination over fitness landscapes [40]. All these processes, interactions, and effects are important within a GA and need to be properly understood in order to achieve good performance in highly difficult problems.

In our study, we take a model of a parallel varying-mutation GA (GA-SRM) to emphasize each of the four processes mentioned above. Experiments are conducted for NK-landscapes with $N = 96$ varying K from 0 to 48. Results by a random bit climber RBC+ are also included for reference. We show that an appropriate selection pressure and postponing drift make GAs quite robust on NK-landscapes. Compared to strictly local search algorithms, we see that even simple GAs with these two features perform better than RBC+ for a broad range of classes of problems ($K \geq 4$). We also show that the interaction of parallel varying-mutation with crossover improves further the reliability of the GA for $12 < K < 32$. Contrary to intuition, we find that for small K a mutation-only evolutionary algorithm is very effective and crossover may be omitted; but the relative importance of crossover interacting with varying-mutation increases with K performing better than mutation alone for $K > 12$.

This work gives good insights on how to improve GAs by understanding better the behavior that arises from the interaction of all major processes involved in the algorithm. It is also a valuable guide to practitioners on how to configure properly the GA to achieve better results as the nonlinearity and complexity of the problem increases.

3.2   Epistasis and NK-Landscapes

Epistasis is a term used in biology to describe a range of non-additive phenomena due to the nonlinear inter-dependence of gene values, in which the expression of one gene masks the genotypic effect of another. In the context of GA this terminology is used to describe nonlinearities in fitness functions due to changes in the values of interacting bits. An implication of epistatic interactions among bits is that the fitness function develops conflicting constraints [41]. That is, a mutation in one bit may improve its own contribution to fitness, but it may decrease the contributions of other bits with which it interacts. Hence, epistatic interactions increase the difficulty in trying to optimize all bits simultaneously. Kauffman [40] designed a family of problems, the NK-landscapes, to explore how epistasis is linked to the ruggedness of search spaces. Here epistasis corresponds to the degree of interaction between genes (bits), and ruggedness is related to local optima, their number, and their density. In essence, NK-landscapes are stochastically generated fitness functions on bit strings, parameterized with N bits and K epistatic interactions between bits. More formally, an NK-landscape is a function $f : \mathcal{B}^N \rightarrow \Re$ where $\mathcal{B} = \{0, 1\}$, N is the bit string length, and K is the number of bits in the string that epistatically interact with each bit.

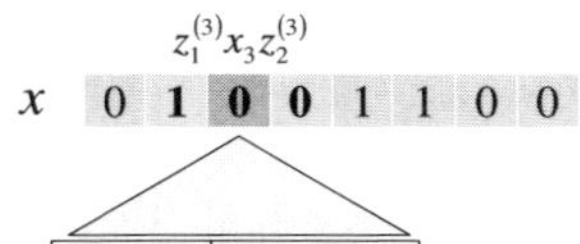

$$z_1^{(3)} x_3 z_2^{(3)}$$

x 0 1 **0** 0 1 1 0 0

$z_1 x_3 z_2$	f_3
000	0.83
001	0.34
010	0.68
011	0.10
100	**0.67**
101	0.24
110	0.60
111	0.64

Fig. 3.1. Example of fitness function $f_3(x_3, z_1^{(3)}, z_2^{(3)})$ associated to bit x_3 in which x_3 epistatically interacts with its left and right neighbouring bits, $z_1^{(3)} = x_2$ and $z_2^{(3)} = x_4$ ($N = 8$, $K = 2$)

Kauffman's original NK-landscape [40] can be expressed as an average of N functions as follows

$$f(\boldsymbol{x}) = \frac{1}{N} \sum_{i=1}^{N} f_i(x_i, z_1^{(i)}, z_2^{(i)}, \cdots, z_K^{(i)}) \qquad (3.1)$$

where $f_i : \mathcal{B}^{K+1} \to \Re$ gives the fitness contribution of bit x_i, and $z_1^{(i)}, z_2^{(i)}, \cdots$, $z_K^{(i)}$ are the K bits interacting with bit x_i in the string $\boldsymbol{x}$. That is, there is one fitness function associated to each bit in the string. NK-landscapes are stochastically generated and usually the fitness contribution f_i of bit x_i is a number between $[0.0, 1.0]$ drawn from a uniform distribution. Figure 3.1 shows an example of the fitness function $f_3(x_3, z_1^{(3)}, z_2^{(3)})$ associated to bit x_3 for $N = 8, K = 2$, in which x_3 epistatically interacts with its left and right neighbouring bits, $z_1^{(3)} = x_2$ and $z_2^{(3)} = x_4$, respectively.

For a giving N, we can tune the ruggedness of the fitness function by varying K. In the limits, $K = 0$ corresponds to a model in which there are no epistatic interactions and the fitness contribution from each bit value is simply additive, which yields a single peaked smooth fitness landscape. On the opposite extreme, $K = N-1$ corresponds to a model in which each bit value is epistatically affected by all other bits, yielding a maximally rugged fully random fitness landscape. Varying K from 0 to $N - 1$ gives a family of increasingly rugged multi- peaked landscapes.

Besides defining N and K, it is also possible to arrange the epistatic pattern between bit x_i and K other interacting bits. That is, the distribution of the K bits among the N. Kauffman investigated NK-landscapes with two kinds of epistatic patterns: (i) *nearest neighbour*, in which a bit interacts with its $K/2$ left and $K/2$ right adjacent bits, and (ii) *random*, in which a bit interacts with K other randomly chosen bits in the chromosome. By varying N, K, and the distribution of K among the N, we can study the effects of the size of the search space, intensity of epistatic interactions, and epistatic pattern on the performance of genetic algorithms.

The reader is referred to Altenberg [41], Heckendorn et al. [42], and Smith and Smith [43,44] for extensions of NK-landscapes to a more general framework of tunable random landscapes.

3.3 Related Works

NK-landscapes have been the center of several theoretical and empirical studies both for the statistical properties of the generated landscapes and for their *GA-hardness* [45,46,47,48,49]. Most previous works that focus on the performance of GAs on NK-landscapes have limited their study to small landscapes, typically $10 \leq N \leq 48$, and observed the behavior of GAs only for few generations.

Recently, studies are being conducted on larger landscapes, expending more evaluations, and the performance of GAs are being benchmarked against local search and hill climbers [50,42,51]. Merz and Freisleben [50] studied the performance of evolutionary algorithms on high-dimensional landscapes with N up to 1024 bits, but with low epistasis ($K = \{2, 4, 11\}$) relative to N. Works by Heckendorn et al. [42] and Mathias et al. [51] are closely related to our work that focuses on the effectiveness and scalability of GAs on NK-landscapes varying the epistasis among bits from low to high values. Heckendorn et al. [42] analyzed the epistatic features of *embedded landscapes* showing that for NK-landscapes all the schema information is random if K is sufficiently large, predicting that a standard GA would have no advantage over a strictly local search algorithm. To verify this, Heckendorn et al. [42] empirically compared the performance of a random bit climber (RBC+), a simple GA, and the enhanced CHC (Eshelman's adaptive GA that combines a conservative selection strategy with a highly disruptive crossover operator) [52] known to be robust in a wide variety of problems. Experiments were conducted for $N = 100$ varying K from 0 to 65. A striking result of this study was the overall better performance of the random bit climber RBC+. Motivated by [42], Mathias et al. [51] provided an exhaustive experimental examination of the performance of similar algorithms, including also Davis' RBC [53]. A main conclusion of this study is that over the range $19 \leq N \leq 100$ there is a niche for the enhanced CHC in the region of $N > 30$ for $K = 1$ and $N > 60$ for $1 \leq K < 12$. Yet, this niche is very narrow compared to the broad region where RBC and RBC+ show superiority.

Thought works by Heckendorn et al. [42] and Mathias et al. [51] have mostly focused on recombination or recombination enhanced GAs (like CHC), the conclusions achieved by these works have been extended to the overall performance of GAs. These, unfortunately, have given the wrong impression that GAs are doomed to perform badly on epistatic and k-satisfiability problems (kSAT) [42].

3.4   The Algorithms

3.4.1   GA-SRM

GA-SRM [54, 55] is a generational, non-elitist, genetic algorithm that achieves good exploration-exploitation balances by applying varying-mutation parallel to recombination and including a truncated selection mechanism enhanced with a fitness duplicates prevention method.

Namely, GA-SRM applies a varying-mutation operator parallel (concurrent) to the conventional crossover & background mutation operator. The varying-mutation operator is called Self-Reproduction with Mutation (SRM). SRM creates λ_{SRM} offspring by mutation alone with a high mutation rate that varies adaptively during the generations, as explained below. The crossover & background mutation operator (CM) creates λ_{CM} offspring by conventional crossover and successive mutation with a fixed, small, mutation rate $p_m^{(CM)}$ per bit. SRM parallel (concurrent) to CM avoids interferences between high mutations and crossover, increasing the levels of cooperation to introduce beneficial mutations and favorable recombinations.

Offspring created by SRM and CM coexist and compete to reproduce by way of (μ, λ) proportional selection [57]. This selection method is extinctive in nature. That is, the number of parents μ is smaller than the total number of created offspring $\lambda = \lambda_{SRM} + \lambda_{CM}$. Selection pressure can be easily adjusted by changing the values of μ [57].

GA-SRM enhances selection by preventing fitness duplicates. If several individuals have exactly the same fitness then one of them is chosen at random and kept. The other equal fitness individuals are eliminated from the population. The fitness duplicates elimination is carried out before extinctive selection is performed. Preventing duplicates removes an unwanted source of selective bias [58], postpones drift, and allows a fair competition between offspring created by CM and SRM.

Mutation rate within SRM is dynamically adjusted every time a normalized mutants survival ratio γ falls under a threshold τ. The normalized mutant survival ratio is specified by

$$\gamma = \frac{\mu_{SRM}}{\lambda_{SRM}} \cdot \frac{\lambda}{\mu} \tag{3.2}$$

where μ_{SRM} is the number of individuals created by SRM present in the parent population $P(t)$ after extinctive selection. The block diagram of GA-SRM is depicted in Figure 3.2.

Higher mutations imply that more than one bit will be mutated in each individual and several strategies are possible for choosing the bits that will undergo mutation [54, 56]. One mutation strategy is called Adaptive Dynamic-Probability (ADP). ADP subjects to mutation every bit of the chromosome with probability $p_m^{(SRM)}$ per bit, reducing it by a coefficient β each time γ falls under the threshold τ. Mutation rate $p_m^{(SRM)}$ varies from an initial high rate α to $1/N$ per bit, where N is the bit string length. Other mutation strategy is called Adaptive

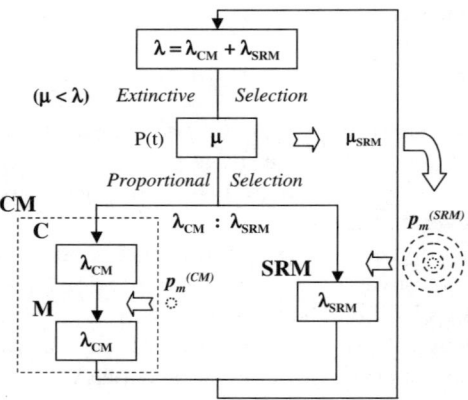

Fig. 3.2. Block diagram of GA-SRM

Dynamic-Segment. ADS directs mutation only to the bits within a segment of the chromosome using constant high mutation probability $p_m^{(SRM)} = \alpha$ per bit, reducing the mutation segment size ℓ by a coefficient β every time γ falls under τ. The mutation segment ℓ varies from N to $1/\alpha$. Both ADP and ADS flip in average the same number of bits. Unless indicated otherwise, mutation strategy used for SRM in this work is ADP.

3.4.2 RBC+

RBC+ [42] is a variant of a random bit climber (RBC) defined by Davis [53]. Both are local search algorithms that use a single bit neighbourhood. We implement RBC+ following indications given in [42, 51, 53].

RBC+ begins with a random string of length N. A random permutation of the string positions is created and bits are complemented (i.e. flipped) one at the time following the order indicated by the random permutation. Each time a bit is complemented the string is re-evaluated. All changes that result in equally good or better solutions are kept and accounted as an accepted change. After testing all N positions indicated by the random permutation, if accepted changes were detected a new permutation of string positions is generated and testing continues. If no accepted changes were detected a local optimum has been found, in which case RBC+ opts for a "soft-restart". That is, a random bit is complemented, the change is accepted regardless of the resulting fitness, a new permutation of string positions is generated, and testing continues. These soft-restarts are allowed until $5 \times N$ changes are accepted (including the bit changes that constituted the soft restarts). When RBC+ has exhausted the possibility of a soft-restart it opts for a "hard-restart" generating a new random bit string. This process continues until a given total number of evaluations have been expended. The difference between RBC and RBC+ is the inclusion of soft-restarts in the latter.

3.5 Configurations for Performance Observation

In order to observe and compare the effect on performance of selection, parallel varying-mutation, and recombination, we use the following algorithms: (i) A simple canonical GA with proportional selection and crossover & background mutation (CM), denoted as cGA; (ii) a simple GA with (μ, λ) proportional selection and CM, denoted as $GA(\mu, \lambda)$; (iii) a GA that uses (μ, λ) proportional selection, CM, and parallel varying-mutation SRM, denoted as $GA\text{-}SRM(\mu, \lambda)$; and (iv) a version of $GA\text{-}SRM(\mu, \lambda)$ with no crossover ($p_c = 0.0$), denoted as $M\text{-}SRM(\mu, \lambda)$. To observe the effect of drift the algorithms are used with the fitness duplicates elimination feature either on or off. The superscript ^{ed} attached to the name of a GA indicates that the elimination of duplicates feature is on, i.e. cGA^{ed}, $GA^{ed}(\mu, \lambda)$, $GA\text{-}SRM^{ed}(\mu, \lambda)$, and $M\text{-}SRM^{ed}(\mu, \lambda)$.

The offspring population is set to 200 and all GAs use linear scaling, two-point crossover as the recombination operator and the standard bit-flipping method as the mutation operator after crossover. Linear scaling is implemented as indicated in [59] (p.79), where the coefficients a and b that implement the linear transformation $f' = af + b$ are chosen to enforce equality of the raw (f) and scaled (f') average fitness values and cause the maximum scaled fitness to be twice the average fitness. The number of evaluations is set to 2×10^6 for both GAs and RBC+. The genetic operators parameters used for the GAs are summarized in Table 3.1. In the case of GA-SRM, from our experiments we observed that values of β in the range $1.1 < \beta \leq 2$ produce similar good results. Setting $\beta > 2$ caused a too fast reduction of mutation rate and the algorithm attained inferior results. Fixing $\beta = 2$ we sampled the threshold τ for all combinations of N and K. Results presented here use the best sampled τ.

We conduct our study on NK-landscapes with $N = 96$ bits, varying the number of epistatic interactions from $K = 0$ to $K = 48$ in increments of 4. For each combination of N and K, 50 different problems are randomly generated. We use landscapes with *random* epistatic patterns among bits and consider circular genotypes to avoid boundary effects. The figures plot the mean value of the best solution observed over the 50 problems starting each time with the same initial population. The vertical bars overlaying the mean curves represent 95% confidence intervals. Note that we maximize all problems.

Table 3.1. Genetic algorithms's parameters

Parameter	cGA	$GA(\mu, \lambda)$	$GA\text{-}SRM(\mu, \lambda)$ ADP	$M\text{-}SRM(\mu, \lambda)$ ADP
Selection	Proportional	(μ, λ) Prop.	(μ, λ) Prop.	(μ, λ) Prop.
p_c	0.6	0.6	0.6	0.0
$p_m^{(CM)}$	$1/N$	$1/N$	$1/N$	$1/N$
$p_m^{(SRM)}$	–	–	$[\alpha = 0.5, 1/N]$ $\beta = 2, \ell = N$	$[\alpha = 0.5, 1/N]$ $\beta = 2, \ell = N$
$\lambda_{CM} : \lambda_{SRM}$	–	–	$1 : 1$	$1 : 1$

3.6 Performance Observation, Insights on GA Behavior, and Discussion

3.6.1 Selection

First, we recall some properties of NK-landscapes and observe the effect of selection on the performance of a simple GA. Figure 3.3 plots results by cGA(200) and GA with $(\mu, \lambda)=\{(100,200), (60,200), (30,200)\}$. Results by the random bit climber RBC+ are also included for comparison.

As indicated in Section 3.2, $K = 0$ corresponds to an additive genetic model (no epistatic interactions) that yields a single peaked smooth fitness landscape with its optimum located around 0.66 due to the central limit theorem. All algorithms are expected to find the global optimum in this kind of landscape, which can be seen in Figure 3.3 for $K = 0$.

On the opposite extreme, $K = N - 1$ corresponds to a maximally rugged fully random fitness landscape in which each bit is epistatically affected by all other bits. Results for $K = N - 1$ are not included in this work, but all algorithms are expected to perform equally badly, not better than random search, due to the random nature of the landscape.

As K increases the landscapes change from smooth through a family of increasingly rugged landscapes to fully uncorrelated landscapes. In [40], sampling the landscapes by one-mutant adaptive walks, it was observed that low levels of epistatic interactions increase the number of peaks and seem to bend the landscape yielding optima higher than the $K = 0$ landscape. Further increasing K would make fitness landscapes more multi-peaked, but it would cause a *complexity catastrophe* [40], i.e. the attainable optima fall toward the mean fitness of the landscape. Recent works, using exhaustive search for $N \leq 25$, confirm

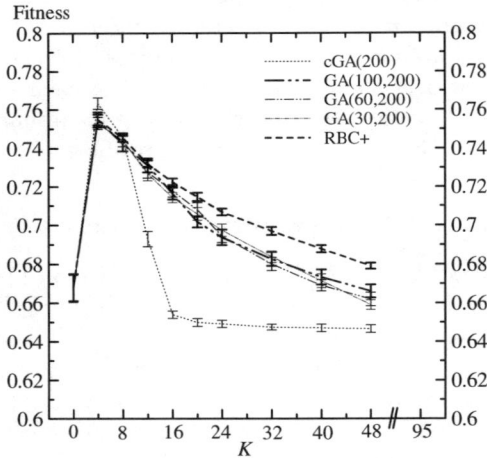

Fig. 3.3. Higher selection pressure

that there is a positive correlation between K and the number of peaks [43] and that the global optima of the landscape increase for small K [51]. Note from Figure 3.3 that the highest optima is for small K. Contrary to [40], however, [43] and [51] show that the global optima remains high as K increases from small to large values. Thus, another important observation from Figure 3.3 is that the decreasing best values found by the algorithms when K increases ($K > 4$) indicates that the search performance is in fact worsening, and not that the values of the attainable optima in the NK-landscapes fall toward the mean fitness of the landscape as erroneously predicted in [40].

For $0 < K < N - 1$, different from $K = 0$ and $K = N - 1$, differences in performance by the GAs can be seen. From Figure 3.3 some important observations regarding the performance of the algorithms are as follows. (i) cGA does relatively well for very low epistasis ($K \leq 8$) but its performance falls sharply for medium and high epistasis ($K \geq 12$). Similar behavior by another simple GA has been observed in [42] and [51]. (ii) GA(μ, λ) that includes an stronger selection pressure performs worse than cGA for low epistasis but it outperforms cGA(200) for medium and high epistasis ($K \geq 12$). The behaviors of GA(100,200), GA(60,200), and GA(30,200) are similar. Results by cGA and GA(μ, λ) indicate the importance of an appropriate selection pressure to pull the population to fittest regions of the search space (note that the genetic operators are the same in both algorithms). The selection pressure induced by proportional selection seems to be appropriate only for very small K. As K increases a stronger selection pressure works better. (iii) The overall performance of RBC+ is better than both cGA and GA(μ, λ).

3.6.2 Duplicates Elimination

Second, we observe the effect of genetic drift by setting on the fitness duplicates elimination feature. Genetic drift emerges from the stochastic operators of selection, recombination, and mutation. Genetic drift refers to the change on bit (allele) frequencies due to chance alone, particularly in small populations where an extreme consequence of it is the homogenization of the population. As shown below, in NK-landscapes, fitness duplicates elimination is an effective way of eliminating clones without calculating hamming distances among the individuals.

Figure 3.4 plots results by cGA^{ed}(200) and GA^{ed}(μ, λ) with (μ, λ)= {(100, 200), (60,200), (30,200)}. Results by cGA(200) and RBC+ are also included for comparison. From this figure we can see that eliminating duplicates affects differently the performance of the GAs. (i) It deteriorates even more the performance of cGA. (ii) On the other hand, the combination of higher selection pressure with duplicates elimination produces a striking increase on performance. Note that all GA^{ed}(μ, λ) algorithms achieved higher optima than RBC+ for $4 \leq K \leq 40$. The optima achieved by GA^{ed}(100,200) is lower than RBC+ for $K = 48$. However, note that GA^{ed}(60,200) and GA^{ed}(30,200) achieved higher optima than RBC+. This suggest that for $K = 48$ even the pressure imposed by (μ, λ)=(100,200) is not strong enough.

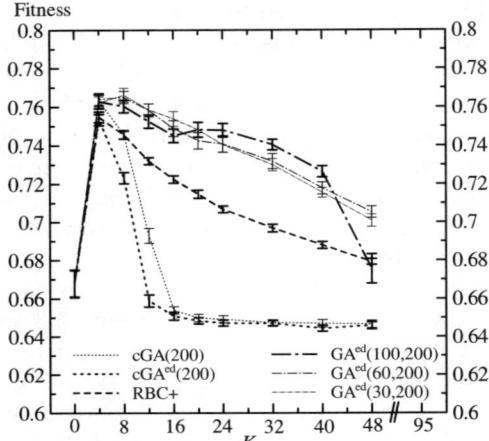

Fig. 3.4. Duplicates elimination

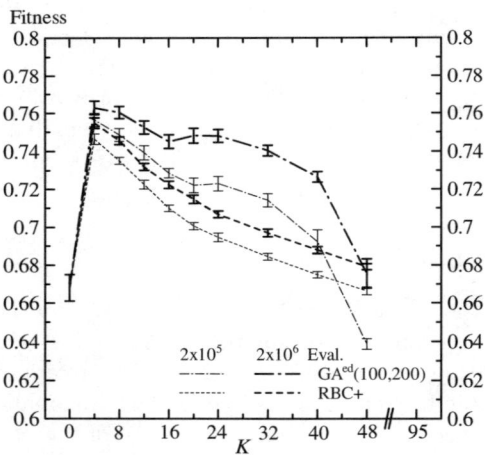

Fig. 3.5. Mean fitness after 2×10^5 and 2×10^6 evaluations

In our study we run the algorithms for 2×10^6 evaluations whereas only 2×10^5 evaluations have been used in previous studies [42, 51]. Figure 3.5 illustrates the optima achieved by $\text{GA}^{ed}(100,200)$ and RBC+ after 2×10^5 and 2×10^6 evaluations. From this figure we can see that allocating more evaluations allows both algorithms to find higher optima, being the rate of improvement by the GA greater than the random bit climber RBC+. Note that for most values of K, even after 2×10^6 evaluations, RBC+ still does not reach the optima achieved by $\text{GA}^{ed}(100,200)$ after 2×10^5 evaluations.

As shown in Section 3.6.1, as K increases a higher selection pressure improves the performance of the simple GA. However, it would also increase the likelihood

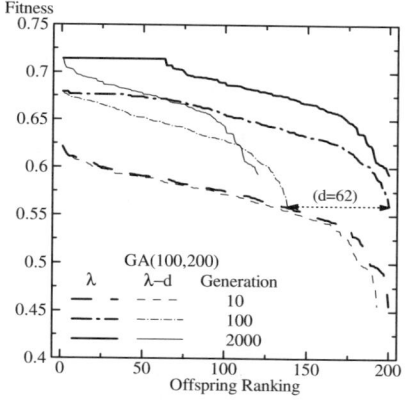

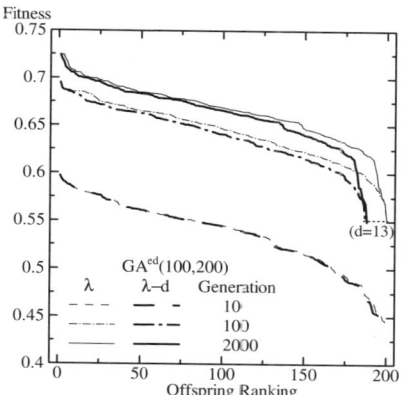

Fig. 3.6. Offspring ranking by GA(μ, λ) ($K = 12$)

Fig. 3.7. Offspring ranking by GA^{ed} (μ, λ) ($K = 12$)

of duplicates. Preventing duplicates distributes more fairly selective pressure in the population, removes an unwanted source of selective bias in the algorithm [58] and postpones genetic drift [60], [61]. GA$^{ed}(\mu, \lambda)$ takes advantage of the higher selection pressure avoiding the unwanted selective bias. The drop in performance by cGA^{ed} compared to cGA suggests that the latter uses the duplicates as a way to increase its selection pressure. In the following we elaborate further on the effects of no eliminating duplicates.

Figure 3.6 illustrates with thick lines the ranking of the λ offspring by the GA(100,200), which does not eliminate duplicates, for generations 10, 100 and 2000 during one run of the algorithm ($K = 12$). The offspring with highest fitness is giving rank 1. The horizontal segments in those lines indicate the presence of duplicates. This figure also illustrates with thin lines what the rank of the $\lambda - d$ not duplicates offspring would be, where d is the number of duplicates. From this figure we can see that if duplicates are not eliminated they accumulate rapidly. In this example, the number of duplicates at generation 10 is 7, which increases to 62 at generation 100 and to 80 at generation 2000. Figure 3.7 presents a similar plot for GA^{ed}(100,200), which does eliminate duplicates. In this case the thick (thin) lines indicate the ranking of the offspring after (before) duplicates elimination. From this figure we can see that, compared to GA(100,200), eliminating duplicates at each generation prevent them from increasing their number. Note that at generation 100 there are only 13 duplicates and that at generation 2000 their number remains similar (this effect can be observed with more detail in Figure 3.9 for various values of K). The plots in Figure 3.6 can be interpreted as the cumulative effect of drift on the population composition, whereas the plots in Figure 3.7 as the instantaneous effect of it. An important conclusion from both figures is that by eliminating duplicates we can increase substantially the likelihood that the algorithm will explore a larger number of different candidate solutions, which increases the possibility of finding higher optima.

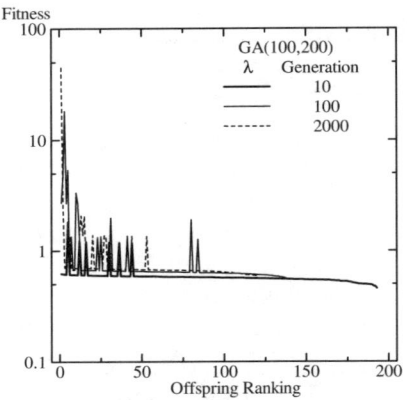

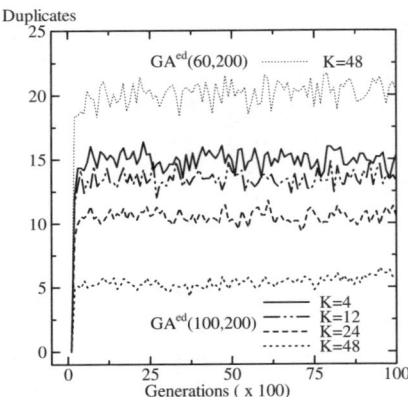

Fig. 3.8. Accumulated fitness of the unique genotype offspring, $(K = 12)$

Fig. 3.9. Number of eliminated duplicates

Another important aspect of duplicates is related to selection. In the case of algorithms that do not eliminate duplicates, such GA(100,200), selection of parents will be based on the ranking of the all λ offspring as shown by the thicker lines in Figure 3.6, which contains ensembles of clones (i.e. individuals with the same genotype besides having the same fitness). From a genotype uniqueness point of view, each ensemble of clones represents one individual. However, the selection mechanism will assign a selection probability to each clone, the same for all clones within an ensemble, as if they were different individuals. As a consequence, the chances of selecting a given genotype are multiplied by the number of clones of that genotype present in the offspring population. To illustrate this better, Figure 3.8 plots the fitness of unique genotypes when duplicates are not eliminated. Here, an ensemble of clones is treated as one individual and its fitness is the accumulated fitness of the clones in the ensemble. In the figure, ensembles of clones can be clearly recognized by the peaks in the curves. This figure clearly indicates that low fitness genotypes (due to the duplicates effect) can end up with higher selective advantage than high fitness unique genotypes. It should be noted that this unwanted selective bias, which is not based in actual fitness, cannot be avoided by fitness scaling mechanisms, ranking procedures, or even extinctive (truncated) deterministic selection schemes (such (μ, λ) Selection).

In the case of eliminating duplicates, selection of parents will be based on the ranking of the $\lambda - d$ unique genotype offspring as shown by the thicker lines in Figure 3.7. In this case, selection will depend exclusively on the fitness of the individuals.

Figure 3.9 illustrates the number of fitness duplicates over the generations eliminated by $GA^{ed}(100,200)$ for landscapes with values of $K = \{4, 12, 24, 48\}$ and the number of duplicates eliminated by $GA^{ed}(60,200)$ for $K = 48$. Most of these fitness duplicates were actual clones. For $K = 4$, as an example, 148,993 fitness duplicates were created during the 10,000 generations (average in the 50 runs). Out of these, 99.88% corresponded to actual clones. Similar percentages were observed for other values of K.

3.6.3   Parallel Varying-Mutation

Third, we observe the effect of parallel varying-mutation. Figure 3.10 plots results by GA-SRM^{ed}(100;200). Results by cGA(200), GA^{ed}(100,200) and RBC+ are also included for comparison. From this figure the following observations are relevant. (i) The inclusion of parallel varying-mutation can improve further convergence reliability. Note that GA-SRM^{ed}(100,200) with ADP achieves higher optima than GA^{ed}(100,200) for $4 \leq K < 32$. For $K = 32$ and $K = 40$, however, GA-SRM^{ed}(100,200) is not better than GA^{ed}(100,200), which indicates that varying-mutation is not working properly at this values of K. (ii) The optima achieved by GA^{ed}(100,200) is lower than RBC+ for $K = 48$, which seems to be caused by a lack of appropriate selection pressure as mentioned in Section 3.6.2. However, for the same selection pressure, GA-SRM^{ed}(100,200) achieved higher optima than RBC+. This nicely shows that the effectiveness of a given selection pressure is not only correlated to the complexity of the landscape, but also to the effectiveness of the operators searching in that landscape. In the case of GA-SRM^{ed}(100,200), the inclusion of varying-mutation increases the effectiveness of the operators, hiding selection deficiencies. It would be better to correct selection pressure and try to use mutation to improve further the search.

3.6.4   No Crossover

Next, we observe the effect of (not) using recombination. Figure 3.11 plots results by M-SRM^{ed}(100,200), which is a GA-SRM^{ed}(100,200) with crossover turned off. Results by GA-SRM^{ed}(100,200), GA^{ed}(100,200), and RBC+ are also included for comparison. From this figure the following observations are important. (i) For $K \leq 12$ the mutation-only algorithm M-SRM^{ed}(100,200) performs similar or better than GA-SRM^{ed}(100,200) that includes crossover. For some instances of other combinatorial optimization problems it has also been shown that a mutation-only evolutionary algorithm can produce similar or better results with higher efficiency than a GA that includes crossover, see for example [62]. For $K \geq 16$, GA-SRM that includes both crossover and parallel varying-mutation achieves higher optima; note that the difference between GA-SRM and M-SRM increases with K. (ii) Similar to GA-SRM, the mutation-only algorithm M-SRM achieves higher optima than RBC+ for $K \geq 4$, which illustrates the potential of evolutionary algorithms, population based, with or without recombination, over strictly local search algorithms in a broad range of classes of problems. This is in accordance with theoretical studies of first hitting time of population-based evolutionary algorithms. He and Yao [63] have shown that in some cases the average computation time and the first hitting probability of an evolutionary algorithm can be improved by introducing a population.

The behavior of M-SRM compared with GA-SRM is at first glance counterintuitive and deserves further explanation. Results in [42] imply that the usefulness

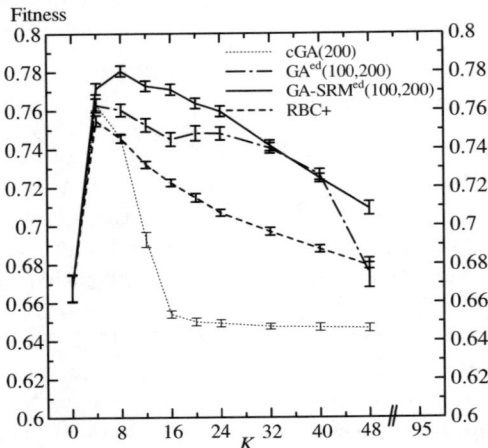

Fig. 3.10. Parallel varying-mutation

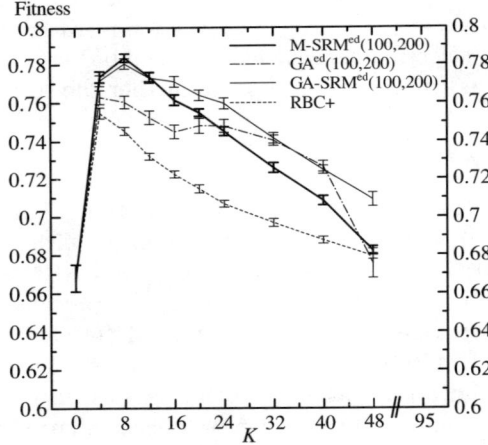

Fig. 3.11. No recombination, M-SRM^{ed}(100,200)

of recombination would decrease with K. However, Figure 3.11 seems to imply exactly the opposite. A sensible explanation for these apparently opposing results comes from the structure of the underlying fitness landscape. Kauffman [40] has shown that clustering of high peaks can arise as a feature of landscapes structure without obvious modularity, as is the case of NK-landscapes with K epistatic inputs to each site chosen at random among sites. In this case, recombination would be a useful search strategy because the location of high peaks carry information about the location of the other high peaks [40].

In the case of small K the problems are the easier and a mutation-only EA proves to be very effective, although the landscape is more structured than for high K. As K increases the structure of the landscape fades away, decreasing also the effectiveness of recombination. However, what figure Figure 3.11 is showing is that the decay of mutation alone seems to be faster than the decay of recombination interacting with varying-mutation as the complexity of the landscape increases with K and its structure fades. In other words, the relative importance of recombination interacting with varying-mutation increases with K respect to mutation alone. In this work we used two-point crossover as the recombination operator. It would be interesting to investigate the effects of other recombination operators, such as one-point and uniform crossover.

3.7 Epitasis Pattern and Mutation Strategy

Finally, the effect of the mutation strategy used in parallel varying-mutation is investigated on landscapes with different patterns of epistasis when the fitness duplicates feature is on. Figure 3.12 and Figure 3.13 plots results for NK-landscapes with *random* and *nearest neighbour* pattern of epistasis by GA-SRM^{ed}(100,200) with either ADS or ADP. Results by the other algorithms are also included for comparison. From the figures, we can see that mutation strategy seems not to have effect in landscapes with *random* pattern of epistasis when the fitness duplicates feature is on. Note that performance by GA-SRM^{ed}(100,200) with ADS or ADP is very similar for any value of K. However, for *nearest neighbour* it is observed that for $4 \leq K < 32$ the strategy that mutates within a continue mutation segment, i.e. ADS, performs better than the strategy that mutates any bit of the chromosome, i.e. ADP.

Looking at Figure 3.12 and Figure 3.13 it can be seen that the behavior of RBC+ is similar in both kinds of landscapes with the exception of $K = 4$. However, note that the behavior of cGA is more robust in landscapes

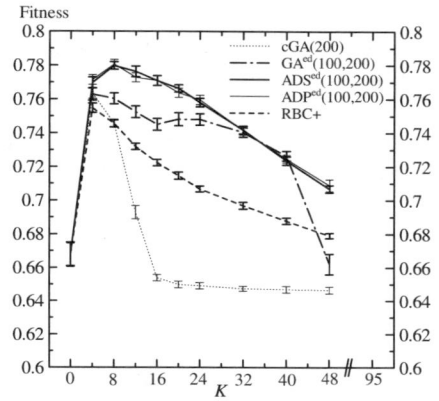

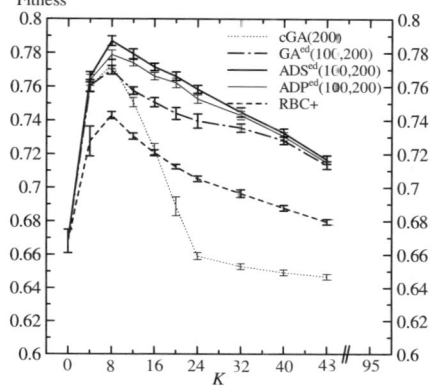

Fig. 3.12. Random epistasis   **Fig. 3.13.** Nearest neighbour epistasis

with *nearest neighbour* than *random* pattern of epistasis. Similarly, it can be seen that higher fitness is achieved by GA-SRM on *nearest neighbour* than on *random* landscapes. These results are in accordance with Altenberg [47], who suggests that GA-friendlier representations could be evolved in order to facilitate good performance. Precisely, the better performance by ADS shows that a mutation strategy appropriate to the representation could further improve performance.

3.8 Conclusions

We have examined the effects of selection, drift, mutation, and recombination on NK-landscapes with $N = 96$ bits, varying K from 0 to 48 on landscapes with *random* patterns of epistasis. We showed that GAs can be robust algorithms on NK-landscapes postponing drift by eliminating fitness duplicates and using a selection pressure stronger than a canonical GA. Even simple GAs with these two features performed better than the single bit climber RBC+ for a broad range of classes of problems ($K \geq 4$). We also showed that the interaction of parallel varying-mutation with crossover (GA-SRM) improves further the reliability of the GA. Contrary to intuition we found that a mutation-only EA can perform as well as GA-SRM that includes crossover for small values of K, where crossover is supposed to be advantageous. However, we found that the decay of mutation alone seems to be faster than the decay of crossover interacting with varying-mutation as the complexity of the landscape increases with K and its structure fades. In other words, the relative importance of recombination interacting with varying-mutation increased with K respect to mutation alone. Finally, we showed that performance can be improved on landscapes with *nearest neighbour* patterns of epistasis by using a segment mutation strategy.

NK-landscapes are useful for testing the GA's overall performance on a broad range of sub-classes of epistatic problems and also for testing each one of the major processes involved in a GA. This work gives good insights on how to improve GAs by understanding better the behavior that arises from the interaction of all major processes involved in the algorithm. It is also a valuable guide to practitioners on how to configure properly the GA to achieve better results as the non-linearity and complexity of the problem increases.

In this work we have focused on single objective fitness landscapes. At present, there is work in progress on the performance of evolutionary algorithms on multi-objective epistatic landscapes. We will report these results in the near future. As future works, we would like to extend our study including a comparison with algorithms that incorporate linkage learning. Also, we should look deeper into the relationship among selection pressure, the range for varying-mutation, and K to incorporate heuristics that can increase the adaptability of GAs according to the complexity of the problem.

4

Engineering Stochastic Local Search Algorithms: A Case Study in Estimation-Based Local Search for the Probabilistic Travelling Salesman Problem

Prasanna Balaprakash, Mauro Birattari, and Thomas Stützle

IRIDIA, CoDE, Université Libre de Bruxelles, Brussels, Belgium
{pbalapra,mbiro,stuetzle}@ulb.ac.be

Summary. In this article, we describe the steps that have been followed in the development of a high performing stochastic local search algorithm for the probabilistic travelling salesman problem, a paradigmatic combinatorial stochastic optimization problem. In fact, we have followed a bottom-up algorithm engineering process that starts from basic algorithms (here, iterative improvement) and adds complexity step-by-step. An extensive experimental campaign has given insight into the advantages and disadvantages of the prototype algorithms obtained at the various steps and directed the further algorithm development. The final stochastic local search algorithm was shown to substantially outperform the previous best algorithms known for this problem. Besides the systematic engineering process for the development of stochastic local search algorithms followed here, the main reason for the high performance of our final algorithm is the innovative adoption of techniques for the estimation of the cost of neighboring solutions using delta evaluation.

Keywords: Stochastic Local Search, Algorithm Engineering, Stochastic Optimization Problems, Probabilistic Travelling Salesman Problem, Estimation-Based Local Search.

4.1 Introduction

Stochastic local search (SLS) algorithms are powerful tools for tackling computationally hard decision and optimization problems that arise in many application areas. The field of SLS is rather vast and there exists a large variety of algorithmic techniques and strategies. They range from simple constructive and iterative improvement algorithms to general purpose SLS methods such as iterated local search, tabu search, and ant colony optimization.

A frequently used approach for the development of SLS algorithms appears to be that the algorithm designer takes his favorite general-purpose SLS method. Then, this method is adapted in a sometimes more, often less organized way to the problem under consideration, guided by the researcher's intuitions and previous experiences.

There is an increasing awareness that the development of SLS algorithms should be done in a more structured way, ideally in an engineering style

C. Cotta and J. van Hemert (Eds.): Recent Advances in Evol. Comp., SCI 153, pp. 53–66, 2008.
springerlink.com   © Springer-Verlag Berlin Heidelberg 2008

following a well-defined process and applying a set of well established procedures. One possibility is to follow a bottom-up engineering process that starts from basic algorithms and adds complexity step-by-step. Such a process could be organized as follows. First, start by analyzing the problem under concern and gaining insights into its structure and possibly previous work on it. Second, develop basic constructive and iterative improvement algorithms and analyze them experimentally. Third, integrate these basic heuristics into simple SLS methods to improve upon their performance. Fourth, if deemed necessary, add more advanced concepts to extend the SLS algorithm by, for example, consider populations. Clearly, such an approach is to be considered an iterative one where insights gained through experiments at one level may lead to further refinements at the same or at previous levels.

In this chapter, we illustrate the steps that we have followed in the development of new state-of-the-art algorithms for the PROBABILISTIC TRAVELLING SALESMAN PROBLEM (PTSP). In fact, we were following a bottom-up approach, which in this case was mainly focused on the implementation and refinement of a very effective iterative improvement algorithm. As it will be shown later, this process was supported by a comprehensive experimental analysis, the usage of the automatic tuning of some algorithm parameters, an efficient implementation of supporting data structures, and an integration of the iterative improvement algorithm into an iterated local search.

The high-level steps of the adopted bottom-up approach materialize in the following main elements. The main underlying ideas contributing to the success of this research were (i) the inclusion of speed-up techniques known from the deterministic TSP to the PTSP local search algorithms, and (ii) the use of empirical estimation in the evaluation of local search moves. These ideas have been implemented into a new estimation-based iterative improvement algorithm for the PTSP. Experimental results with the final iterated local search algorithm show that we actually have obtained a new state-of-the-art algorithm that outperforms the previous best algorithm for this problem.

4.2   The Probabilistic Travelling Salesman Problem

The PTSP [70] is a paradigmatic example of routing problems under uncertainty. It is similar to the TSP with the only difference that each node has a probability of requiring being visited. The *a priori* optimization approach [71] for the PTSP consists in finding an *a priori* solution that visits all the nodes and that minimizes the expected cost of *a posteriori* solutions. The *a priori* solution must be found prior to knowing which nodes are to be visited. The associated *a posteriori* solution is computed *after* knowing which nodes need to be visited. It is obtained by skipping the nodes that do not require being visited and visiting the others in the order in which they appear in the *a priori* solution. An illustration is given in Figure 4.1. Formally, a PTSP instance is defined on a complete graph $G = (V, A, C, P)$, where $V = \{1, 2, \ldots, n\}$ is a set of nodes, $A = \{\langle i, j \rangle : i, j \in V, i \neq j\}$ is the set of edges that completely connects the

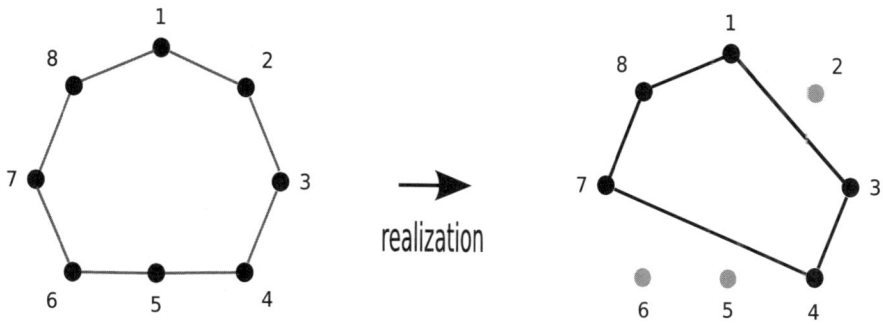

Fig. 4.1. An *a priori* solution for a PTSP instance with 8 nodes. The nodes in the *a priori* solution are visited in the order 1, 2, 3, 4, 5, 6, 7, 8, and 1. Assume that a realization of ω prescribes that nodes 1, 3, 4, 7, and 8 are to be visited. The *a posteriori* solution visits then nodes in the order in which they appear in the *a priori* solution and skips nodes 2, 5, and 6 since they do not require being visited.

nodes, $C = \{c_{ij} : \langle i, j \rangle \in A\}$ is a set of costs associated with edges, and $P = \{p_i : i \in V\}$ is a set of probabilities that for each node i specifies its probability p_i of requiring being visited. Hence, for the PTSP the stochastic element of the problem is defined by a random variable ω that is distributed according to an n-variate Bernoulli distribution and a realization of ω is a binary vector of size n where a '1' in position i indicates that node i requires being visited and a '0' indicates that it does not. We assume that the costs are symmetric. The goal in the PTSP is to find an *a priori* solution that minimizes the expected cost of the *a posteriori* solution, where the expectation is computed with respect to the given n-variate Bernoulli distribution.

4.3   Iterative Improvement Algorithms for the PTSP

Iterative improvement algorithms start from some initial solution and repeatedly try to move from a current solution x to a lower cost neighboring one. An iterative improvement search terminates in a local optimum, that is, a solution that does not have any improving neighbor. In the PTSP literature, mainly *2-exchange* and *node-insertion* neighbourhood structures were considered (see Figure 4.2 for examples).

Crucial to the performance of many iterative improvement algorithms is the possibility of performing *delta evaluation*, that is, of computing the cost of a neighbor by only considering the cost contribution of the solution components in which the two solutions differ. For example, in the case of a *2-exchange* move that deletes edges $\langle a, b \rangle$ and $\langle c, d \rangle$ and adds edges $\langle a, c \rangle$ and $\langle b, d \rangle$, the cost difference is given by $c_{a,c} + c_{b,d} - c_{a,b} - c_{c,d}$.

2-p-opt and **1-shift**, the current state-of-the-art iterative improvement algorithms for the PTSP, use analytical computations, that is, closed-form expressions based on heavy mathematical derivations, for correctly taking into account

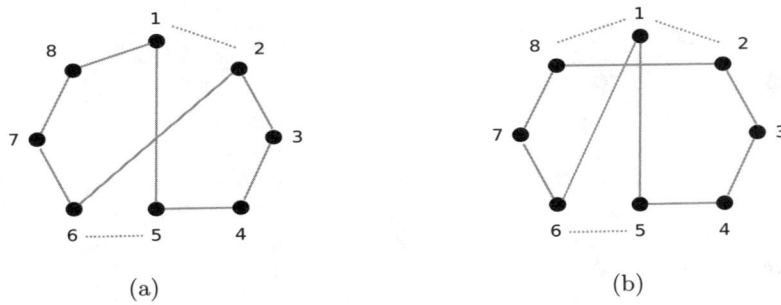

Fig. 4.2. Plot 2(a) shows a *2-exchange* move that is obtained by deleting two edges $\langle 1, 2 \rangle$ and $\langle 5, 6 \rangle$ of the solution and by replacing them with $\langle 1, 5 \rangle$ and $\langle 2, 6 \rangle$. Plot 2(b) shows a *node-insertion* move obtained by deleting node 1 from its current position in the solution and inserting it between nodes 5 and 6.

the random variable ω in such a *delta evaluation* [72, 73, 74]. Unfortunately, to allow the re-use of already computed expressions and, thus, to make the *delta evaluation* more efficient, 2-p-opt and 1-shift require to scan the neighbourhood in a fixed lexicographic order.

4.4 Engineering an Iterative Improvement Algorithm for the PTSP

We now present the main steps that we followed for engineering an iterative improvement algorithm for the PTSP. It is based on two main hypotheses.

Hypothesis 1: Exploiting known TSP speed-up techniques can increase computation speed.

Hypothesis 2: We can use estimations of the costs for the *delta evaluation* to make the local search faster.

4.4.1 Introducing TSP Speed-Up Techniques

Iterative improvement algorithms for the TSP strongly exploit neighbourhood reduction techniques such as fixed radius search, candidate lists, and don't look bits [75, 76]. These techniques allow the local search to focus on the most relevant part for obtaining improvements by pruning a large part of the neighbourhood. While reducing very strongly the number of neighbors, the exploitation of these techniques does not allow to scan the neighbourhood in the lexicographic order, which is required to use the speed-ups of the *delta evaluation* as it is used in 2-p-opt and 1-shift.

We implemented an iterative improvement algorithm using the three above mentioned TSP speed-up techniques. For doing so, we have to compute the cost difference between two neighboring solutions from scratch. This is done by using the closed-form expressions proposed for the *2-exchange* and the *node-insertion* neighbourhood structures [74]. In fact, we implemented an iterative improvement algorithm based on the *2.5-exchange neighbourhood* [77]. The *2.5-exchange neighbourhood* is well-known in TSP solving and it combines the *node-insertion neighbourhood* and the *2-exchange neighbourhood* structure into a hybrid one. We call the resulting algorithm 2.5-opt-ACs, where AC and s stand for *analytical computation* and *speedup*, respectively.

2.5-opt-ACs was experimentally compared to 2-p-opt and 1-shift. Our analysis is based on PTSP instances that we obtained from TSP instances generated with the DIMACS instance generator [78]. They are homogeneous PTSP instances, where all nodes of an instance have a same probability p of appearing in a realization. We carried out experiments on clustered instances of 300 nodes, where the nodes are arranged in a number of clusters, in a $10^6 \times 10^6$ square. We considered the following probability levels: $[0.1, 0.9]$ with a step size of 0.1. All algorithms were implemented in C and the source codes were compiled with gcc, version 3.3. Experiments were carried out on AMD Opteron^{TM}244 1.75 GHz processors with 1 MB L2-Cache and 2 GB RAM, running under the Rocks Cluster GNU/Linux. Each iterative improvement algorithm is run until it reaches a local optimum. (In the following we present only some of the most important results; more details are given in [79].)

The results given in Figure 4.3, which illustrate the development of the average cost obtained, show that 2.5-opt-ACs *dominates* 2-p-opt and 1-shift. Irrespective of the probability value, 2.5-opt-ACs reaches local optima about four times faster than 2-p-opt. Compared to 1-shift, the same holds when $p \geq 0.5$; for small p, however, the speed difference between 2.5-opt-ACs and 1-shift is small. Concerning the average cost of local optima found, 2.5-opt-ACs is between 2% and 5% better than 2-p-opt. The same trend is true when compared to 1-shift, except for $p \leq 0.3$, where the difference between 2.5-opt-ACs and 1-shift is small. All the observed cost differences are statistically significant, as shown by the *t*-test; an exception is for $p \leq 0.2$, where the difference between 2.5-opt-ACs and 1-shift is not significant.

4.4.2   Estimation-Based Local Search

Empirical estimation is a technique for estimating the expectation through Monte Carlo simulation. Empirical estimation can also be applied to estimate the cost for the *delta evaluation*. This has the advantage that one can use any neighbourhood structure without requiring complex mathematical derivations. In particular, the *2.5-exchange neighbourhood* can easily be used. Clearly, an estimation-based *delta evaluation* procedure also does not impose constraints on the order of exploring the neighbourhood. Thus, it is easy to integrate the TSP neighbourhood reduction techniques. Given these advantages, our hypothesis is that the estimation-based *delta evaluation* procedure can lead to a very fast and highly effective iterative improvement algorithms.

In empirical estimation, an unbiased estimator of $F(x)$ for a solution x can be computed using M independent realizations of the random variable $\boldsymbol{\omega}$ and

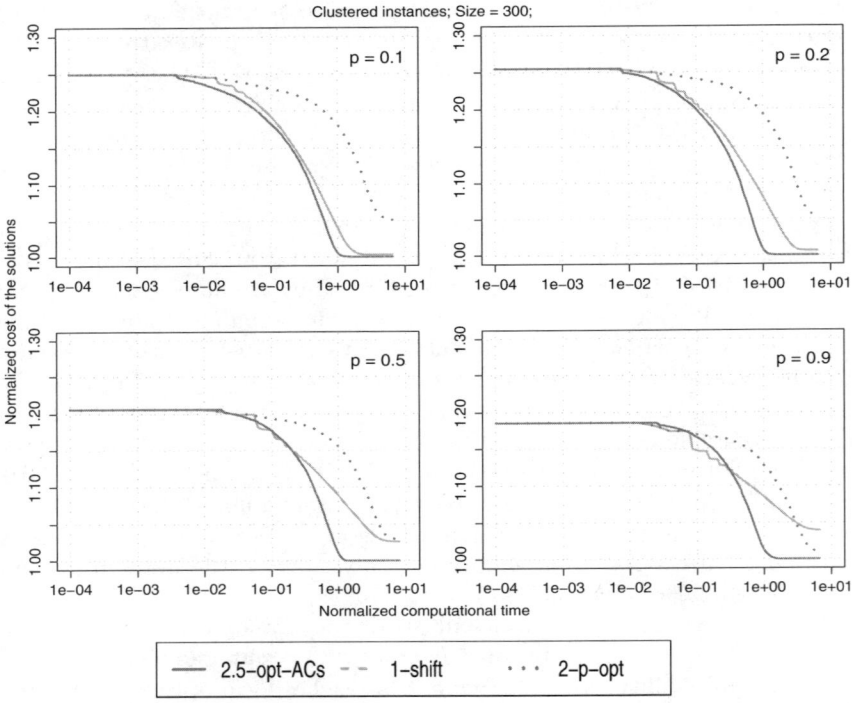

Fig. 4.3. Experimental results on clustered homogeneous PTSP instances of size 300. The plots represent the development of the average cost of the solutions obtained by 2-p-opt and 1-shift normalized by the one obtained by 2.5-opt-ACs. Each algorithm is stopped when it reaches a local optimum.

the resulting costs of these *a posteriori* solutions [80, 79]. The cost difference between a solution x and a neighboring solution x' can then be estimated by $\hat{F}_M(x') - \hat{F}_M(x)$, which is given by:

$$\hat{F}_M(x') - \hat{F}_M(x) = \frac{1}{M} \sum_{r=1}^{M} \Big(f(x', \omega_r) - f(x, \omega_r) \Big). \tag{4.1}$$

We use the same M realizations for all steps of the iterative improvement algorithms. Alternatively, one may sample for each comparison M new realizations; however, this was proven to be not effective in our experiments (for details see [79]).

For estimating the cost differences between two neighboring *a priori* solutions by a realization ω, one needs to identify the edges that are not common in the two *a posteriori* solutions. This is done as follows. For every edge $\langle i, j \rangle$ that is deleted from x, one needs to find the corresponding edge $\langle i^*, j^* \rangle$ that is deleted in the *a posteriori* solution of x. This so-called *a posteriori edge* is obtained

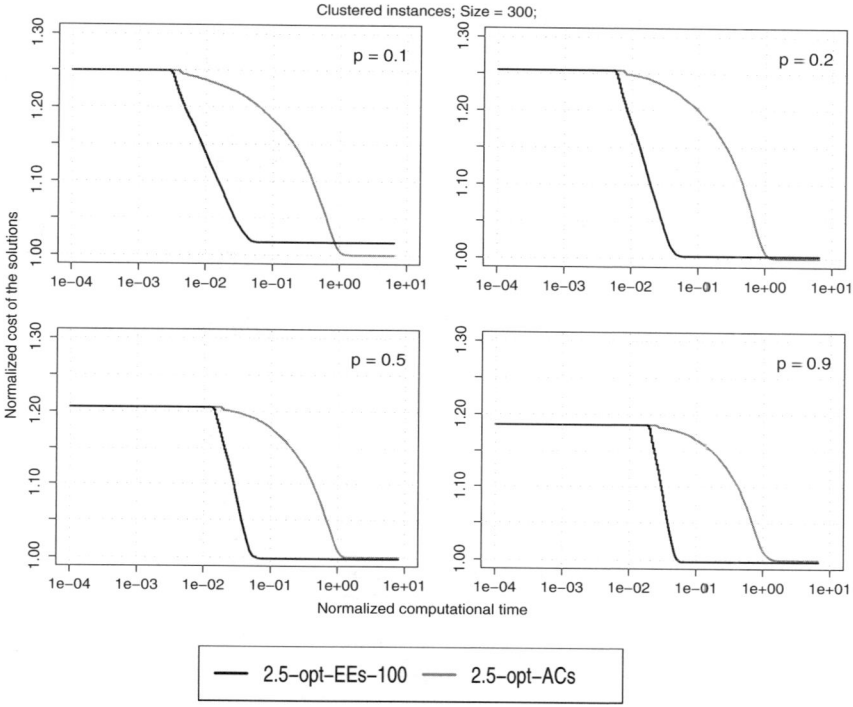

Fig. 4.4. Experimental results on clustered homogeneous PTSP instances of size 300. The plots represent the average solution cost obtained by `2.5-opt-EEs-100` normalized by the one obtained by `2.5-opt-ACs`. Each algorithm is stopped upon reaching a local optimum.

as follows. If node i requires being visited, we have $i^* = i$; otherwise, i^* is the first predecessor of i in x such that $\omega[i^*] = 1$, that is, the first predecessor that requires being visited. If node j requires being visited, then $j^* = j$; otherwise, j^* is the first successor of j such that $\omega[j^*] = 1$. In a *2-exchange* move that deletes the edges $\langle a, b \rangle$ and $\langle c, d \rangle$ from x and replaces them by $\langle a, c \rangle$ and $\langle b, d \rangle$, hence, first the corresponding *a posteriori edges* $\langle a^*, b^* \rangle$ and $\langle c^*, d^* \rangle$ for a given realization ω are to be identified. The cost difference between the two *a posteriori* solutions is then given by $c_{a^*,c^*} + c_{b^*,d^*} - c_{a^*,b^*} - c_{c^*,d^*}$. and Eq 4.1 then simply sums the cost differences for each of the M realizations. This procedure can be directly extended to *node-insertion* moves. Furthermore, the algorithm adopts the neighbourhood reduction techniques *fixed-radius search, candidate lists*, and *don't look bits* [81, 77, 75]. We call the resulting first-improvement algorithm `2.5-opt-EEs`. (See [79] for more details).

As a default, we use 100 realizations in `2.5-opt-EEs`, which is indicated by denoting this version as `2.5-opt-EEs-100`. Next, we compare `2.5-opt-EEs--100` with `2.5-opt-ACs`; these two algorithms differ only in the *delta evaluation*

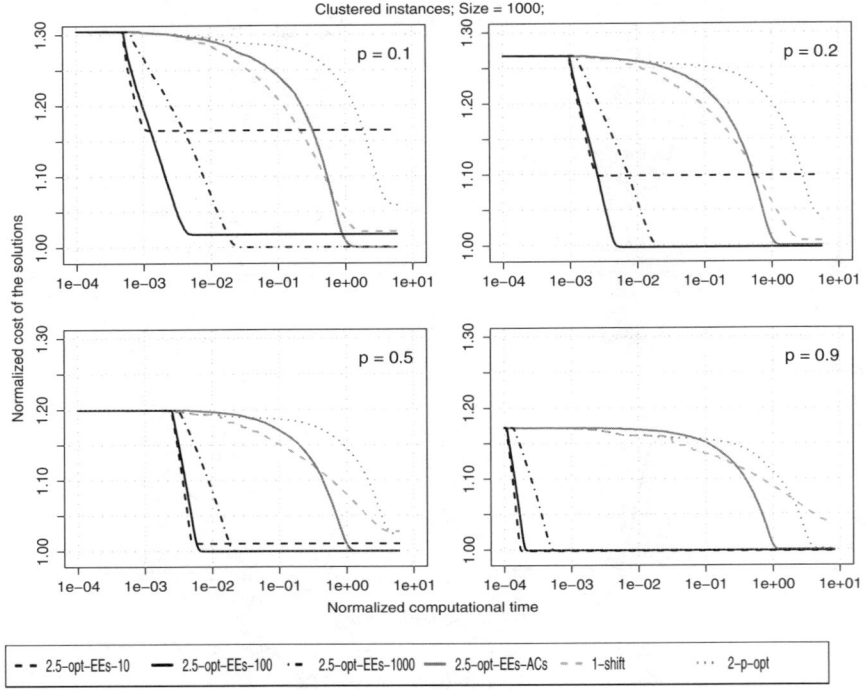

Fig. 4.5. Experimental results on clustered homogeneous PTSP instances of size 1000. The plots give the solution cost obtained by 2.5-opt-EEs-10, 2.5-opt-EEs--100, 2.5-opt-EEs-1000, 1-shift, and 2-p-opt normalized by the one obtained by 2.5-opt-ACs. Each algorithm stops upon reaching a local optimum.

procedure: *empirical estimation* versus *analytical computation*. Figure 4.4 gives the experimental results. Both algorithms reach similar average costs with the only exception of $p = 0.1$, where 2.5-opt-EEs-100 returns local optima with an average cost that is approximately 2% higher than that of 2.5-opt-ACs. However, 2.5-opt-EEs-100 is much faster than 2.5-opt-ACs; it reaches local optima, irrespective of the probability value, approximately 1.5 orders of magnitude faster.

The poorer solution cost of 2.5-opt-EEs-100 for $p = 0.1$ can be attributed to the number of realizations used to estimate the cost difference between two solutions. Since the variance of the cost difference estimator is very high at low probability levels, the adoption of 100 realizations is not sufficient to obtain a good estimate of the cost difference. We therefore added experiments to examine the impact of the number of realizations considered on the performance of 2.5-opt-EEs. For this purpose, we consider samples of size 10, 100, and 1000 and we denote the algorithms 2.5-opt-EEs-10, 2.5-opt-EEs-100, and 2.5-opt-EEs-1000. We considered PTSP instances with 1000 nodes, which are generated in the same way as described before, to study the performance of the algorithms when applied to large instances. (For a PTSP instance size of 1000, 2.5-opt-ACs, 2-p-opt, and 1-shift suffer from numerical problems and they need to resort to an arbitrary precision arithmetic for $p > 0.5$ [79], which makes them even slower.) The results are given in Figure 4.5.

As conjectured, the use of a large number of realizations, in our case $M = 1000$, is effective with respect to the cost of the local optima for low probability values. Even though larger sample sizes incur more computation time, the total search time is very short compared to the *analytical computation* algorithms. On the other hand, the use of few realizations, in our case $M = 10$, does not lead to a further very significant reduction of computation time: concerning the average computation time, 2.5-opt-EEs-10 is faster than 2.5-opt-EEs--100 approximately by a factor of two, while 2.5-opt-EEs-1000 is slower than 2.5-opt-EEs-100 by a factor of four. Nonetheless, an important observation is that 2.5-opt-EEs-1000 is approximately 1.5 orders of magnitude faster than 2.5-opt-ACs. Concerning the average solution cost, 2.5-opt-EEs-1000 is similar to 2.5-opt-EEs-100 and 2.5-opt-ACs with the exception of $p = 0.1$, where the average cost of the local optima obtained by 2.5-opt-EEs-1000 is approximately 3% lower than that of 2.5-opt-EEs-100 and comparable with the one of 2.5-opt-ACs. 2.5-opt-EEs-10 reaches clearly much worse costs than the other estimation-based algorithms; only for high probability values it appears to be competitive.

With respect to the instance size, the trends concerning the relative performance of 2.5-opt-EEs-100 and 2.5-opt-ACs are similar as for instances of size 300. However, the differences between the computation times of the two algorithms are larger and 2.5-opt-EEs-100 reaches, irrespective of the value of p, local optima approximately 2.3 orders of magnitude faster than 2.5-opt-ACs. Similarly, for the comparison between 2.5-opt-ACs and 1-shift and 2-p-opt, respectively, the results for instance size 1000 are analogous to those for instance size 300.

4.4.3 Improving the Estimation-Based Local Search

The results of the previous section clearly show that the performance of 2.5-opt-EEs depends on the number of realizations and the probabilities associated with the nodes. In particular, for low probability values, 2.5-opt-EEs is less effective with few realizations. This insight led to our third hypothesis:

Hypothesis 3: 2.5-opt-EEs can be improved by choosing an appropriate sample size and a special treatment of the low probability cases.

Two main ideas were developed to improve 2.5-opt-EEs. The first is to use an *adaptive sampling procedure* that selects the appropriate number of realizations with respect to the variance of the cost estimator. In fact, as shown in [82], the variance of the cost of the *a posteriori* solutions depends strongly on the probabilities associated with the nodes: the smaller the probability values, the higher the variance. The increased variance could be handled by increasing the sample size since averaging over a large number of realizations reduces the variance of the estimator. However, for high probability values using a large number of realizations is a waste of time. For addressing this issue, we adopted an adaptive sampling procedure that makes use of *Student's t-test* in the following way: Given two neighboring *a priori* solutions, the cost difference between their corresponding *a posteriori* solutions is sequentially computed on a number of realizations. Once the *t-test* rejects the null hypothesis of zero cost difference, the computation is stopped; if the null hypothesis cannot be rejected, the computation is stopped after a maximum of M realizations, where M is a parameter. The sign of the estimator determines the solution of lower cost. The *estimation-based* iterative improvement algorithm that adds adaptive sampling to 2.5-opt-EEs will be called 2.5-opt-EEas.

The second main idea is to adopt the *importance sampling* technique for reducing the variance of the estimator when dealing with highly stochastic problem instances: given two neighboring *a priori* solutions, this technique considers, instead of realizations from the given distribution ω, realizations from another distribution ω^*—the so-called biased distribution. ω^* forces the nodes involved in the cost difference computation to occur more frequently. The resulting cost difference between two *a posteriori* solutions for each realization is corrected for the adoption of the biased distribution and the correction is given by the likelihood ratio of the original distribution with respect to the biased distribution. We denote 2.5-opt-EEais the algorithm that adds to 2.5-opt-EEas the above described importance sampling procedure. Note that the adoption of the importance sampling technique in 2.5-opt-EEas requires additional parameters for defining the biased distribution. These parameters have been tuned by Iterative F-Race [83]. We refer the reader to [82] for a more detailed description of 2.5-opt-EEais and its tuning.

Next, we compared the performance of 2.5-opt-EEas and 2.5-opt-EEais to 2.5-opt-EEs, which does not use an adaptive sample size and importance sampling. In the case of 2.5-opt-EEs, we again consider samples of size 10, 100, and 1000. The clustered instances of 1000 nodes are used for the experiments with the probability levels [0.050, 0.200] with a step size of 0.025 and [0.3, 0.9] with a step size of 0.1. Results of the comparison are given in Figure 4.6, where 2.5-opt-EEs-1000 is taken as a reference. (We only highlight the most important results; more details are given in [82].)

The computational results show that, especially for low probabilities, 2.5-opt-EEais is more effective than the other algorithms. For what concerns the comparison of 2.5-opt-EEais and 2.5-opt-EEas, the results show that the adoption of importance sampling allows the former to achieve high quality solutions for very low probabilities, that is, for $p < 0.2$—the average cost of the local optima obtained by 2.5-opt-EEais is between 1% and 3% less than that of 2.5-opt-EEas. The observed differences are significant according to the paired *t*-test. For $p \geq 0.3$, the average solution cost and the computation time of 2.5-opt-EEais are comparable to the ones of 2.5-opt-EEas.

Concerning the comparison to 2.5-opt-EEs, the following results are most noteworthy. 2.5-opt-EEais reaches an average solution similar to that of

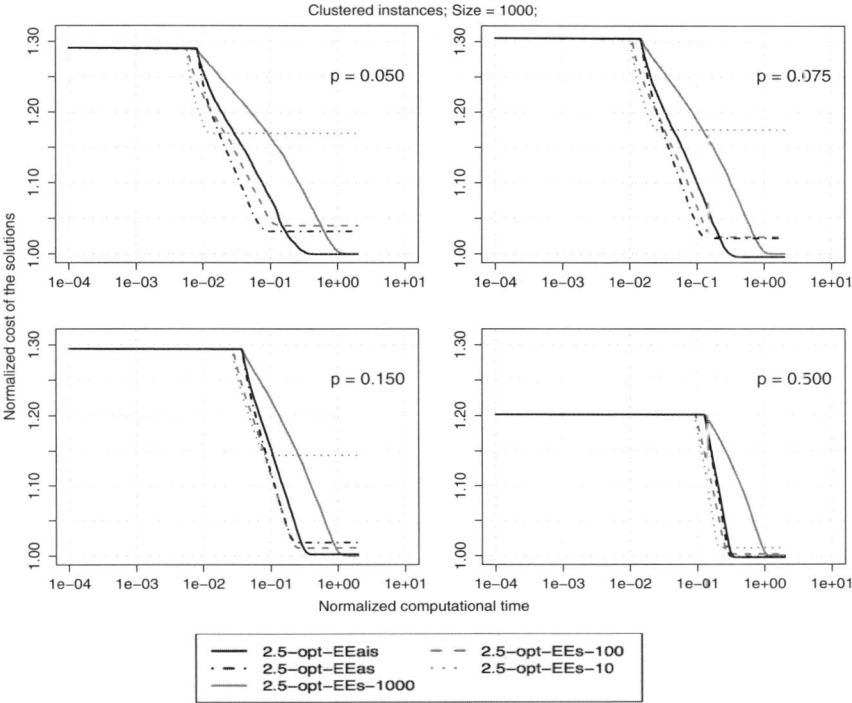

Fig. 4.6. Experimental results on clustered homogeneous PTSP instances of size 1000. The plots represent the cost of the solutions obtained by `2.5-opt-EEas`, `2.5-opt-EEais`, `2.5-opt-EEs-10`, and `2.5-opt-EEs-100` normalized by the one obtained by `2.5-opt-EEs-1000`. Each algorithm is stopped when it reaches a local optimum.

`2.5-opt-EEs-1000`, but it does so approximately four times faster. Compared to `2.5-opt-EEs-100`, `2.5-opt-EEais` reaches for low probability values, $p \leq 0.2$, an average cost that is between 1% and 3% lower (the differences are statistically significant according to the paired t-test), while for $p \geq 0.3$ the two algorithms are comparable. Taking into account both the computation time and the cost of the solutions obtained, we can see that `2.5-opt-EEais` emerges as a clear winner among the considered *estimation-based* algorithms.

Finally, we compared `2.5-opt-EEais` with `2.5-opt-ACs`. In order to avoid over-tuning [84], we generated 100 new instances for each probability level. The results are shown in Figure 4.7. The computational results show that `2.5-opt--EEais` is very competitive. Regarding the time required to reach local optima, irrespective of the probability levels, `2.5-opt-EEais` is approximately 2 orders of magnitude faster than `2.5-opt-ACs`. The average cost of local optima obtained by `2.5-opt-EEais` is comparable to the one of `2.5-opt-ACs`.

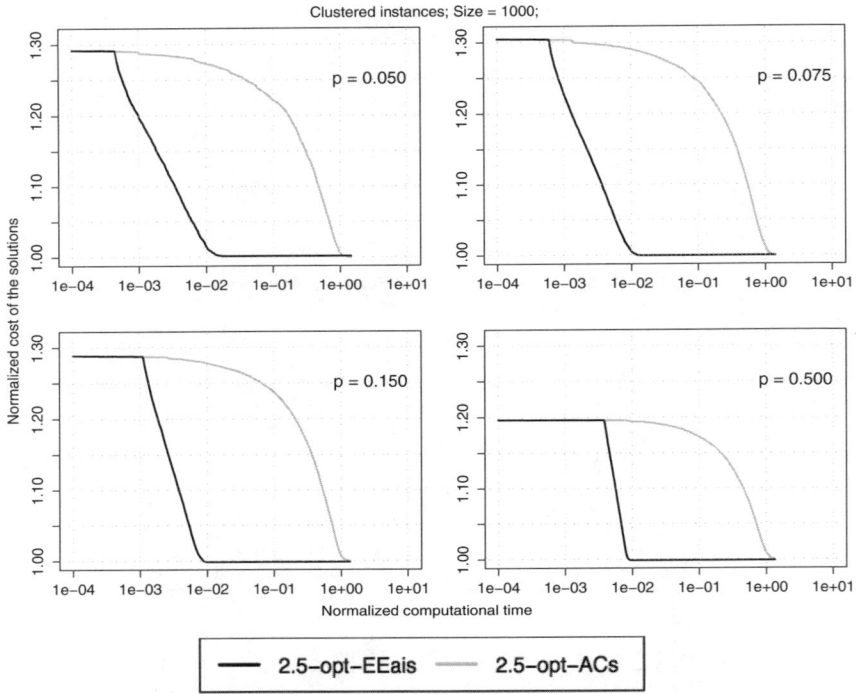

Fig. 4.7. Experimental results on clustered homogeneous PTSP instances of size 1000. The plots represent the solution cost obtained by `2.5-opt-EEais` normalized by the one obtained by `2.5-opt-ACs`. Each algorithm stops upon reaching a local optimum.

4.5   Estimation Based Iterated Local Search

As a final step, we tried to derive a new state-of-the-art SLS algorithm for the PTSP. Clearly, the significant performance gains obtained by `2.5-opt-EEais` over the previous state-of-the-art iterative improvement algorithms should make this feasible. Given the very high performance of iterated local search (ILS) [85] algorithms for the TSP, we decided to adopt this general-purpose SLS method also for the PTSP. That is, we implement our fourth hypothesis:

Hypothesis 4: `2.5-opt-EEais` is a good candidate procedure to derive a new state-of-the-art SLS algorithm for the PTSP.

Our ILS algorithm is a straightforward adaptation of ILS algorithms for the TSP. It starts from a nearest neighbor tour and uses `2.5-opt-EEais` as the subsidiary local search algorithm. The acceptance criterion compares two local optima by using the *t-test* with up to a maximum of n realizations. If no statistically significant difference is detected, the solution with lower cost estimate is accepted. The perturbation consists of a number of simple moves. In particular,

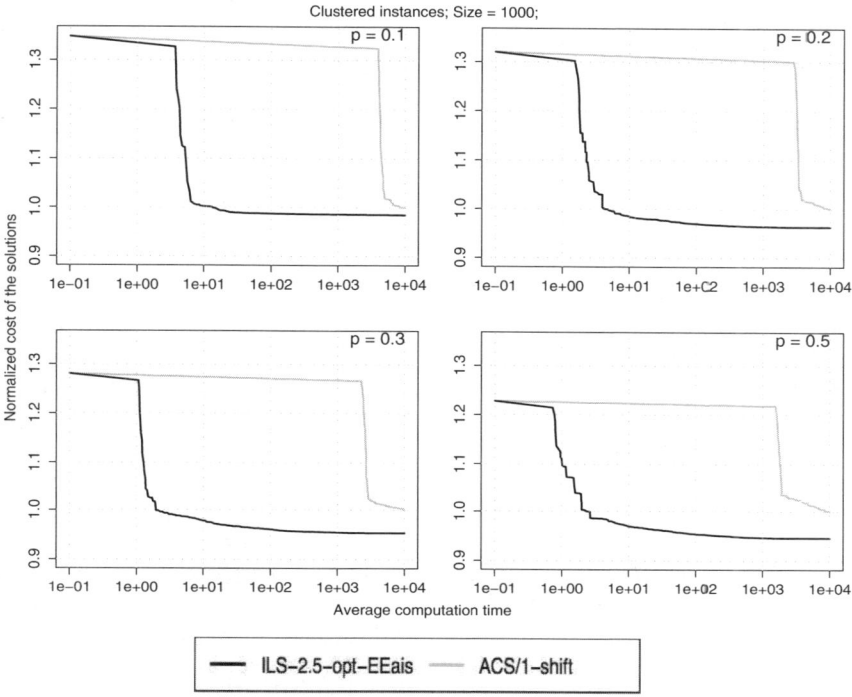

Fig. 4.8. Experimental results on clustered homogeneous PTSP instances of size 1000. Each algorithm is allowed to run for 10 000 seconds. The plots give the solution cost obtained by ILS-2.5-opt-EEais normalized by the one obtained by ACS/1-shift.

we perform two random double-bridge moves and perturbe the position of $ps\%$ of the nodes, where ps is a parameter. This change of the position is done by picking uniformly at random $ps\%$ of nodes, removing them from the tour and then re-inserting them again according to the farthest insertion heuristic. This composite perturbation is motivated by the change of the usefulness of edge exchange moves and insertion moves, as indicated by the crossover in the relative performance of 2-p-opt and 1-shift. We denote the final algorithm by ILS-2.5-opt-EEais.

We first tuned the parameter ps of ILS-2.5-opt-EEais using Iterative F-Race, resulting in a value of $ps = 6$. Next, we compared its performance to ACS/1-shift [86], an ant colony optimization algorithm [87] that uses 1-shift as the underlying local search and was shown to be a state-of-the-art stochastic local search algorithm for the PTSP. For this task, we adopted the parameter settings of ACS/1-shift that were fine-tuned in [86]. We compared the two algorithms on clustered instances with 1000 nodes using 10 000 seconds as a stopping criterion.

The experimental results in Figure 4.8 show that `ILS-2.5-opt-EEais` outperforms `ACS/1-shift` both in terms of final solution quality and computation time to reach a specific bound on the solution quality. In fact, the final average solution cost of `ILS-2.5-opt-EEais` is between 1% and 5% lower that that of `ACS/1-shift`; all observed differences are statistically significant according to a paired t-test with $\alpha = 0.05$.

4.6   Conclusions

In this chapter, we presented a case study in engineering an effective SLS algorithm for the PTSP. Our approach has used a bottom-up process. It had the particularity that it focused very strongly on the development and refinement of the underlying iterative improvement algorithm. We do not claim that this strong focus on this element of the process is always necessary. In fact, this process required, in this particular case, a significant number of new ideas. However, we strongly believe that this bottom-up approach is a potentially very successful way of deriving very high performing algorithms.

SLS engineering is relatively a new area of research in SLS algorithms and it is receiving considerable attention in recent years. Therefore, it has a number of avenues open for further contributions. One of the main focus of research in SLS engineering is to develop a framework of principled procedures for SLS design, implementation, analysis, and in-depth experimental studies. Moreover, SLS engineering needs a tight integration with tools that support the algorithm development process such as software frameworks, statistical tools, experimental design, automated tuning, search space analysis, and efficient data structures. Given researchers' and practitioners' quest for high performing algorithms, we strongly believe that SLS engineering is going to play a major role in designing SLS algorithms.

Acknowledgements

The authors thank Leonora Bianchi for providing the source code of `2-p-opt` and `1-shift`. This research has been supported by COMP²SYS, a Marie Curie Training Site funded by the Commission of the European Community, and by the ANTS project, an *Action de Recherche Concertée* funded by the French Community of Belgium. The authors acknowledge support from the fund for scientific research F.R.S.-FNRS of the French Community of Belgium.

Part II

Hybrid Approaches

5

A Lagrangian Decomposition/Evolutionary Algorithm Hybrid for the Knapsack Constrained Maximum Spanning Tree Problem

Sandro Pirkwieser[1], Günther R. Raidl[1], and Jakob Puchinger[2]

[1] Institute of Computer Graphics and Algorithms
   Vienna University of Technology, Vienna, Austria
   {pirkwieser,raidl}@ads.tuwien.ac.at
[2] NICTA Victoria Laboratory
   The University of Melbourne, Australia
   jakobp@csse.unimelb.edu.au

Summary. We present a Lagrangian decomposition approach for the Knapsack Constrained Maximum Spanning Tree problem yielding upper bounds as well as heuristic solutions. This method is further combined with an evolutionary algorithm to a sequential hybrid approach. Thorough experimental investigations, including a comparison to a previously suggested simpler Lagrangian relaxation based method, document the advantages of our approach. Most of the upper bounds derived by Lagrangian decomposition are optimal, and when additionally applying local search (LS) and combining it with the evolutionary algorithm, large and supposedly hard instances can be either solved to provable optimality or with a very small remaining gap in reasonable time.

Keywords: Knapsack Constrained Maximum Spanning Tree, Lagrangian Decomposition, Evolutionary Algorithm, Hybrid Approach, Volume Algorithm, Local Search.

5.1 Introduction

The *Knapsack Constrained Maximum Spanning Tree* (KCMST) problem arises in practice in situations where the aim is to design a profitable communication network under a strict limit on total costs, e.g. for cable laying or similar resource constraints.

We are given an undirected connected graph $G = (V, E)$ with node set V and edge set $E \subseteq V \times V$ representing all possible connections. Each edge $e \in E$ has associated a weight $w_e \in \mathbb{Z}_+$ (corresponding to costs) and a profit $p_e \in \mathbb{Z}_+$. In addition, a weight limit (capacity) $c > 0$ is specified. A feasible KCMST is a spanning tree $G_T = (V, T)$, $T \subseteq E$ on G, i.e. a cycle-free subgraph connecting all nodes, whose weight $\sum_{e \in T} w_e$ does not exceed c. The objective is to find a KCMST with maximum total profit $\sum_{e \in T} p_e$.

C. Cotta and J. van Hemert (Eds.): Recent Advances in Evol. Comp., SCI 153, pp. 69–85, 2008.
springerlink.com © Springer-Verlag Berlin Heidelberg 2008

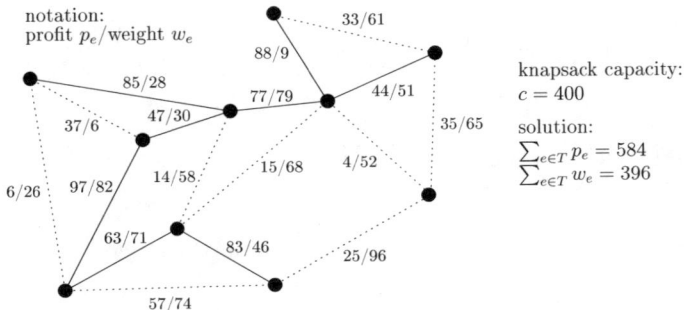

Fig. 5.1. Exemplary KCMST instance and its solution

More formally, we can introduce binary variables x_e, $\forall e \in E$, indicating which edges are part of the solution, i.e. $x_e = 1 \leftrightarrow e \in T$ and $x_e = 0$ otherwise, and write the KCMST problem as:

$$\max \ p(x) = \sum_{e \in E} p_e x_e \qquad (5.1)$$

s. t. x represents a spanning tree on G $\qquad\qquad$ (5.2)

$$\sum_{e \in E} w_e x_e \le c \qquad (5.3)$$

$$x_e \in \{0, 1\} \qquad\qquad \forall e \in E \qquad (5.4)$$

Obviously, the problem represents a combination of the classical *minimum spanning tree* (MST) problem (with changed sign in the objective function) and the classical 0–1 knapsack problem due to constraint (5.3). Yamada et al. [88] gave a proof for the KCMST problem's $\mathcal{NP}$-hardness. An exemplary instance and its solution are shown in Fig. 5.1.

After summarizing previous work for this problem in the next section, we present a Lagrangian decomposition approach in Section 5.3. It is able to yield tight upper bounds as well as lower bounds corresponding to feasible heuristic solutions. The latter are gained via a Lagrangian heuristic including local search. Section 5.4 describes an evolutionary algorithm for the KCMST problem utilizing the edge-set representation. Section 5.5 explains how this evolutionary algorithm can be effectively combined with the Lagrangian decomposition approach in a sequential manner. Computational results are presented in Section 5.6. The results document the excellent performance of the whole hybrid system, which is able to solve many test instances with planar graphs of up to 12000 nodes and complete graphs up to 300 nodes to provable optimality or with a very small gap in reasonable time.

This article extends our previous conference contribution [89] in various ways: more algorithmic details are presented, in particular concerning the volume algorithm for solving the Lagrangian dual; a new comparison of the Lagrangian decomposition with a previously proposed simpler Lagrangian relaxation is performed; and substantially more computational results for a larger variety of differently structured test instances are included.

5.2 Previous Work

In the literature, the KCMST problem is known under several different names and as minimization and maximization variants. As the minimization problem can trivially be transformed into a maximization variant, we ignore this difference in the following. Aggarwal et al. [90] were the first describing this problem and called it *MST problem subject to a side constraint*. They proved its $\mathcal{NP}$-hardness and proposed a branch-and-bound approach for solving it. Jörnsten and Migdalas [91] (*MST network subject to a budget constraint*) describe a *Lagrangian relaxation* (LR) in which the knapsack constraint (5.3) is relaxed, yielding a simple minimum spanning tree problem which can be solved efficiently. They further document the superiority of Lagrangian decomposition, and subsequently solving each subproblem to optimality, for generating valid bounds. An approximation algorithm also based on LR and a method to reduce the problem size is suggested in [92] (*constrained MST problem*). The later articles from Xue [93] (*weight-constrained MST*) and Jüttner [94] (*constrained minimum cost spanning tree problem*) deal with two similar primal-dual algorithms. Recently, Yamada et al. [88] (KCMST problem) also described a LR approach, which yields feasible heuristic solutions, too. These are further improved by a 2-opt local search. In order to determine provable optimal solutions for instances of restricted size, the LR is embedded in a branch-and-bound framework. While the approach is able to optimally solve instances with up to 1000 nodes and 2800 edges when edge weights and profits are uncorrelated, performance degrades substantially in the correlated case. Our Lagrangian decomposition approach was introduced in the first author's master thesis [95]. Finally, the recent master thesis of Henn [96] gives an overview on previous work, introduces a way to reduce the problem size and presents another exact branch-and-bound scheme.

Generally, LR is a commonly used technique from the area of mathematical programming to determine upper bounds for maximization problems. Though the solutions obtained are in general infeasible for the original problem, they can lend themselves to create feasible solutions and thus to derive lower bounds, too. For a general introduction to LR, see [120, 121, 122].

Since LR plays a fundamental role in the mentioned previous work, we briefly present its straight-forward application to the KCMST problem. We denote it as KCMST-LR(λ):

$$\max \; p(x) = \sum_{e \in E} x_e (p_e - \lambda w_e) + \lambda c \tag{5.5}$$

s. t. x represents a spanning tree $\tag{5.6}$

$\quad x_e \in \{0, 1\} \qquad\qquad\qquad \forall e \in E \tag{5.7}$

In order to find a best suited Lagrangian multiplier $\lambda \geq 0$ for the relaxed weight constraint, one has to solve the Lagrangian dual problem:

$$\min_{\lambda \geq 0} v(\text{KCMST-LR}(\lambda)), \qquad (5.8)$$

where the objective value of the optimal solution of KCMST-LR(λ) is denoted by $v(\text{KCMST-LR}(\lambda))$.

5.3   Lagrangian Decomposition for the KCMST Problem

Lagrangian decomposition (LD) is a special variant of LR that can be meaningful when there is evidence of two or possibly more intertwined subproblems, and each of them can be efficiently solved on its own by specialized algorithms.

As the KCMST problem is a natural combination of the maximum spanning tree problem and the 0–1 knapsack problem, we apply LD with the aim of such a partitioning. For this purpose, we duplicate variables x_e, $\forall e \in E$, by introducing new, corresponding variables y_e and including linking constraints, leading to the following equivalent reformulation:

$$\max \ p(x) = \sum_{e \in E} p_e x_e \qquad (5.9)$$

$$\text{s. t.} \ x \text{ represents a spanning tree} \qquad (5.10)$$

$$\sum_{e \in E} w_e y_e \leq c \qquad (5.11)$$

$$x_e = y_e \qquad\qquad \forall e \in E \qquad (5.12)$$

$$x_e, y_e \in \{0, 1\} \qquad\qquad \forall e \in E \qquad (5.13)$$

The next step is to relax the linking constraints (5.12) in a Lagrangian fashion using Lagrangian multipliers $\lambda_e \in \mathbb{R}$, $\forall e \in E$. By doing so we obtain the Lagrangian decomposition of the original problem, denoted by KCMST-LD(λ):

$$\max \ p(x) = \sum_{e \in E} p_e x_e - \sum_{e \in E} \lambda_e (x_e - y_e) \qquad (5.14)$$

$$\text{s. t.} \ x \text{ represents a spanning tree} \qquad (5.15)$$

$$\sum_{e \in E} w_e y_e \leq c \qquad (5.16)$$

$$x_e, y_e \in \{0, 1\} \qquad\qquad \forall e \in E \qquad (5.17)$$

Stating KCMST-LD(λ) in a more compact way and emphasizing the now independent subproblems yields

$$(\text{MST}) \ \max \ \{(p - \lambda)^T x \mid x \,\hat{=}\, \text{a spanning tree}, \ x \in \{0, 1\}^E\} \ + \quad (5.18)$$

$$(\text{KP}) \ \max \ \{\lambda^T y \mid w^T y \leq c, \ y \in \{0, 1\}^E\}. \qquad (5.19)$$

For a particular λ, the maximum spanning tree (MST) subproblem (5.18) can be efficiently solved by standard algorithms. In our implementation we apply Kruskal's algorithm [123] based on a union-find data structure when the underlying graph is sparse and Prim's algorithm [124] utilizing a pairing heap with dynamic insertion [125] for dense graphs. The 0–1 knapsack subproblem (5.19) is known to be weakly $\mathcal{NP}$-hard, and practically highly efficient dynamic programming approaches exist [126]; we apply the COMBO algorithm [127].

It follows from LR theory that for any choice of Lagrangian multipliers λ, the optimal solution value to KCMST-LD(λ), denoted by v(KCMST-LD(λ)), is always at least as large as the optimal solution value of the original KCMST problem, i.e., KCMST-LD(λ) provides a valid upper bound. To obtain the tightest (smallest) upper bound, we have to solve the Lagrangian dual problem:

$$\min_{\lambda \in \mathbb{R}^E} v(\text{KCMST-LD}(\lambda)). \tag{5.20}$$

5.3.1 Solving the Lagrangian Dual Problem

The dual problem (5.20) is piecewise linear and convex, and standard algorithms like an iterative subgradient approach can be applied for (approximately) solving it. More specifically, we use the *volume algorithm* [128] which has been reported to outperform standard subgradient methods in many cases including set covering, set partitioning, max cut, and Steiner tree problems [129,130]. Our preliminary tests on the KCMST problem also indicated its superiority over a standard subgradient algorithm [95]. The volume algorithm's name is inspired by the fact that primal solutions are considered and that their values come from approximating the volumes below the active faces of the dual problem. See Algorithm 5.1 for a pseudocode description.

The derived upper and lower bounds are stored in variables z_{UB} and z_{LB}, respectively. The primal vectors of the two subproblems, which represent an approximation to a primal solution, are denoted by $\mathbf{x}_\text{P}$ and $\mathbf{y}_\text{F}$, the Lagrangian multiplier vector is $\boldsymbol{\lambda}$.

At the beginning in line 1 an initial solution is created by solving the MST problem using edge values $v_e = p_e/w_e$, if this fails $v_e = 1/w_e$. In this way, either we derive a feasible solution or the problem instance is infeasible. In line 1 the Lagrangian multipliers are initialized to $\lambda_e = 0.5p_e$. We remark that this as well as some other specific settings in the volume algorithm may influence the final solution quality significantly. Our choices are based on preliminary tests partly documented in [95] and the primary intention to find a relatively simple and generally robust configuration. The primal vectors are initialized in line 1. The target value T is always estimated by $T := 0.95z_{\text{LB}}$ with the exception $T := 0.95T$ if $z_{\text{UB}} < 1.05T$. Parameter f is initialized with 0.1 and multiplied by 0.67 after 20 consecutive *red* iterations (i.e. no better upper bound was found) when $f > 10^{-8}$ and is multiplied by 1.1 after a *green* iteration (i.e. a better upper bound was found and $\mathbf{v}^t \cdot (\mathbf{x}^t - \mathbf{y}^t) \geq 0$) when $f < 1$. These two parameters influence the step size, which determines the amount of change of the Lagrangian multipliers. Factor α controls the update of the primal vectors. It is initialized with 0.01 and periodically checked after every 100 iterations: if the upper bound decreased less than 1% and $\alpha > 10^{-5}$ then $\alpha := 0.85\alpha$. These initializations are done in line 1 and the update in line 1. The volume algorithm terminates

---

Algorithm 5.1. Volume Algorithm applied to KCMST

Result: best lower bound z_{LB}, best upper bound z_{UB} and best solution found sol_{best}

1 $(sol, p(sol)) \leftarrow \texttt{getInitialSolution()}$;
2 $sol_{\text{best}} \leftarrow sol$;
3 $z_{\text{LB}} \leftarrow p(sol)$;
4 choose initial values for $\boldsymbol{\lambda}$;
5 $(z_{\text{MST}}^0, \mathbf{x}^0) \leftarrow \texttt{solve\_MST}(\mathbf{p} - \boldsymbol{\lambda})$; // see (5.18)
6 $(z_{\text{KP}}^0, \mathbf{y}^0) \leftarrow \texttt{solve\_KP}(\boldsymbol{\lambda})$; // see (5.19)
7 $z_{\text{UB}} = z_{\text{MST}}^0 + z_{\text{KP}}^0$;
8 $(\mathbf{x}_{\text{P}}, \mathbf{y}_{\text{P}}) = (\mathbf{x}^0, \mathbf{y}^0)$; // initialize primal values
9 initialize T, f and α accordingly;
10 $t = 0$; // iteration counter
11 $steps = 0$;
12 **while** $z_{\text{LB}} \neq \lfloor z_{\text{UB}} \rfloor$ *and* $steps \neq steps_{\text{max}}$ **do**
13 $\quad$ $t = t + 1$;
14 $\quad$ $\mathbf{v}^t = \mathbf{x}_{\text{P}} - \mathbf{y}_{\text{P}}$; // determine actual subgradients
15 $\quad$ $s = f(z_{\text{UB}} - T)/\|\mathbf{v}^t\|^2$; // determine step size
16 $\quad$ $\boldsymbol{\lambda}^t = \boldsymbol{\lambda} + s\mathbf{v}^t$; // determine actual multipliers
17 $\quad$ $(z_{\text{MST}}^t, \mathbf{x}^t) \leftarrow \texttt{solve\_MST}(\mathbf{p} - \boldsymbol{\lambda}^t)$;
18 $\quad$ $(z_{\text{KP}}^t, \mathbf{y}^t) \leftarrow \texttt{solve\_KP}(\boldsymbol{\lambda}^t)$;
19 $\quad$ $z^t = z_{\text{MST}}^t + z_{\text{KP}}^t$; // actual upper bound
20 $\quad$ $\texttt{LagrangianHeuristic}(\mathbf{x}^t)$; // see Section 5.3.3
$\quad$ // update z_{LB} and sol_{best}
21 $\quad$ $(\mathbf{x}_{\text{P}}, \mathbf{y}_{\text{P}}) = \alpha(\mathbf{x}^t, \mathbf{y}^t) + (1 - \alpha)(\mathbf{x}_{\text{P}}, \mathbf{y}_{\text{P}})$; // update primal values
22 $\quad$ **if** $z^t < z_{\text{UB}}$ **then** // better (lower) upper bound found
23 $\quad\quad$ **if** $z^t < \lfloor z_{\text{UB}} \rfloor$ **then**
24 $\quad\quad\quad$ $steps = 0$;
25 $\quad\quad$ **else**
26 $\quad\quad\quad$ $steps = steps + 1$;
27 $\quad\quad$ $z_{\text{UB}} = z^t$; // update best upper bound
28 $\quad\quad$ $\boldsymbol{\lambda} = \boldsymbol{\lambda}^t$; // update multipliers
29 $\quad$ **else** // no improvement, *red* iteration
30 $\quad\quad$ $steps = steps + 1$;
31 $\quad$ update T, f and α accordingly;

---

when either the lower and upper bounds become identical and, thus, an optimal solution has been reached, or when the upper bound did not improve over the last 300 iterations, i.e. $steps_{\text{max}}$ is set to 300 in line 1. All these update rules are similar to those used in [130].

In each iteration the current subgradients $\mathbf{v}^t$, the step size s, and the new multipliers $\boldsymbol{\lambda}^t$ are determined. Using these multipliers both subproblems are solved and the upper bound z^t is calculated. Furthermore, a Lagrangian heuristic, described in Section 5.3.3, is applied to the solution of the MST subproblem, if necessary updating the lower bound and the best solution so far. Afterwards the primal values are updated using α; they are a convex combination of the preceding dual solutions $\mathbf{x}^0, \mathbf{y}^0$ to $\mathbf{x}^t$ and $\mathbf{y}^t$. Only in case a better (i.e. lower) upper bound is found, the multipliers are set to the new values, and *steps* is reset to 0 iff $z^t < \lfloor z_{\mathrm{UB}} \rfloor$.

5.3.2 Strength of the Lagrangian Decomposition

According to integer linear programming theory, LR always yields a bound that is at least as good as the one obtained by the corresponding linear programming (LP) relaxation, providing the Lagrangian dual problem is solved to optimality. The LR's bound can be substantially better when the relaxed problem does not fulfill the integrality property , i.e., the solution to the LP relaxation of the relaxed problem – KCMST-LD(λ) in our case – is in general not integer.

To see whether or not this condition is fulfilled here, we have to consider both independent subproblems. For the MST problem, compact models having the integrality property exist, see e.g. [131]. For the knapsack problem, however, the integrality property is not fulfilled. Thus, we may expect to obtain bounds that are better than those from the linear programming relaxation of KCMST.

In comparison, in the LR approach from [88, 94] the knapsack constraint is relaxed and only the MST problem remains. This approach therefore fulfills the integrality property and, thus, is in general weaker than our LD.

We further remark that the proposed LD can in principle be strengthened by adding the cardinality constraint $\sum_{e \in E} y_e = |V| - 1$ to the knapsack subproblem. The resulting cardinality constrained or exact k-item knapsack problem is still only weakly $\mathcal{NP}$-hard, and pseudo-polynomial algorithms based on dynamic programming are known for it [126]. Our investigations indicate, however, that the computational demand required for solving this refined formulation is in practice substantially higher and does not pay off the typically only small improvement of the obtained bound [95].

5.3.3 Deriving Lower Bounds

In some iterations of the volume algorithm, the obtained spanning tree is feasible with respect to the knapsack constraint and can be directly used as a lower bound. Hence, we have already a trivial Lagrangian heuristic . In order to further improve such solutions this heuristic is strengthened by consecutively applying a local search based on the following edge exchange move.

1. Select an edge $(u, v) \in E \setminus T$ to be considered for inclusion.
2. Determine the path $P \subseteq T$ connecting nodes u and v in the current tree. Including e in T would yield the cycle $P \cup \{(u, v)\}$.
3. Identify a least profitable edge $\tilde{e} \in P$ that may be replaced by (u, v) without violating the knapsack constraint:

$$\tilde{e} = \mathrm{minarg} \ \{ p_e \mid e \in E \wedge w(T) - w_e + w_{(u,v)} \leq c \}, \qquad (5.21)$$

where $w(T) = \sum_{e \in T} w_e$. In case of ties, an edge with largest weight is chosen.

4. If replacing $\tilde{e}$ by (u,v) improves the solution, i.e. $p_{\tilde{e}} < p_{(u,v)} \vee (p_{\tilde{e}} = p_{(u,v)} \wedge w_{\tilde{e}} > w_{(u,v)})$, perform this exchange.

For selecting edge (u,v) in step 1 we consider two possibilities:

Random selection: Randomly select an edge from $E \setminus T$.

Greedy selection: At the beginning of the local search, all edges are sorted according to decreasing $p'_e = p_e - \lambda_e$, the reduced profits used to solve the MST subproblem. Then, in every iteration of local search, the next less profitable edge not active in the current solution is selected. This results in a greedy search where every edge is considered at most once.

Since Lagrangian multipliers are supposed to be of better quality in later phases of the optimization process, local search is only applied when the ratio of the incumbent lower and upper bounds is larger than a certain threshold τ. Local search stops after ρ consecutive non-improving iterations have been performed.

5.4   A Suitable Evolutionary Algorithm

Evolutionary algorithms (EAs) have often proven to be well suited for finding good approximate solutions to hard network design problems. In particular for constrained spanning tree problems, a large variety of EAs applying very different representations and variation operators have been described, see e.g. [132] for an overview.

Here, we apply an EA based on a direct edge-set representation for heuristically solving the KCMST problem, since this encoding and its corresponding variation operators are known to provide strong locality and heritability. Furthermore, variation operators can efficiently be performed in time that depends (almost) only linearly on the number of nodes. In fact, our EA closely follows the description of the EA for the degree constrained minimum spanning tree problem in [132]. Only the initialization and variation operators are adapted to conform with the knapsack constraint.

The general framework is steady-state, i.e. in each iteration one feasible offspring solution is created by means of recombination, mutation, and eventually local improvement, and it replaces the worst solution in the population. Duplicates are not allowed in the population; they are always immediately discarded. The EA's operators work as follows.

Initialization. To obtain a diversified initial population, a random spanning tree construction based on Kruskal's algorithm is used. Edges are selected with a bias towards those with high profits. The specifically applied technique corresponds to that in [132]. In case a generated solution is infeasible with respect to the knapsack constraint, it is stochastically repaired by iteratively selecting a not yet included edge at random, adding it to the tree, and removing an edge with highest weight from the induced cycle.

Recombination. An offspring is derived from two selected parental solutions in such a way that the new solution candidate always exclusively consists of inherited edges: In a first step all edges contained in both parents are immediately adopted. The remaining parental edges are merged into a single candidate list. From this list, we iteratively select edges by binary tournaments with replacement favoring high-profit edges. Selected edges are included in the solution if they do not introduce a cycle; otherwise, they are discarded. The process is repeated until a complete spanning tree is obtained. Finally, its validity with respect to the knapsack constraint is checked. An infeasible solution is repaired in the same way as during initialization, but only considering parental edges for inclusion.

Mutation. We perform mutation by inserting a randomly selected new edge and removing another edge from the introduced cycle. The choice of the edge to be included is biased towards high-profit edges by utilizing a normally-distributed rank-based selection as described in [132]. The edge to be removed from the induced cycle is chosen at random among those edges whose removal would retain a feasible solution.

Local Search. With a certain probability, a newly derived candidate solution is further improved by the local search procedure described in Section 5.3.3.

5.5 Hybrid Lagrangian Evolutionary Algorithm

Preliminary tests clearly indicated that the EA cannot compete with the performance of LD in terms of running time and solution quality. However, following similar ideas as described in [130] for the price-collecting Steiner tree problem, we can successfully apply the EA for finding better final solutions after performing LD. Hereby, the EA is adapted to exploit a variety of (intermediate) results from LD. In detail, the following steps are performed after LD has terminated and before the EA is executed:

1. If the profit of the best feasible solution obtained by LD corresponds to the determined upper bound, we already have an optimal solution. No further actions are required.

2. For the selection of edges during initialization, recombination, and mutation of the EA, original edge profits p_e are replaced by reduced profits $p'_e = p_e - \lambda_e$. In this way, Lagrangian dual variables are exploited, and the heuristic search emphasizes the inclusion of edges that turned out to be beneficial in LD.

3. The edge set to be considered by the EA is reduced from E to a subset E' containing only those edges that appeared in any of the feasible solutions encountered by LD. For this purpose, LD is extended to mark these edges.

4. The best feasible solution obtained by LD is included in the EA's initial population.

5. Finally, the upper bound obtained by LD is passed to the EA and exploited by it as an additional stopping criterion: When a solution with a corresponding total profit is found, it is optimal and the EA terminates.

An outline of the collaboration is given in Fig. 5.2.

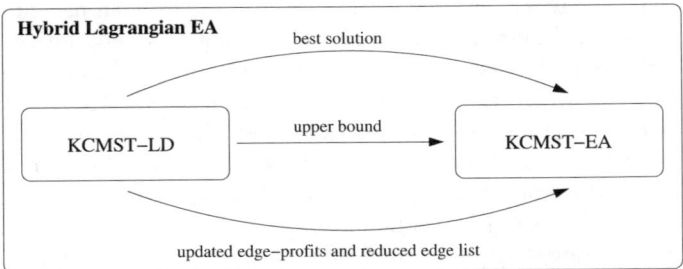

Fig. 5.2. Information exchange in the hybrid approach

5.6   Computational Results

The described algorithms have been tested on a large variety of different problem instances, and comparisons regarding the strength of the Lagrangian dual have been performed in particular with the previous LR based primal-dual method of [94]. This section includes several representative results; further details can be found in [95]. All experiments were performed on a 2.2 GHz AMD Athlon 64 PC with 2 GB RAM.

We show and compare results for the Lagrangian relaxation (LR) based on [94], our Lagrangian decomposition with the simple primal heuristic (LD) and optionally local search (LD+LS), and the combination of LD and the EA (LD+LS+EA).

5.6.1   Test Instances

Unfortunately, no test instances from previously published algorithms for the KCMST problem are publicly available or could be obtained from the authors. As in [88], we consider instances based on complete graphs $K_{|V|,\gamma}$ and planar graphs $P_{|V|,|E|,\gamma}$. Parameter γ represents the type of correlation between profits and weights:

uncorrelated ("u"): p_e and w_e, $e \in E$, are independently chosen from the integer interval $[1, 100]$;

weakly correlated ("w"): w_e is chosen as before, and $p_e := \lfloor 0.8w_e + v_e \rfloor$, where v_e is randomly selected from $[1, 20]$;

strongly correlated ("s"): w_e is chosen as before, and $p_e := \lfloor 0.9w_e + 10 \rfloor$.

For details on the methods used to construct the planar graphs, we refer to [88, 95]. Since we could not obtain the original instances, we created them in the same way by our own. In addition we constructed larger maximal planar graphs $P_{|V|,\gamma}$, i.e. graphs that cannot be augmented by any further edge without violating planarity (for $|V| > 2 : |E| = 3|V| - 6$). In case of complete graphs, the knapsack capacity is $c = 20|V| - 20$, in case of (maximal) planar graphs $c = 35|V|$.

In particular for larger strongly correlated instances, we recognized that they are often easier to solve due to the relatively small number of possible profit and weight values and the resulting high probability for edges having assigned exactly the same profit/weight values. For example in case of our largest $P_{8000,s}$ instances, there are 23994 edges but only 100 different profit/weight combinations. In the expected case this leads to ≈ 240 edges sharing each possible profit/weight value pair. Therefore, we also created maximal planar graphs from a profit (weight) interval of $[1, 1000]$ and correspondingly scaled the correlations and the knapsack capacity. We denote these refined instances as $P^*_{|V|,\gamma}$

We further created particularly challenging test instances according to the description in [96]. They are based on random and complete graphs and the following special profit/weight correlations.

outliers ("o"): p_e and w_e lie with probability 0.9 in $[1001, 2000]$ and with probability 0.1 in $[1, 1000]$.

weakly correlated ("w_2"): p_e are uniformly distributed in $[1, 1000]$ and $w_e = \min\{1000, X + 0.5p_e\}$ with X chosen randomly in $[1, 1000]$.

strongly correlated ("s_2"): p_e are uniformly distributed in $[1, 1000]$ and $w_e = p_e + 20 + \beta$ with β uniformly in $[-20, 20]$.

To determine capacity c, the weight of the profit-maximal tree W_1 (in case of several such trees, the one having the least weight is chosen) and the weight of the weight-minimal tree W_2 are computed. Then c is derived in one of the following ways: $c = (W_1 + W_2)/4$ (low limit "l"), $c = (W_1 + W_2)/2$ (medium limit "m"), or $c = 3(W_1 + W_2)/4$ (high limit "h"). The variant used is given as additional subscript δ in the instance class name.

For each considered type, size, correlation, and capacity combination, 10 independent instances had been created.

5.6.2 Parameter Settings

In addition to the settings already described in Section 5.3.1 we are using the following setup for computing the results presented here.

For the optional local search, greedy edge selection is used for random and complete graphs with an application threshold set to $\tau = 0.99$ and random edge selection with $\tau = 0.995$ for the maximal planar graphs. In all cases $\rho = 100$ is used as maximum number of iterations without improvement. Heuristically derived solutions are not used for updating the target value T, thus the local search does not directly influence the volume algorithm.

For the EA, the population size is 100, binary tournament selection is used, and recombination and mutation are always applied. For the biasing towards edges with higher profits, parameters α and β (see [132]) are both set to 1.5. Local search is applied with a probability of 20% for each new candidate solution in the same manner as described before, except with $\rho = 50$. The maximum number of iterations is 10000 for (maximal) planar graphs and 30000 for random and complete graphs. In case of maximal planar graphs the edge set reduction was applied.

5.6.3   Comparing LR and LD

To see the performance differences between Lagrangian decomposition and the simpler Lagrangian relaxation, we compared our algorithm to a re-implementation of the method described in [94]. We made this choice since preliminary tests revealed that this method combines the good upper bounds of the bisection method in [88] and the good lower bounds of the primal-dual algorithm in [93] and, thus, outperforms both. Results on planar, complete, and random graphs are shown in Table 5.1; average values over 10 different instances are printed. Column $t[s]$ states the CPU-time in seconds, *iter* are the number of iterations. The table further lists relative errors of the achieved lower bounds $\%\text{-}gap_\mathrm{L} = \frac{p^*-\underline{p}}{p^*} \cdot 100\%$ and those of the upper bound $\%\text{-}gap_\mathrm{U} = \frac{\overline{p}-p^*}{p^*} \cdot 100\%$, with $\underline{p}$ and $\overline{p}$ being the derived lower and upper bounds, respectively, and the optimal solution value p^* was determined by an exact approach[1].

Most importantly, we can see that LD achieves in almost all cases substantially smaller gaps than LR and is never worse. In fact, LD's $\%\text{-}gap_\mathrm{U}$ is never larger

Table 5.1. Comparison between Lagrangian relaxation and decomposition

Instance	Jüttner [94] LR				Our approach LD			
	$t[s]$	*iter*	$\%\text{-}gap_\mathrm{L}$	$\%\text{-}gap_\mathrm{U}$	$t[s]$	*iter*	$\%\text{-}gap_\mathrm{L}$	$\%\text{-}gap_\mathrm{U}$
$P_{50,127,u}$	<0.01	6	0.4046	0.1349	0.05	805	0.0140	0.0478
$P_{50,127,w}$	<0.01	6	1.0079	0.0485	0.09	704	0.0097	0.0291
$P_{50,127,s}$	<0.01	4	4.3953	0	0.11	741	0.0487	0
$P_{100,260,u}$	<0.01	7	0.2035	0.0249	0.10	726	0	0.0055
$P_{100,260,w}$	<0.01	7	1.8282	0.0144	0.12	730	0.0072	0.0072
$P_{100,260,s}$	<0.01	5	4.5438	0	0.16	746	0.0121	0
$K_{20,u}$	<0.01	6	0.5369	0.2684	0.04	708	0.0061	0.0732
$K_{20,w}$	<0.01	5	2.6822	0.1293	0.06	628	0.0485	0.0162
$K_{20,s}$	<0.01	4	13.5186	0	0.08	723	0.0378	0
$K_{40,u}$	<0.01	6	0.1935	0.0164	0.11	680	0.0055	0.0055
$K_{40,w}$	<0.01	7	1.5371	0	0.23	721	0	0
$K_{40,s}$	<0.01	4	5.6600	0	0.26	964	0.0459	0
$K_{100,u}$	0.02	7	0.0454	0.0010	0.80	970	0	0.0010
$K_{100,w}$	0.02	7	2.9257	0	1.32	978	0.0058	0
$K_{100,s}$	0.01	4	5.7794	0	2.10	1529	0.0866	0
$R_{100,1238,o,l}$	<0.01	8	0.2208	0.0429	5.79	2443	0.0039	0.0096
$R_{100,1238,o,m}$	<0.01	8	0.0563	0.0068	0.87	1069	0.0012	0.0016
$R_{100,1238,o,h}$	<0.01	6	5.8593	0.0007	0.34	784	0	0.0002
$R_{100,1238,w_2,l}$	<0.01	9	0.5505	0.0413	2.72	1591	0.0036	0.0126
$R_{100,1238,w_2,m}$	<0.01	9	0.1772	0.0143	1.11	1024	0.0050	0.0051
$R_{100,1238,w_2,h}$	<0.01	8	0.0315	0.0065	0.48	865	0.0010	0.0005
$R_{100,1238,s_2,l}$	<0.01	8	1.9856	0.0035	3.66	1063	0.0106	0.0020
$R_{100,1238,s_2,m}$	<0.01	7	2.1569	0.0008	3.57	973	0.0045	0.0004
$R_{100,1238,s_2,h}$	<0.01	8	0.3101	0.0005	3.22	979	0.0027	0.0003
avg. values	<0.01	6	2.3587	0.0314	1.14	964	0.0150	0.0090

---

[1] We also implemented a yet unpublished exact branch-and-cut algorithm, which is able to solve these instances to proven optimality.

than 0.073% and %-gap_L is always below 0.087%, whereas the maxima of LR are $\approx 0.27\%$ and even $\approx 13.5\%$, respectively. Thus, in the course of solving LD much more high-quality feasible solutions are derived. As already observed in [88], strongly correlated instances are typically harder than uncorrelated ones, and Henn [96] also considered those with low capacity limit to be more challenging.

Sometimes LD is able to solve the instances to optimality but cannot prove their optimality since the upper bounds were not tight enough. In general, we can conclude that LD already delivers excellent bounds in short time.

5.6.4   LD Combined with LS and EA

In order to investigate the performance of the proposed LD+LS and the LD+LS+EA hybrid, we turn to the larger maximal planar graphs, for which Table 5.2 presents results. Average values over 10 instances and 10 runs per instance (for the stochastic algorithms) are reported. If appropriate we state in the last row the average of these values over all instances.

We state again $t[s]$ and *iter*, but also the average lower bounds LB, i.e. the objective values of the best feasible solutions. Upper bounds (UB) are expressed in terms of the relative gap to these lower bounds: $gap = (UB - LB)/LB$; corresponding standard deviations are listed in columns σ_{gap}. Columns %-*Opt* show percentage of instances for which the gap is zero and, thus, optimality has been proven.

For LD+LS+EA, the table also lists the overall time $t[s]$, LB, corresponding gap information, the percentage of overall optimal solutions %-*Opt*, and additionally the average number of EA iterations $iter_{EA}$, the relative amount of edges discarded after performing LD $red = (|E| - |E'|)/|E| \cdot 100\%$, and the percentage of optimal solutions %-Opt_{EA}, among %-*Opt*, found by the EA.

The solutions obtained by LD are already quite good and gaps are in general small. Applying the local search (LD+LS) always improves the average lower bound and in some cases helps to find more provably optimal solutions, which in turn reduces the number of iterations of the volume algorithm. The hybrid approach (LD+LS+EA) further boosts the average solution quality in almost all cases and substantially increases the numbers of solutions for which optimality could be proven. As expected, the finer-grained $P^*_{|V|,\gamma}$ instances with larger profit and weight ranges are for all algorithms significantly harder to solve than the coarse-grained $P_{|V|,\gamma}$ instances. The possible edge-set reduction decreases with increasing correlation and range. We remark that these large graphs are much harder to solve than the ones used in [88], thus the results are very satisfying; for LD+LS+EA, the gap is always less than 0.00023%.

Tests on random and complete graphs are shown in Table 5.3. The general results are quite similar than before, i.e. the local search as well as the EA are both consistently improving the quality. Preliminary tests suggested not to reduce the edge-sets on these type of instances; otherwise too many improving edges are missing. In comparison to the results presented by Henn [96], our approach was also highly successful on the challenging instances with outlier correlation (instances $R_{|V|,|E|,o,\delta}$ and $K_{|V|,o,\delta}$). In particular, LD+LS+EA was

Table 5.2. Results of Lagrangian decomposition and hybrid algorithms on maximal planar graphs

Instance	LD						LD+LS						LD+LS+EA							
	$t[s]$	iter	LB	gap $[\cdot10^{-5}]$	σ_{gap} $[\cdot10^{-5}]$	%-Opt	$t[s]$	iter	LB	gap $[\cdot10^{-5}]$	σ_{gap} $[\cdot10^{-5}]$	%-Opt	$t[s]$	red	$iter_{EA}$	LB	gap $[\cdot10^{-5}]$	σ_{gap} $[\cdot10^{-5}]$	%-Opt	%-Opt$_{EA}$
$P_{2000,u}$	1.48	791	147799.50	0.0683	0.2049	90	2.28	782	147799.55	0.0342	0.1489	95	2.90	41.21	150	147799.60	0	0	100	5
$P_{2000,w}$	1.52	853	85570.50	0.3519	0.7513	80	2.38	844	85570.63	0.1994	0.5261	86	4.26	42.61	457	85570.78	0.0235	0.1643	98	12
$P_{2000,s}$	2.12	1030	82521.70	1.9389	2.3118	40	2.66	868	82523.30	0	0	100	2.66	21.99	0	82523.30	0	0	100	0
$P_{4000,u}$	3.35	859	294872.00	0.0340	0.1019	90	5.59	841	294872.03	0.0238	0.0866	93	8.64	40.17	316	294872.10	0	0	100	7
$P_{4000,w}$	4.19	1053	170956.70	0.8195	0.9155	40	6.15	978	170957.79	0.1813	0.306	72	14.66	43.82	842	170958.06	0.0234	0.1147	96	24
$P_{4000,s}$	4.71	1066	165049.80	1.0300	0.8590	30	5.99	915	165051.44	0.0364	0.1439	94	9.95	19.92	410	165051.48	0.0121	0.0848	98	4
$P_{6000,u}$	5.66	912	441977.80	0.0680	0.1038	70	9.33	886	441977.96	0.0317	0.0786	86	15.41	40.25	339	441978.10	0	0	100	14
$P_{6000,w}$	6.55	1022	256317.40	0.3904	0.4621	50	9.25	964	256318.09	0.1210	0.2452	76	24.47	45.14	909	256318.36	0.0156	0.0764	96	20
$P_{6000,s}$	8.14	1157	247587.90	1.7368	1.3032	60	10.44	996	247592.04	0.0646	0.1481	84	33.73	19.94	1401	247592.09	0.0444	0.1264	89	5
$P_{8000,u}$	8.32	960	589446.50	0.1017	0.1357	60	13.81	918	589446.89	0.0356	0.077	81	28.44	39.98	595	589447.09	0.0017	0.0168	99	18
$P_{8000,w}$	9.78	1107	341902.50	0.5555	0.5139	30	14.18	1037	341903.85	0.1609	0.2124	58	48.40	44.82	1384	341904.37	0.0088	0.0499	97	39
$P_{8000,s}$	10.88	1125	330117.10	1.5147	1.3065	20	14.20	990	330121.86	0.0727	0.1294	76	57.00	17.99	1727	330121.96	0.0424	0.1051	86	10
$P_{2000,u}$	3.90	1189	1475226.30	0.2432	0.3076	20	5.74	1189	1475226.59	0.2236	0.2751	20	34.85	40.22	8000	1475226.93	0.2007	0.2466	20	0
$P_{2000,w}$	3.87	1151	854766.40	0.4685	0.5092	10	5.32	1151	854767.24	0.3698	0.3156	10	38.34	39.67	8906	854768.45	0.2281	0.1500	12	2
$P^*_{2000,s}$	4.80	1263	829462.20	0.7475	0.5877	10	6.17	1122	829468.13	0.0326	0.0535	73	16.61	4.98	2597	829468.15	0.0301	0.0522	75	2
$P_{4000,u}$	7.39	1230	2942682.80	0.1565	0.1262	30	11.17	1230	2942683.21	0.1425	0.1175	30	74.33	40.53	7000	2942684.72	0.0911	0.0812	30	0
$P_{4000,w}$	6.92	1179	1708837.70	0.6200	0.4678	20	9.89	1179	1708840.68	0.4457	0.3306	20	80.85	39.36	7852	1708845.20	0.1813	0.1454	23	3
$P_{4000,s}$	8.15	1288	1659023.60	0.5063	0.3548	0	11.92	1183	1659031.56	0.0265	0.0299	56	54.24	4.53	4320	1659031.59	0.0247	0.0296	59	3
$P^*_{6000,u}$	9.45	1178	4409459.00	0.0727	0.0827	30	15.22	1178	4409459.28	0.0664	0.0789	30	122.01	39.88	7000	4409460.15	0.0465	0.0445	30	0
$P^*_{6000,w}$	10.41	1256	2561360.20	0.4529	0.3598	0	15.40	1256	2561365.54	0.2444	0.1299	0	166.25	38.06	9683	2561369.06	0.1069	0.0799	9	9
$P_{6000,s}$	12.15	1335	2488583.60	0.4983	0.7021	10	18.10	1186	2488595.69	0.0125	0.0194	70	67.86	4.06	2967	2488595.70	0.0121	0.0193	71	1
$P_{8000,u}$	13.83	1290	5884437.60	0.0952	0.0556	0	22.51	1290	5884438.27	0.0838	0.0433	0	232.95	40.12	9667	5884439.85	0.0569	0.0354	5	5
$P^*_{8000,w}$	12.58	1189	3417468.90	0.1112	0.1093	40	18.76	1183	3417469.29	0.0998	0.1038	42	136.08	40.59	5414	3417471.21	0.0436	0.0565	49	7
$P^*_{8000,s}$	15.43	1298	3318159.40	0.4460	0.3934	20	25.90	1219	3318173.87	0.0099	0.0142	67	105.05	3.90	3246	3318173.89	0.0093	0.0139	69	2
avg. values	7.31	1115	–	0.5428	0.5427	33.7	10.93	1057	–	0.1132	0.1505	59.1	57.49	–	–	–	0.0501	0.0705	67.1	8.0

Table 5.3. Results of Lagrangian decomposition and hybrid algorithms on random and complete graphs with range 1000

Instance	LD						LD+LS						LD+LS+EA							
	$t[s]$	$iter$	LB	gap $[\cdot 10^{-5}]$	σ_{gap} $[\cdot 10^{-5}]$	%-Opt	$t[s]$	$iter$	LB	gap $[\cdot 10^{-5}]$	σ_{gap} $[\cdot 10^{-5}]$	%-Opt	$t[s]$	$(red)^*$	$iter_{EA}$	LB	gap $[\cdot 10^{-5}]$	σ_{gap} $[\cdot 10^{-5}]$	%-Opt	%-Opt$_{EA}$
R300,11213,o,l	9.53	1737	542839.40	1.7477	1.8326	10	11.72	1737	542840.60	1.5271	1.5937	10	29.99	92.93	27000	542843.63	0.9706	0.6928	10	0
R300,11213,o,m	7.10	1536	580716.50	0.2583	0.2464	30	8.89	1506	580716.60	0.2411	0.2576	40	21.43	91.63	18000	580716.64	0.2342	0.2477	40	0
R300,11213,o,h	3.57	1260	591409.00	0.1690	0.2507	50	5.11	1259	591409.30	0.1183	0.1320	50	13.73	91.02	12285	591409.54	0.0778	0.1132	64	14
R300,11213,s_2,l	24.58	1563	77466.60	8.5209	5.6046	20	24.45	1409	77473.00	0.2581	0.5161	80	24.69	80.64	336	77473.20	0	0	100	20
R300,11213,s_2,m	15.37	1351	155244.80	5.4064	5.1165	0	14.77	1051	155253.20	0	0	100	14.73	81.54	2222	155253.20	0	0	100	0
R300,11213,s_2,h	16.52	1332	232877.70	6.5305	5.2668	10	16.74	1238	232892.50	0.1718	0.2847	70	18.34	85.28	2222	232892.89	0.0043	0.0428	99	29
R300,22425,o,l	26.39	3324	568771.90	6.8383	6.1475	10	32.10	3324	568788.80	3.8714	4.3327	10	52.08	95.24	26700	568796.00	2.6042	3.3654	11	1
R300,22425,o,m	14.70	1943	588410.30	0.2210	0.2020	30	18.83	1943	588410.50	0.1870	0.1605	30	33.05	95.46	18078	588410.80	0.1360	0.1272	40	10
R300,22425,o,h	7.28	1358	594373.50	0.0168	0.0505	90	10.10	1358	594373.50	0.0168	0.0505	90	12.40	94.54	3000	594373.50	0.0168	0.0505	90	0
R300,22425,s_2,l	44.08	2059	77445.70	12.2628	9.0170	10	42.58	1793	77455.20	0	0	100	42.58	86.26	0	77455.20	0	0	100	0
R300,22425,s_2,m	29.69	1687	154940.30	7.8185	8.9007	10	28.81	1392	154952.40	0	0	100	28.81	93.71	0	154952.40	0	0	100	0
R300,22425,s_2,h	34.63	1964	232424.80	16.2741	12.5659	10	36.55	1885	232461.90	0.3013	0.3874	50	44.59	89.39	10682	232462.37	0.0990	0.1811	77	27
K300,o,l	247.29	19163	582646.00	4.0334	7.1749	10	316.33	19163	582660.30	1.5789	1.4435	10	333.98	97.50	27000	582663.46	1.0366	0.8511	10	0
K300,o,m	40.44	2909	592797.70	0.1856	0.1401	30	45.96	2864	592797.90	0.1518	0.1401	40	55.19	97.70	10212	592798.50	0.0506	0.0773	70	30
K300,o,h	30.13	2373	596076.40	0.0503	0.1074	80	35.49	2371	596076.50	0.0336	0.0671	80	36.13	96.94	1239	596076.70	0	0	100	20
K300,s_2,l	63.20	2495	77225.70	28.6269	20.8442	0	60.80	2195	77247.80	0	0	100	60.80	93.07	0	77247.80	0	0	100	0
K300,s_2,m	62.25	2704	154445.00	12.4958	8.3394	0	59.11	2404	154464.30	0	0	100	59.11	94.48	0	154464.30	0	0	100	0
K300,s_2,h	76.60	3396	231665.00	15.9285	18.7408	0	78.10	3142	231701.90	0	0	100	78.10	92.77	0	231701.90	0	0	100	0
avg. values	41.85	3008	–	7.0769	6.1415	22.0	47.02	2890	–	0.4698	0.5203	64.0	53.3183	–	–	–	0.2905	0.3193	72.8	8.3

*No edge-set reduction applied, only states possible amount.

able to solve larger instances (300 instead of 200 nodes) to proven optimality or with a very small gap than could be tackled by Henn's branch-and-bound. We further solved nearly all strongly correlated graph instances to optimality (80 out of 90 with local search, and on average 86 out of 90 with the hybrid algorithm), which also documents that the derived upper bounds are in fact almost always optimal. In case of these graphs particularly the local search was highly effective. The remaining gap of LD+LS+EA is never worse than 0.0026%. In particular for $R_{300,22425,s_2,l}$ and $R_{300,22425,s_2,h}$ instances, our algorithm needed substantially less CPU time than [96][2].

5.7   Conclusions

We presented a Lagrangian decomposition approach for the $\mathcal{NP}$-hard KCMST problem to derive upper bounds as well as heuristic solutions. Experimental results on large and diverse graphs revealed that the upper bounds are extremely tight, in fact most of the time even optimal. Heuristic solutions can be significantly improved by applying a local search, and many instances can be solved to provable optimality already in this way.

For the remaining instances a sequential combination of LD with an evolutionary algorithm has been described. The EA makes use of the edge-set encoding and corresponding problem-specific operators and exploits results from LD in several ways. In particular, the graph can be shrunk by only considering edges also appearing in heuristic solutions of LD, Lagrangian dual variables are exploited by using final reduced costs for biasing the selection of edges in the EA's operators, and the best solution obtained from LD is provided to the EA as seed in the initial population.

Computational results document the effectiveness of the hybrid approach. The EA always improves the quality and sometimes is able to close the gap and provide proven optimal solutions in many of the remaining difficult cases. Hereby, the increase in running time one has to pay is mostly only moderate.

The logical next step we want to pursue is to embed the LD or even the hybrid LD/EA into an exact branch-and-bound algorithm, similar to the one in [88] which makes use of the simple Lagrangian relaxation. Another possibility would be to employ the EA in an intertwined way with an exact method. This would permit us to compare the results with other exact methods in a more direct way.

In general, we believe that such combinations of Lagrangian relaxation and metaheuristics like evolutionary algorithms are highly promising for many combinatorial optimization tasks. Future work therefore includes the consideration of further problems, but also the closer investigation of other forms of collaboration between Lagrangian relaxation based methods and metaheuristics, including intertwined and parallel models.

---

[2] They used a roughly comparable test environment, a 2x 86_64 AMD Opteron workstation.

Acknowledgements

The Institute of Computer Graphics and Algorithms is supported by the European RTN ADONET under grant 504438, by the Austrian Exchange Service, Acciones Integradas Austria/Spain, under grant 13/2006 and by the Austrian Science Fund (FWF) under contract number P20342-N13.

NICTA is funded by the Australian Government's Backing Australia's Ability initiative, in part through the Australian Research Council.

6

A Hybrid Optimization Framework for Cutting and Packing Problems

Case Study on Constrained 2D Non-guillotine Cutting

Napoleão Nepomuceno, Plácido Pinheiro, and André L.V. Coelho

Universidade de Fortaleza, Mestrado em Informática Aplicada
Av. Washington Soares 1321, Sala J-30, Fortaleza, CE, Brazil, 60811-905
napoleao@edu.unifor.br, placido@unifor.br, acoelho@unifor.br

Summary. This work presents a hybrid optimization framework for tackling cutting and packing problems, which is based upon a particular combination scheme between heuristic and exact methods. A metaheuristic engine works as a generator of reduced instances for the original optimization problem, which are formulated as mathematical programming models. These instances, in turn, are solved by an exact optimization technique (solver), and the performance measures accomplished by the respective models are interpreted as score (fitness) values by the metaheuristic, thus guiding its search process. As a means to assess the potentialities behind the novel approach, we provide an instantiation of the framework for dealing specifically with the constrained two-dimensional non-guillotine cutting problem. Computational experiments performed over standard benchmark problems are reported and discussed here, evidencing the effectiveness of the novel approach.

Keywords: Hybrid Methods, Combinatorial Optimization, Cutting and Packing.

6.1 Introduction

The inherent difficulty and the enormous practical relevance of solving hard real-world problems in the field of combinatorial optimization have led to a large research effort in the development of heuristic methods aimed at finding good approximate solutions. And, arguably, one of the most important outcomes of such effort has been realized through the characterization of metaheuristics [108], that is, heuristic archetypes whose conceptual components show flexibility enough for being tailored and/or extended to cope better with the intricacies arising in different classes of applications.

Well-known examples of metaheuristic models include [100]: the several evolutionary computation dialects (e.g., genetic algorithms, evolutionary programming and evolution strategies) and their extensions (e.g., cultural algorithms, memetic algorithms, scatter search, and differential evolution); the recently-proposed swarm intelligence models (like particle swarm and ant colony optimization), as well as other bio-inspired approaches (like neural and immune network models); and trajectory-based methods (e.g., simulated annealing,

C. Cotta and J. van Hemert (Eds.): Recent Advances in Evol. Comp., SCI 153, pp. 87–99, 2008.

GRASP, and tabu and variable neighborhood searches). Usually, such meta-heuristic algorithms have succeeded in providing a proper balance between the efficiency and effectiveness criteria while tackling many non-trivial optimization problems, even though it is common-sense knowledge that they can not bear any sort of guarantee of the optimality of the solutions they find.

However, it is also quite consensual that there is still much room for the conceptualization of more sophisticated solution methods in this research area. In this regard, in recent years, it has become ever more evident that a skilled combination of concepts stemming from different metaheuristics can be a very promising strategy one should recur to when having to confront complicated optimization tasks. The hybridization of metaheuristics with other operations research techniques has been shown great appeal as well, as they typically represent complementary perspectives over the problem solving process as a whole. In general, combinations of components coming from different meta-heuristics and/or from more conventional exact methods into a unique opti-mization framework have been referred to by the label of "hybrid metaheuristics" [99, 115].

In this work, we present an optimization framework for tackling cutting and packing problems, which is based upon a methodology that complies directly with the aforementioned hybridization precept. This methodology involves a par-ticular combination scheme between heuristic and exact methods, which is sum-marized as follows. A metaheuristic engine – currently implemented in the form of a genetic algorithm (GA) – works as a generator of reduced instances for the original optimization problem, formulated as mathematical programming (viz., integer linear programming (ILP)) models. These instances, in turn, are solved by an exact optimization technique (solver), and the performance measures accom-plished by the respective models are interpreted as score (fitness) values by the metaheuristic, thus guiding its search process. As a means to assess the poten-tialities behind the novel approach, we provide an instantiation of the framework for dealing specifically with the constrained two-dimensional non-guillotine cut-ting problem. Computational experiments performed over standard benchmark problems are reported and discussed here, evidencing the effectiveness of the novel approach.

The remainder of the paper is organized as follows. In Section 6.2, we con-vey more information regarding some relatively novel hybrid metaheuristic approaches in optimization, and also discuss the constrained two-dimensional non-guillotine cutting problem. In Section 6.3, the hybrid framework is intro-duced in a step-by-step basis. In Section 6.4, the framework is instantiated for treating the specific problem we have described before. Then, in Section 6.5, we present and discuss some simulation results we have achieved by experiment-ing with the benchmark problem instances. This is done in order to assess the effectiveness of the novel approach by contrasting its performance with that achieved by alternative approaches. In Section 6.6, some final remarks with in-dication for future work conclude the paper.

6.2   Related Work

In this section, we first provide a brief outline of the recent advances in the field of hybrid metaheuristics. Then, we concentrate on the description of the constrained two-dimensional non-guillotine cutting problem.

6.2.1   Outline of Some Hybrid Metaheuristic Approaches

Over the last years, interest in hybrid metaheuristics has risen considerably among researchers in combinatorial optimization. Combinations of methods such as simulated annealing, tabu search and evolutionary algorithms have yielded very powerful search methods [117]. This can be evidenced by the diversity of works about this topic found in the literature.

In [111], for instance, a hybrid approach that combines simulated annealing with local search heuristics is introduced to solve the travelling salesman problem. In [110], the author presented a hybrid method that applies simulated annealing to improve the population obtained by a genetic algorithm. In [102], a local search algorithm, which utilizes problem-specific knowledge, is incorporated into the genetic operators of a GA instance to solve the multi-constraint knapsack problem. In [112], various kinds of local search are embedded in an evolutionary algorithm for locally improving candidate solutions obtained from variation operators.

It is only rather recently that hybrid algorithms which take ideas from both exact and heuristic local search techniques have been proposed. These techniques are traditionally seen as pertaining to two distinct branches of research toward the effective solution of combinatorial optimization problems, each one having particular advantages and disadvantages. Therefore, it appears to be a straightforward idea to try to combine these two distinct techniques into more powerful algorithms [105].

There have been very different attempts to combine strategies and methods from these two scientific streams. Some of these hybrids mainly aim at providing optimal solutions in shorter time, while others primarily focus on getting better heuristic solutions [114]. For instance, in order to reduce the search space, in [118], the authors combine tabu search with an exact method for solving the 0-1 multidimensional knapsack problem. By other means, in [103], a two-phase hybrid method is proposed where high quality tours for the travelling salesman problem are generated, and the sub-problem induced by the set of previous tours is solved exactly on the restricted graph. Recently, some promising concepts such as local branching [107] make use of a general-purpose commercial solver as a black-box tool to explore effectively suitable solution subspaces defined and controlled at a strategic level by a simple external branching framework.

In [117], the author has recently presented a taxonomy of hybrid metaheuristic components, which distinguishes the hybridization into two levels. In the low-level scheme, the result is a functional composition of a single optimization method. In this hybrid class, a given function of a metaheuristic is replaced by another metaheuristic. Conversely, in the high-level scheme, the different meta-heuristics are self-contained, and there is no direct relationship to the internal workings of the others. By other means, in [114], an alternative classification of existing approaches combining exact and metaheuristic algorithms for combinatorial optimization is presented, whereby the following two main categories are distinguished: (i) collaborative combinations; and (ii) integrative combinations. By collaboration, it means that the constituent algorithms exchange information to each other, but none is part of the other. The algorithms may be executed sequentially, intertwined, or in parallel. By integration, it means that one technique is a subordinate component of the other. Thus, there is the distinction between a master algorithm – which can be either an exact or a metaheuristic algorithm – and a slave algorithm. An alternative classification of hybridization methods can be found in [115].

6.2.2 The Constrained Two-Dimensional Non-guillotine Cutting Problem

The constrained two-dimensional non-guillotine cutting problem consists of cutting rectangular pieces from a single large rectangular object, as shown in Fig. 6.1. Each piece is of fixed orientation and must be cut with its edges parallel to the edges of the object. The number of pieces of each type that are cut must lie within prescribed limits and, in addition, the cuts may not go from one end to another. Each piece has an associated value and the objective is to maximise the total value of the pieces cut. This problem has been shown to be NP-Complete [104], meaning that it is impossible to find their optimal solution by resorting to an enumerative, brute-force approach alone, except in trivial cases.

In [104], two combinatory methods that generate constrained cutting patterns by successive horizontal and vertical builds of ordered rectangles are investigated

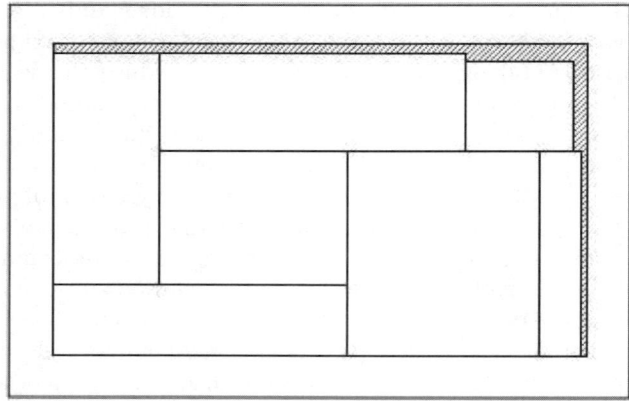

Fig. 6.1. A two-dimensional non-guillotine cutting pattern

to tackle this problem. Each of the algorithms uses a parameter to bound the maximum waste they may create. Error bounds measure how close the pattern wastes are to the waste of the optimal solution. These algorithms are fast and can yield efficient solutions when applied to small problems.

In [97], a tree search approach is presented based upon the lagrangean relaxation of a 0-1 integer linear programming formulation of the problem to derive an upper bound on the optimal solution. The formulation makes use of variables that relate to whether or not a piece of a particular type is cut with its bottom-left hand corner at a certain position. Subgradient optimization is used to optimize the bound derived from the lagrangean relaxation. Problem reduction tests derived from both the original problem and the lagrangean relaxation are given. This procedure is oriented to the solving of moderately-sized problems.

By other means, in [109], a new exact tree-search procedure is introduced. The algorithm limits the size of the tree search by using a bound derived from a lagrangean relaxation of the problem. Subgradient optimization is used to optimize this bound. Reduction tests derived from both the original problem and the lagrangean relaxation version produce substantial computational gains. The computational performance of the algorithm indicates that it is an effective procedure capable of optimally solving practical two-dimensional cutting problems of medium size.

In [106], the authors combined the use of a data structure for characterizing feasible packings with new classes of lower bounds, as well as other heuristics, in order to develop a two-level tree search algorithm for solving high-dimensional packing problems to optimality. In that approach, projections of cut pieces are made onto both the horizontal and vertical edges of the stock rectangle. Each such projection is translated into a graph, where the nodes in the graph are the cut pieces and an edge joins two nodes if the projections of the corresponding cut pieces overlap. The authors showed that a cutting pattern is feasible if and only if the projection graphs have certain properties, and also that problems of considerable size can be solved to optimality in reasonable time with this approach.

In a recent paper [98], Beasley introduced a new non-linear formulation of the constrained two-dimensional non-guillotine cutting problem. Based upon this formulation, a population heuristic is conceived, which explicitly works with a population of solutions and combine them together in some way to generate new solutions. Computational results for that heuristic approach applied to typical test problems taken from the literature are conveyed by the authors and are used as reference here for the purpose of comparison.

6.3 Hybrid Framework

The proposed framework prescribes the integration of two distinct conceptual components, according to the setting illustrated in Fig. 6.2. The first component is named as the Generator of Reduced Instances (GRI), which is in charge of producing reduced instances of the original problem. In theory, any metaheuristic technique can be recruited for such a purpose, provided that the reductions carried out always respect the constraints imposed by the problem at hand. This is necessary to ensure that an optimal solution found for any novel problem instance would also be a feasible solution to the original problem. Conversely, the role of the second element, referred to as the Solver of Reduced Instances (SRI), is to interpret and solve any of the generated problem instances coming out of the GRI. The optimal objective function values achieved with the solving of these sub-problems will serve as feedback to the GRI, in a manner as to guide its search process, although other performance measures (such as the computational costs involved or number of unfulfilled constraints, in case of constrained optimization) could be somehow employed as well. Any exact optimization technique (solver) could be, in thesis, a valid candidate to act as SRI.

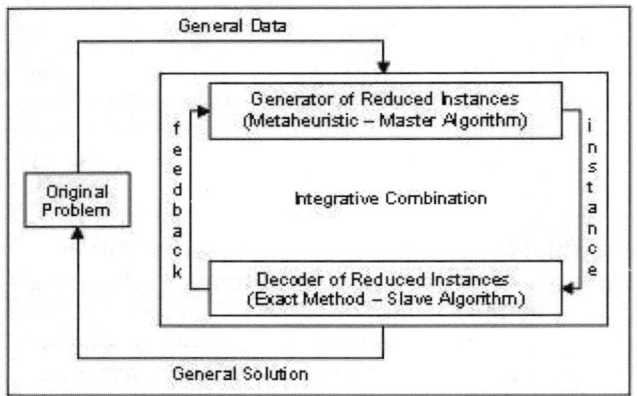

Fig. 6.2. The hybrid framework under investigation

According to the aforementioned classification proposed in [114], the methodology falls into the category of integrative combinations. The quality of the solutions to the instances generated by the metaheuristic is determined when the sub-problems are solved by the exact method, and the best solution obtained throughout the whole metaheuristic process is deemed to be the final solution to the original problem. However, there are cases where it is not possible to solve, or even generate, the reduced problem instance, due to its complexity and/or to other computational limitations of the solver environment. Thus, roughly speaking, the hybrid framework operates with the pragmatic perspective of providing effective solutions to complicated optimization tasks by searching for derived sub-problems that can still be directly handled by the exact methods one has currently at his/her disposal. Fig. 6.3 represents, hypothetically, the search space of the GRI metaheuristic, as well as demonstrates the distribution of sub-problems according to its dimension (size), solution quality, and the applicability of the SRI. The solution space of the framework is restricted to the sub-problems where the SRI is applicable.

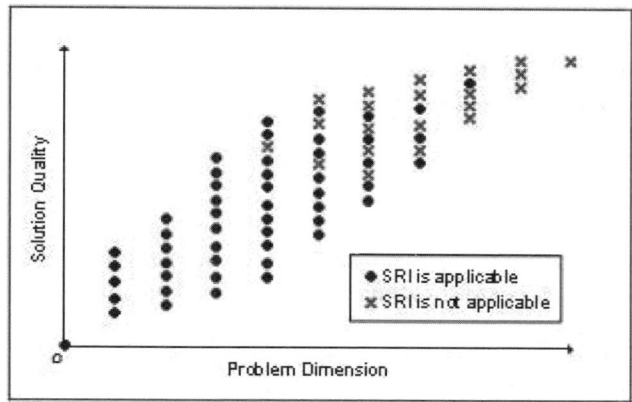

Fig. 6.3. Hypothetical search space of the GRI metaheuristic

Basically, the hybrid methodology, as informally described so far, roughly comprehends a sequence of three distinctive steps, discussed in the following.

Mathematical Formulation of the Problem

In this stage, the aspects to be incorporated in the formal model and the suppositions that can be made relative to the problem itself need to be specified. The desired objectives, the decision variables, and the considered constraints, all need to be made explicit in the mathematical formulation adopted.

Identification of the Reducible Structures

In principle, it would be possible to setup the entire model and solve the original problem using an exact method. However, the number of variables and constraints tends to grow extraordinarily in practical cases. In this step, the task is to identify the main entities that contribute somehow to the increase in the problem dimension. Thus, the idea is to generate and solve only sub-problems in an efficient way.

Specification of the Metaheuristic Sub-problem Generator

Finally, the choice of the metaheuristic technique to act as GRI should be done. This issue, like the previous one, deserves greater attention by the framework user, since it is the GRI component that is in charge to discover the reduced version of the original problem that could still provide the more adequate solution to it.

In the following, we present an instantiation of the proposed hybrid framework to cope specifically with the constrained two-dimensional non-guillotine cutting problem. A previous successful application of this approach for another cutting and packing problem can be found in [113].

6.4   Case Study

Here, the three steps of the methodology are revisited, each one providing detailed information pertaining exclusively to the constrained two-dimensional non-guillotine cutting problem.

Mathematical Formulation

To formulate the constrained two-dimensional non-guillotine cutting problem, we resort to the ILP model proposed in [97]. Consider a set of items grouped into m types. For each item type i, characterized by its length, width (l_i, w_i), and value v_i, there is an associated number of items b_i. Consider also a large object that has (L, W) as its length and width dimensions respectively. The items should be cut orthogonally from the object. Each 0-1 variable x_{ipq} alludes to the decision of whether to cut or not an item of type i at the coordinate (p, q).

$$x_{ipq} = \begin{cases} 1, \text{if an item of type } i \text{ is allocated at position } (p,q) \\ 0, \text{otherwise} . \end{cases} \quad (6.1)$$

It can be assumed that p and q belong, respectively, to the following discretization sets:

$$P = \{p \mid p = \sum_{i=1}^{m} \alpha_i l_i, p \leq L - \min\{l_i, i = 1, \ldots, m\}, \alpha_i \geq 0, \alpha_i \in \mathbb{Z}\} , \quad (6.2)$$

$$Q = \{q \mid q = \sum_{i=1}^{m} \beta_i w_i, q \leq W - \min\{w_i, i = 1, \ldots, m\}, \beta_i \geq 0, \beta_i \in \mathbb{Z}\} . \quad (6.3)$$

To avoid interposition of items, the incidence matrix is defined as a_{ipqrs}:

$$a_{ipqrs} = \begin{cases} 1, \text{if } p \leq r \leq p + l_i - 1 \text{ and } q \leq s \leq q + w_i - 1 \\ 0, \text{otherwise} . \end{cases} \quad (6.4)$$

which should be computed a priori for each type i $(i = 1, \ldots, m)$, for each coordinate (p, q) and for each coordinate (r, s). Other important set definitions come as follows:

$$P_i = \{p \mid p \in P, p \leq L - l_i\} , \quad (6.5)$$

$$Q_i = \{q \mid q \in Q, q \leq W - w_i\} . \quad (6.6)$$

Finally, the constrained two-dimensional non-guillotine cutting problem can be formulated as:

$$\max \quad \sum_{i=1}^{m} \sum_{p \in P_i} \sum_{q \in Q_i} v_i x_{ipq}$$

$$\text{s.t.} \quad \sum_{i=1}^{m} \sum_{p \in P_i} \sum_{q \in Q_i} a_{ipqrs} x_{ipq} \leq 1, \forall r \in P, s \in Q$$

$$\sum_{p \in P_i} \sum_{q \in Q_i} x_{ipq} \leq b_i, i = 1, \ldots, m$$

with $\quad x_{ipq} \in \{0,1\}, i = 1, \ldots, m, \forall p \in P, q \in Q$

Identification of the Reducible Structures

The number of cutting patterns of the model under investigation can be defined by the amount of types m and by the cardinality of the discretization sets, namely $|P|$ and $|Q|$. Several reduced instances of the original problem can then be eventually generated by excluding some of the types and/or some positions demarcated by the discretization sets.

Specification of the Metaheuristic Sub-problem Generator

A modified genetic algorithm has been implemented to account for the GRI task. Each of its individuals is represented by a binary chromosome, which encodes a particular set of discretization points to the aforementioned ILP model. In this encoding, each gene (bit) represents one possible element of discretization along a particular dimension, as shown in Fig. 6.4. Assigning a null value to a bit means that its associated element of discretization will not be used for the construction of the reduced problem; by flipping the value to one, the element will take part in the reduced problem instance. As mentioned earlier, each sub-problem should be solved by an exact method playing the role of the SRI. The optimal solution value attained for this problem instance is assigned as the fitness value of its corresponding GA individual.

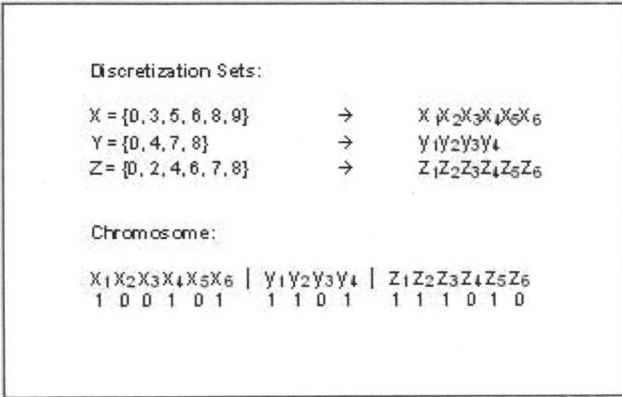

Fig. 6.4. The chromosome representation

For the initial population, n chromosomes are generated in the following manner: One of the items is randomly chosen and, then, only the elements proportional to the respective dimensions of this item will be considered for being selected to take part in the problem instance. Standard selection/variation operators are applied iteratively over the individuals across generations. More precisely, individuals are selected for mating and for replacement with the help of the roulette wheel selection mechanism. Furthermore, the single-point crossover is adopted with rate τ_c, whereas the simple mutation operator is applied with different rates for each chromosome segment. These mutation rates vary according to the length of the respective discretization set: $\tau_{mP} = 1/|P|$ and $\tau_{mQ} = 1/|Q|$. A maximum number of iterations τ_g, a prefixed interval of convergence c, as well as a maximum limit on the execution time of the whole framework t, are used in concert as stopping criteria.

6.5   Computational Experiments

To evaluate the potentialities behind the proposed hybrid framework (HF), a series of experiments have been carried out by our group, and some of the results achieved so far are discussed here. We have performed experiments over 21 typical test problems taken from the literature for which the optimal solutions are already known. These problems are: *Beasley01-Beasley12*, taken from [97]; *HadChr03* and *HadChr11*, taken from [109]; *Wang01*, taken from [119]; *ChrWhi03*, taken from [101]; and *FekSch01-FekSch05*, taken from [106].

For conducting these experiments, the GRI metaheuristic was developed in *Delphi* language, whereas an implementation of the overall hybrid framework has been executed and tested on an Intel Pentium 4 3.2 GHz platform with 1GB of RAM memory available. For implementing the SRI component, we have made use of a LINGO 8.0 DLL [116] as well. The control parameters of the GA engine were set as follows, after performing some experimental calibration: $n = 100$ individuals, $\tau_c = 0.5$, $\tau_g = 50$ iterations, $c = 10$ iterations, and $t = 7,200$ seconds. Moreover, as the optimal solutions for these test problems are known in advance, we have also decided to abort the framework execution whenever the optimal solution is discovered. In Table 6.1, for each problem instance, we present the best final solution achieved with the population heuristic (PH) proposed in [98], the standalone application of the LINGO solver, and the HF approach, along with the actual optimal solution.

The hybrid framework has presented very satisfactory results. Taking as reference the scores achieved by Beasley [98], the hybrid framework presents better solutions (even though with high costs in terms of computation time for the larger problems), with the average percentage deviation from optimality of 0.37%, prevailing over the score of 1.24% achieved by Beasley. For the smaller test problem instances, the hybrid framework achieved the optimal solution very quickly (see Fig. 6.5 for an example). Moreover, it is clear that the hybrid framework is itself

Table 6.1. Computational Results

Problem Instance	PH Best Solution	PH Time in seconds	Lingo Solution	Lingo Time in seconds	HF Best Solution	HF Time in seconds	Optimal Solution
Beasley01	164	0.02	164	0.80	164	0.06	164
Beasley02	230	0.16	230	20.69	230	0.08	230
Beasley03	247	0.53	247	9.27	247	0.45	247
Beasley04	268	0.01	268	0.48	268	0.08	268
Beasley05	358	0.11	358	5.03	358	0.22	358
Beasley06	289	0.43	289	59.09	289	0.31	289
Beasley07	430	0.01	430	80.50	430	0.17	430
Beasley08	834	3.25	834	143.80	834	1.89	834
Beasley09	924	2.18	n/f	–	924	78.06	924
Beasley10	1452	0.03	1452	1151.83	1452	1.42	1452
Beasley11	1688	0.60	n/f	–	1688	17.84	1688
Beasley12	1801	3.48	n/f	–	1865	18.78	1865
HadChr03	1178	0.03	1178	204.11	1178	0.06	1178
HadChr11	1270	0.04	n/f	–	1270	0.81	1270
Wang01	2721	6.86	n/f	–	2726	3507.61	2726
ChrWhi03	1720	8.63	n/f	–	1860	415.22	1860
FekSch01	27486	19.71	n/f	–	27360	5773.08	27718
FekSch02	21976	13.19	n/f	–	21888	7200.00	22502
FekSch03	23743	11.46	n/f	–	23743	5997.21	24019
FekSch04	31269	32.08	n/f	–	32018	7200.00	32893
FekSch05	26332	83.44	n/f	–	27923	830.53	27923

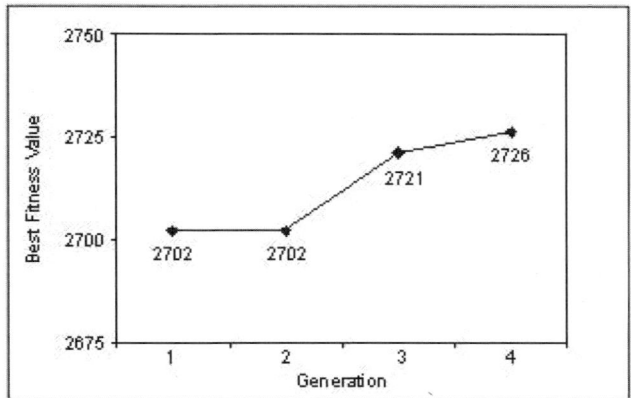

Fig. 6.5. Dynamics of the GA evolutionary process: A step-by-step improvement until reaching the optimum, for instance Wang01

adding value, providing optimal solutions in shorter time than the standalone application of the LINGO solver and presenting feasible solutions for all test problems.

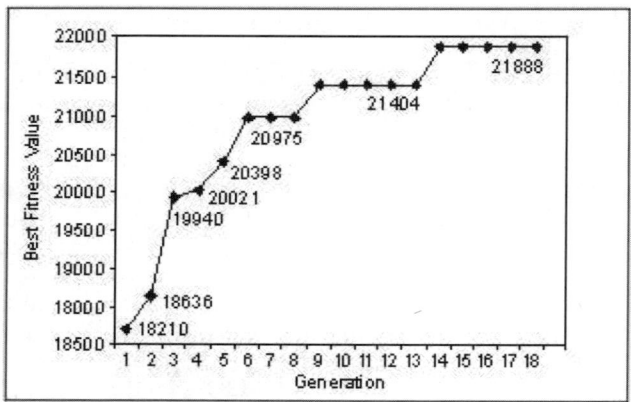

Fig. 6.6. Dynamics of the GA evolutionary process: Difficulties to reach the optimal due to premature convergence, for instance FekSch02

However, there were problems where the whole approach could not succeed as well, like instances *FekSch01-FekSch04* – see Fig. 6.6. Perhaps these problem instances are associated with huge search spaces (due to the particular dimensions of some of the items included), and the GRI component, as currently implemented, could not keep up with the task of exploring well some promising regions of these solution spaces.

6.6   Final Remarks

In this work, we investigate the application of a novel hybrid framework for coping with the constrained two-dimensional non-guillotine cutting problem. The framework is based upon a particular type of hybridization between approximate and exact optimization algorithms. A series of experiments conducted over benchmark instances were also reported and discussed here in a manner as to assess the potentialities of the framework. Overall, the optimization performance achieved with the new methodology have been satisfactory, taking as reference the results achieved by other approaches and taking into account the inherent difficulties associated with the C&P class of problems – indeed, some problem instances could not be directly solved through the application of the exact optimization package alone.

As future work, we plan to investigate the impact of other forms of feedback coming from the SRI component – like the positions effectively used at the solution of the sub-problem –, in a manner as to allow a more tailored redefinition of the sub-problems supplied by the GRI and a better exploration of the solution space. In order to evaluate the generalization level behind this approach, we plan to develop and test other instantiations of the framework for coping with other types of hard combinatorial optimization problems. Likewise, it is in our plans to investigate other types of metaheuristics (like trajectory-based or even hybrid ones to play the role of the GRI component in the framework. Parallel versions of the hybrid framework are also underway.

Acknowledgements

The first and third authors gratefully acknowledge the financial support received from CAPES, through a Ph.D. scholarship, and from CNPq/FUNCAP, through a DCR grant (project number 23661-04).

7

A Hybrid Genetic Algorithm for the DNA Fragment Assembly Problem

Enrique Alba and Gabriel Luque

E.T.S.I. Informática
Campus Teatinos, 29071 Málaga (Spain)
{eat,gabriel}@lcc.uma.es

Summary. In this paper we propose and study the behavior of a new hybrid heuristic algorithm for the DNA fragment assembly problem. The DNA fragment assembly is a problem solved in the early phases of the genome project and thus very important, since the other steps depend on its accuracy. This is an NP-hard combinatorial optimization problem which is growing in importance and complexity as more research centers become involved on sequencing new genomes. Our contribution is a hybrid method that combines a promising heuristic, PALS, with a well-know metaheuristic, a genetic algorithm, obtaining as result a very efficient assembler that allows to find optimal solutions for large instances of this problem.

Keywords: DNA Fragment Assembly, Hybridization, Genetic Algorithm, Problem Aware Local Search.

7.1 Introduction

With the advance of computational science, bioinformatics has become more and more attractive to researchers in the field of computational biology. Genomic data analysis using computational approaches is very popular as well. The primary goal of a genomic project is to determine the complete sequence of the genome and its genetic contents. Thus, a genome project is accomplished in two steps; the first one is the genome sequencing and the second one is the genome annotation (i.e., the process of identifying the boundaries between genes and other features in raw DNA sequence).

In this chapter, we focus on the genome sequencing, which is also known as the DNA fragment assembly problem. The DNA fragment assembly occurs in the very beginning of the process and therefore other steps depend on its accuracy. At present, DNA sequences that are longer than 600 base-pairs (bps) cannot routinely be sequenced accurately. For example, human DNA is about 3.2 billion nucleotides in length and cannot be read at once. Hence, long strands of DNA need to be broken into small fragments for sequencing in a process called *shotgun sequencing*. In this approach, several copies of a portion of DNA are prouced and each copy is broken into many random segments short enough to be sequenced automatically. But this process does not keep neither the ordering of the

C. Cotta and J. van Hemert (Eds.): Recent Advances in Evol. Comp., SCI 153, pp. 101–112, 2008.
springerlink.com

fragments nor the portion from which a particular fragment came. This leads to the DNA fragment assembly problem [133] in which these short sequences have to be reassembled to their (supposed) original form. The automation allows shotgun sequencing to proceed far faster than traditional methods. But comparing all the tiny pieces and matching up the overlaps requires massive computation or carefully designed programs.

The assembly problem is therefore a combinatorial optimization problem that, even in the absence of noise, is NP-hard [134]: given k fragments, there are $2^k k!$ possible combinations. Over the past decade a number of fragment assembly packages have been developed and used to sequence different organisms. The most popular packages are PHRAP, TIGR assembler [135], STROLL [136], CAP3 [137], Celera assembler [138], and EULER [139]. These packages deal with the previously described challenges to different extents, but none of them solves all of them. Each package automates fragment assembly using a variety of algorithms. The most popular techniques are greedy-based procedures while other approaches have tackled the problem with metaheuristics [140]. Recently, Alba and Luque [141] proposed a new promising heuristic for this problem, PALS, in which the search is guided by an estimation of the number of contigs. This estimation is one key point in reducing the computational demands of the technique. We here extend [141] and design a new hybrid method based on that heuristic. This method consists of a genetic algorithm (GA), which uses PALS as a mutation operator. We compare the results of our new approach with the "pure" methods, GA and PALS.

The remainder of this chapter is organized as follows. In the next section, we present background information about the DNA fragment assembly problem. In Section 7.3, the details of the used algorithms are presented. We analyze the results of our experiments in Section 7.5. Last, we end this chapter by giving our final thoughts and conclusions in Section 7.6.

7.2   The DNA Fragment Assembly Problem

In order to determine the function of specific genes, scientists have learned to read the sequence of nucleotides comprising a DNA sequence in a process called DNA sequencing. To do that, multiple exact copies of the original DNA sequence are made. Each copy is then cut into short fragments at random positions. These are the first three steps depicted in Fig. 7.1 and they take place in the laboratory. After the fragment set is obtained, a traditional assemble approach is followed in this order: overlap, layout, and then consensus. To ensure that enough fragments overlap, the reading of fragments continues until a coverage is satisfied. These steps are the last three ones in Fig. 7.1. In what follows, we give a brief description of each of the three phases, namely overlap, layout, and consensus.

Overlap Phase - Find the overlapping fragments. This phase consists in finding the best or longest match between the suffix of one sequence and the prefix of another. In this step, we compare all possible pairs of fragments to

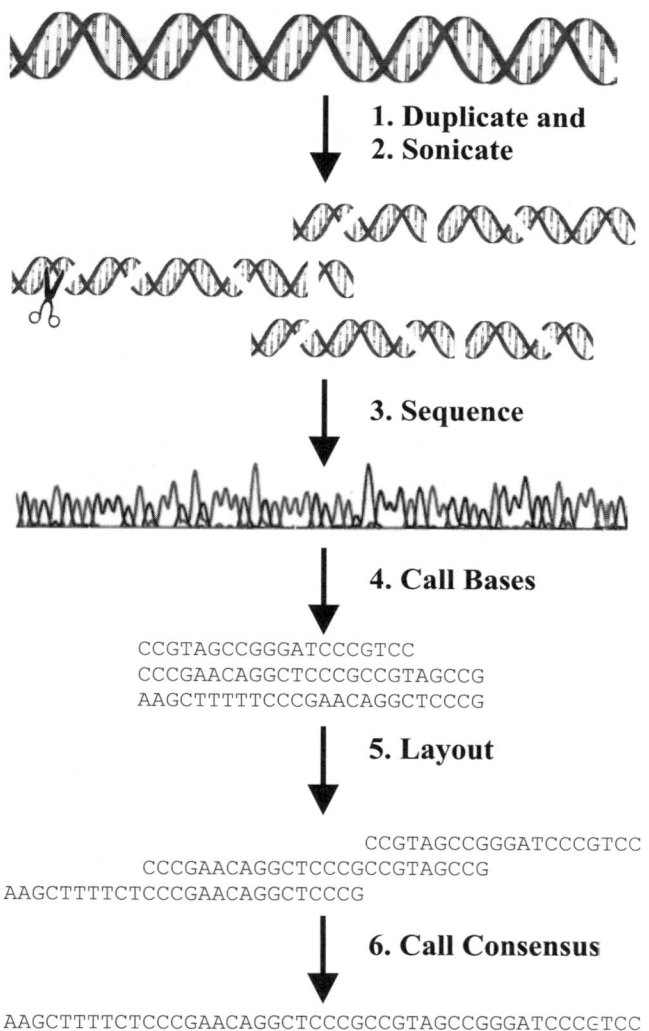

Fig. 7.1. Graphical representation of DNA sequencing and assembly

determine their similarity. Usually, a dynamic programming algorithm applied to semiglobal alignment is used in this step. The intuition behind finding the pairwise overlap is that fragments with a high overlap are very likely next to each other in the target sequence.

Layout Phase - Find the order of fragments based on the computed similarity score. This is the most difficult step because it is hard to tell the true overlap due to the following challenges:

1. Unknown orientation: After the original sequence is cut into many fragments, the orientation is lost. One does not know which strand should be selected. If one fragment does not have any overlap with another, it is still possible that its reverse complement might have such an overlap.
2. Base call errors: There are three types of base call errors: substitution, insertion, and deletion. They occur due to experimental errors in the electrophoresis procedure (the method used in the laboratories to read the DNA sequences). Errors affect the detection of fragment overlaps. Hence, the consensus determination requires multiple alignments in highly coverage regions.
3. Incomplete coverage: It happens when the algorithm is not able to assemble a given set of fragments into a single contig. A contig is a sequence in which the overlap between adjacent fragments is greater or equal to a predefined threshold (cutoff parameter).
4. Repeated regions: "Repeats" are sequences that appear two or more times in the target DNA. Repeated regions have caused problems in many genome-sequencing projects, and none of the current assembly programs can handle them perfectly.
5. Chimeras and contamination: Chimeras arise when two fragments that are not adjacent or overlapping on the target molecule join together into one fragment. Contamination occurs due to the incomplete purification of the fragment from the vector DNA.

After the order is determined, the progressive alignment algorithm is applied to combine all the pairwise alignments obtained in the overlap phase.

Consensus Phase - Derive the DNA sequence from the layout. The most common technique used in this phase is to apply the majority rule in building the consensus.

To measure the quality of a consensus, we can look at the distribution of the coverage. Coverage at a base position is defined as the number of fragments at that position. It is a measure of the redundancy of the fragment data, and it denotes the number of fragments, on average, in which a given nucleotide in the target DNA is expected to appear. It is computed as the number of bases read from fragments over the length of the target DNA [133].

$$Coverage = \frac{\sum_{i=1}^{n} length\ of\ the\ fragment\ i}{target\ sequence\ length} \tag{7.1}$$

where n is the number of fragments. The higher the coverage, the fewer number of the gaps, and the better the result.

7.3 Algorithms

In this section we detail the three proposed algorithms that we have designed to solve the DNA fragment assembly problem. Firstly, we explain the two cannonical methods, a GA and PALS (Problem Aware Local Search), and then we address the hybrid one (GA+PALS).

7.4   Genetic Algorithm

A genetic algorithm [143] is an iterative technique that applies stochastic operators on a pool of individuals (the population). Every individual in the population is the encoded version of a tentative solution. Initially, this population is randomly generated. An evaluation function associates a fitness value to every individual indicating its suitability to the problem.

The genetic algorithm that we have developed for the solution of the DNA fragment assembly follows the basic structured shown in Algorithm 7.1. In next paragraphs we give some details about the most important issues of our GA implementation.

Algorithm 7.1. Basic GA

Generate (P(0))
t := 0
while not Termination_Criterion(P(t)) **do**
    Evaluate(P(t))
    P'(t) := Selection(P(t))
    P'(t) := Recombination(P'(t))
    P'(t) := Mutation(P'(t))
    P(t+1) := Replacement(P(t), P'(t))
    t := t+1
end while
return Best_Solution_Found

For the indivuals, we use a permutation representation with integer number encoding. A permutation of integers represents a sequence of fragment numbers, where successive fragments are later assumed to overlap. The solution in this representation requires a list of fragments assigned with a unique integer ID. For example, 8 fragments will need eight identifiers: 0, 1, 2, 3, 4, 5, 6, 7. The permutation representation requires special operators to make sure that we always get legal (feasible) solutions. In order to maintain a legal solution, the two conditions that must be satisfied are (1) all fragments must be presented in the ordering, and (2) no duplicate fragments are allowed in the ordering. For example, one possible ordering for 4 fragments is 3 0 2 1. It means that fragment 3 is at the first position and fragment 0 is at the second position, and so on.

In the DNA fragment assembly problem, the fitness function measures the multiple sequences alignment quality and finds the best scoring alignment. Parsons, Forrest, and Burks mentioned two different fitness functions [144]. In this work, we use the first one. This function sums the overlap score for adjacent fragments in a given solution. When this fitness function is used, the objective is to maximize such a score. It means that the best individual will have the highest score.

$$F(l) = \sum_{i=0}^{n-2} w_{f[i], f[i+1]}, \tag{7.2}$$

where $w_{i,j}$ is the overlap between fragments i and j. This function have several drawbacks, but despite them and its apperent simplicity it is very difficult to find a better one (see next subsection).

For our experimental runs, we use the order-based crossover (OX) as recombination operator. The order-based crossover operator first copies the fragment ID between two random positions in Parent1 into the offspring's corresponding positions. We then copy the rest of the fragments from Parent2 into the offspring in the relative order presented in Parent2. If the fragment ID is already present in the offspring, then we skip that fragment. The method preserves the feasibility of every string in the population.

Finally, the swap mutation operator is applied to maintain diversity in the population. This operator randomly selects two positions from a permutation and then swaps the two fragment positions. Since this operator does not introduce any duplicate number in the permutation, the solution it produces is always feasible.

7.4.1   Problem Aware Local Search (PALS)

Classical assemblers use fitness functions that favor solutions in which strong overlap occurs between adjacent fragments in the layouts, using equations like Eq. 7.2. However, the actual objective is to obtain an order of the fragments that minimizes the number of contigs, with the goal of reaching one single contig, i.e., a complete DNA sequence composed of all the overlapping fragments. Therefore, the number of contigs is always used as the high-level criterion to judge the whole quality of the results, since it is difficult to capture the dynamics of the problem into other mathematical functions. Contig values are computed by applying a final step of refinement with a greedy heuristic regularly used in this domain [145]. We have even found that in some (extreme) cases it is possible that a solution with a better fitness (using Eq. 7.2) than other one generates a larger number of contigs (worse solution). All this suggests that the fitness (overlapping) should be complemented with the actual number of contigs.

And here we arrive to the true obstacle: the calculation of the number of contigs is a quite time-consuming operations, and this definitely precludes any algorithm to use it. A solution to this problem is the utilization of a method which should not need to know the exact number of contigs, thus being computationally light. Our key contribution is to indirectly estimate the number of contigs by measuring the number of contigs that are created or destroyed when tentative solutions are manipulated in the algorithm. We propose a variation of Lin's 2-opt [146] for the DNA field, which does not only use the overlap among the fragments, but that it also takes into account (in an intelligent manner) the number of contigs that have been created or destroyed. The pseudo-code of our proposed method is shown in Algorithm 7.2.

Our algorithm works on a single solution (an integer permutation) which is generated using the `GenerateInitialSolution` method, and it is iteratively modified by the application of movements in a structured manner. A movement

---

Algorithm 7.2. PALS

$s \leftarrow$ GenerateInitialSolution() {Create the initial solution}
repeat
  $L \leftarrow \emptyset$
  for i = 0 to N **do**
    for j = 0 to N **do**
      $\Delta_c, \Delta_f \leftarrow$ CalculaeDelta(s,i,j) {See Algorithm 7.3}
      if $\Delta_c >= 0$ **then**
        $L \leftarrow L \cup < i, j, \Delta_f, \Delta_c >$ {Add candidate movements to L}
      end if
    end for
  end for
  if $L <> \emptyset$ **then**
    $< i, j, \Delta_f, \Delta_c > \leftarrow$ Select(L) {Select a movement among the candidates}
    ApplyMovement(s,i,j) {Modify the solution}
  end if
until no changes
return: s

---

is a perturbation (**ApplyMovement** method) that, given a solution s, and two positions i and j, reverses the subpermutation between the positions i and j.

The key step in PALS is the calculation of the variation in the overlap (Δ_f) and in the number of contigs (Δ_c) among the current solution and the resulting solution after applying a movement (see Algorithm 7.3). This calculation is computationally light since we do not calculate neither the fitness function nor the number of contigs, but instead we estimate the variation of these values. To do this, we only need to analyze the affected fragments by the tentative movement (i, j, $i - 1$, and $j + 1$), reducing the overlap score of affected fragments of the current solution and adding the one of the modified solution to Δ_f (equations of lines 4-5 of Algorithm 7.3) and testing if some current contig is broken (first two if statements of Algorithm 7.3) or two contigs are merged (last two if statements of Algorithm 7.3) by the movement operator.

In each iteration, PALS makes these calculations for all possible movements, storing the candidate movements in a list L. Our method only considers those candidate movements that do not increase the number of contigs ($\Delta_c \leq 0$). Once it has completed the previous calculations, the method selects a movement of the list L and applies it. The algorithm stops when no more candidate movements are generated.

To complete the definition of our method we must decide how the initial solution is generated (**GenerationInitialSolution** method) and how a movement is selected among all possible candidates (**Select** method). In [141] we tested several alternatives for these operations, and we concluded that the best setting is to start from a random solution and to select the best movement found in each iteration.

Algorithm 7.3. `CalculateDelta(s,i,j)` function

$\Delta_c \leftarrow 0$
$\Delta_f \leftarrow 0$
{Calculate the variation in the overlap}
$\Delta_f = w_{s[i-1]s[j]} + w_{s[i]s[j+1]}$ {Add the overlap of the modified solution}
$\Delta_f = \Delta_f - w_{s[i-1],s[i]} - w_{s[j]s[j+1]}$ {Remove the overlap of the current solution}
{Test if a contig is broken, and if so, it increases the number of contigs}
if $w_{s[i-1]s[i]} > cutoff$ **then**
    $\Delta_c = \Delta_c + 1$
end if
if $w_{s[j]s[j+1]} > cutoff$ **then**
    $\Delta_c = \Delta_c + 1$
end if
{Test if two contig are merged, and if so, it decreases the number of contigs}
if $w_{s[i-1]s[j]} > cutoff$ **then**
    $\Delta_c = \Delta_c - 1$
end if
if $w_{s[i]s[j+1]} > cutoff$ **then**
    $\Delta_c = \Delta_c - 1$
end if
return: Δ_f, Δ_c

7.4.2 Hybrid Approach

Finally, we define here the hybrid algorithm proposed in this article. In its broadest sense, hybridization refers to the inclusion of problem-dependent knowledge in a general search algorithm [147] in one of two ways: *strong hybrids*, where problem-knowledge is included as problem-dependent representation and/or operators, and *weak hybrids*, where several algorithms are combined in some manner to yield the new hybrid algorithm. Therefore, we define a weak hybrid called GA+PALS, where a GA uses PALS as an additional evolutionary operator. The

---

Algorithm 7.4. Hybrid GA+PALS

Generate (P(0))
t := 0
while not Termination_Criterion(P(t)) **do**
    Evaluate(P(t))
    P'(t) := Selection(P(t))
    P'(t) := Recombination(P'(t))
    P'(t) := Mutation(P'(t))
    P'(t) := PALS(P'(t))
    P(t+1) := Replacement(P(t), P'(t))
    t := t+1
end while
return Best_Solution_Found

---

pseudo-code of this approach is shown in Algorithm 7.4. In the main loop of this method after the traditional recombination and mutation operators are applied, several solutions are randomly selected (according to a very low probability) from the current offspring and they are improved using the local search algorithm. The rationale for this sort of hybridization is that, while the GA locates "good" regions of the search space (exploration), PALS allows for exploitation in the best regions found by its partner (neighbourhood intensification). Evidently, the motivation in this case is to see if, by taking the best of these two heuristics (i.e., the genetic algorithm and PALS), we could produce another heuristic which would perform better than any of the two approaches from which it was created (new emergent behaviour).

7.5 Experimental Results

In this section we analyze the behavior of our proposed GA+PALS method. First, the target problem instances used and the parameter setting are presented in Section 7.5.1. In the next subsection, we study the results of the different algorithms presented previously in Section 7.3, and finally, we compare our approaches with other assemblers.

The experiments have been executed on a Intel Pentium IV 2.3GHz with 1GB running SuSE Linux 8.1. Because of the stochastic nature of the algorithms, we performed 30 independent runs of each test to gather meaningful experimental data and apply statistical confidence metrics to validate our results and conclusions.

7.5.1 Target Problem Instances and Parameter Settings

To test and analyze the performance of our algorithm we generated several problem instances with GenFrag [148]. GenFrag takes a known DNA sequence and uses it as a parent strand from which random fragments are generated according to the criteria supplied by the user (mean fragment length and coverage of parent sequence).

We have chosen a large sequence from the NCBI web site[1]: the Neurospora crassa (common bread mold) BAC, with accession number BX842596, which is 77,292 bases long. The instances generated are free from errors of types 4 and 5 (see Section 7.2) and the remainder errors are considered and eliminated during the calculation of the overlap among the fragments.

We must remark that the instance is quite complex. It is often the case that researches use only one or two instances of low-medium sizes (15-30k bases long). We dare to include two large instances (up to 77k bases long) because the efficiency of our technique, that will be shown to be competitive to modern assemblers.

We experimented with coverage 4 and 7. These instances are very hard since they are generated from very long sequences using a small/medium value of

---

[1] http://www.ncbi.nlm.nih.gov/

Table 7.1. Information of datasets. Accession numbers are used as instance names.

Parameters	BX842596	
Coverage	4	7
Mean fragment length	708	703
Number of fragments	442	773
Name	BX842596(4)	BX842596(7)

Table 7.2. Parameters for GA and GA+PALS

Paramter	Value
Independent runs	30
Cutoff	30
Max Evaluations	512000
Popsize	1024
P_c	0.8
P_m	0.2
Selection	Ranking
P_{PALS}	2/Popsize

coverage and a very restrictive cutoff (threshold to join adjacent fragments in the same contig). The combination of these parameters produces a very complex instance. For example, longer target sequences have been solved in the literature [137], however they have a higher coverage which makes them not so difficult. The reason is that the coverage measures the redundancy of the data, and the higher coverage, the easier the problem. The cutoff, which we have set to thirty (a very high value), provides one filter for spurious overlaps introduced by experimental error. Instances with these features have been solved in the past only when target sequences vary from 20k to 50k base pairs [144, 145, 149] while we solve instances up to 70k base pairs.

Table 7.1 presents information about the specific fragments sets we use to test our algorithm.

From these previous analyses, we conclude that the best settings for our problem instances of the fragment assembly problem are the following: the GA uses a population size of 1024 individuals, OX as crossover operator (with probability 0.8), and with an edge swap mutation operator (with probability 0.2). We use the same configuration for the hybrid one, including the PALS as additional evolutionary operator with a very low probability (2/population_size). A summary of the conditions for our experimentation is found in Table 7.2. Values in the tables are average results over 30 independent runs. Since we deal with stochastic algorithms, we have carried out an statistical analysis of the results which consists of the following steps. A Kolmogorov-Smirnov test is performed in order to check whether the variables are normal or not. All the Kolmogorov-Smirnov normality tests in this work were not successful, and therefore we use a non-parametric test: Kruskal-Wallis (with 95% of confidence).

7.5.2 Performance Analysis

We thoroughly compare in this section the results obtained by the hybrid GA+PALS, a GA and the PALS heuristic itself. Additionally, these algorithms are compared versus some of the most popular assemblers in the literature in terms of the final number of assembled contigs in the last part of this section.

We show in tables 7.3 and 7.4 the results obtained by the studied algorithms for instance BX842596 with coverage 4 and 7, respectively. Specifically, we compare the algorithms in terms of the best and average solutions found, the best and average number of contigs in the solution, and the average elapsed wall clock time. The best results are **bolded**. All the obtained results are statistically significant (it is indicated by + symbol in **Test** row).

In these two tables we can see that the proposed hybrid method is the best of the three compared algorithms for the two studied instances in terms of best and average solution found, and the number of contigs in the solution. This means that GA+PALS is very good at finding accurate solutions, which is the main goal of biologists. Conversely, it is much slower than PALS, as it was expected. The GA is the worst algorithm with statistically significant results in all the cases (**Test** row).

An interesting finding is that, even though the GA is the worst algorithm, it becomes the more accurate after hybridizing it with PALS as a local search step. Also, although the run time is clearly penalized (population driven search), it is still highly competitive with the existing assemblers (see Table 7.5) in the literature (which require several hours). Indeed, the hybrid method is not only the most accurate of the compared algorithms, but also very robust in its accuracy (lowest average contigs, see Table 7.3). Additionally, since GAs are very easily parallelizable [150], the utilization of parallel platforms or even grid systems might allow a considerable reduction of the execution time.

Table 7.3. Comparison of the studied algorithms. Instance BX842596(4).

	Best Solution	Average Solution	Best Contigs	Average Contigs	Time (seconds)
PALS	227130	226744.33	4	9.90	**3.87**
GA	92772	88996.60	6	14.29	32.62
GA+PALS	**228810**	**227331.24**	**2**	**4.71**	298.06
Test	•	+	•	+	+

Table 7.4. Comparison of the studied algorithms. Instance BX842596(7).

	Best Solution	Average Solution	Best Contigs	Average Contigs	Time (seconds)
PALS	443512	440779.30	2	7.80	**24.84**
GA	308297	304330.60	4	14.29	185.77
GA+PALS	**445426**	**443234.81**	**2**	**3.69**	1531.31
Test	•	+	•	+	+

Table 7.5. Best final number of contigs for the studied assemblers and for other specialized systems

	PALS	GA	GA+PALS	PMA [145]	CAP3 [137]	Pharp [142]
BX842596(4)	4	6	2	7	6	6
BX842596(7)	2	4	2	2	2	2

The three studied algorithms are compared versus some of the most important assemblers existing in the literature in Table 7.5. These algorithms are a pattern matching algorithm (PMA) [145] and two commercially available packages: CAP3 [137] and Phrap [142]. From this table we can conclude that the proposed GA+PALS is a really competitive tool in the best trade-off for biologists and computer scientists, clearly outperforming all the compared algorithms, and thus representing a new state of the art.

7.6   Conclusions

The DNA fragment assembly is a very complex problem in computational biology. Since the problem is NP-hard, the optimal solution is at impossible to find for real cases, only for very small problem instances. Hence, computational techniques of affordable complexity such as heuristics are needed for it.

We proposed in this chapter a new hybrid algorithm for solving the DNA fragment assembly problem. The algorithm was obtained after hybridizing a GA with the PALS heuristic, a recently proposed algorithm that represents the state of the art for this problem. The resulting algorithm clearly improves the results of PALS, as well as those of an equivalent GA without local search. The drawback is that the resulting GA+PALS considerably increases the computational time with respect to PALS.

In the future we plan to study new designs that will allow to reduce the computational time of the hybrid method. Concretely, we plan to focus on parallel and distributed models. Also, another open research line is the development of new recombination operators that could allow to calculate an estimation of the contigs created or destroyed by the operator as 2-opt permits.

Acknowledgments

The authors are partially supported by the Ministry of Science and Technology and FEDER under contract TIN2005-08818-C04-01 (the OPLINK project).

8

A Memetic-Neural Approach to Discover Resources in P2P Networks

Ferrante Neri, Niko Kotilainen, and Mikko Vapa

Department of Mathematical Information Technology, Agora, University of Jyväskylä, P.O. Box 35 (Agora), FI-40014 University of Jyväskylä, Finland
neferran@cc.jyu.fi, niko.kotilainen@jyu.fi, mikko.vapa@jyu.fi

Summary. This chapter proposes a neural network based approach for solving the resource discovery problem in Peer to Peer (P2P) networks and an Adaptive Global Local Memetic Algorithm (AGLMA) for performing in training of the neural network. The neural network, which is a multi-layer perceptron neural network, allows the P2P nodes to efficiently locate resources desired by the user. The necessity of testing the network in various working conditions, aiming to obtain a robust neural network, introduces noise in the objective function. The AGLMA is a memetic algorithm which employs two local search algorithms adaptively activated by an evolutionary framework. These local searchers, having different features according to the exploration logic and the pivot rule, have the role of exploring decision space from different and complementary perspectives. Furthermore, the AGLMA makes an adaptive noise compensation by means of explicit averaging on the fitness values and a dynamic population sizing which aims to follow the necessity of the optimization process. The numerical results demonstrate that the proposed computational intelligence approach leads to an efficient resource discovery strategy and that the AGLMA outperforms an algorithm classically employed for executing the neural network training.

Keywords: Memetic Algorithms, Neural Networks, P2P Networks, Telecommunication, Noisy Optimization Problems.

8.1 Introduction

During recent years the use of peer-to-peer networks (P2P) has significantly increased. P2P networks are widely used to share files or communicate with each other using Voice over Peer-to-Peer (VoP2P) systems, for example Skype. Due to the large number of users and large files being shared communication load induced to the underlying routers is enormous and thus demand of high performance in peer-to-peer networks is constantly growing.

In order to obtain a proper functioning of a peer-to-peer network a crucial point is to efficiently execute the peer-to-peer resource discovery, meaning the search of information (files, users, devices etc.) within a network of computers connected by Internet. An improper resource discovery mechanism would lead to overwhelming query traffic within the P2P network and consequently to a waste of bandwidth of each single user connected to the network.

C. Cotta and J. van Hemert (Eds.): Recent Advances in Evol. Comp., SCI 153, pp. 113–129, 2008.
springerlink.com

Although several proposals are present in commercial packages (e.g., Gnutella [151] and KaZaA), this problem is still intensively studied in literature. Resource discovery strategies can be divided into two classes: breadth-first search and depth-first search. Breadth-First Search (BFS) strategies forward a query to multiple neighbors at the same time whereas Depth-First Search (DFS) strategies forward only to one neighbor. In both strategies, the choice of those neighbors receiving the query is carried out by heuristic methods. These heuristics might be stochastic e.g. random selection [152], or based on deterministic rules [153].

BFS strategies have been used in Gnutella [151], where the query is forwarded to all neighbors and the forwarding is controlled by a time-to-live parameter. This parameter is defined as the amount of hops required to forward the query. Two nodes are said to be n hops apart if the shortest path between them has length n [153]. The main disadvantage of the Gnutella's mechanism is that it generates a massive traffic of query messages when the time-to-live parameter is high thus leading to a consumption of an unacceptable amount of bandwidth.

In order to reduce query traffic, Lv et al. [152] proposed the *Expanding Ring*. This strategy establishes that the time-to-live parameter is gradually increased until enough resources have been found. Although use of the *Expanding Ring* is beneficial in terms of query reduction, it introduces some delay to resource discovery and thus implies a longer waiting time for the user. Kalogeraki et al. [154] proposed a Modified Random Breadth-First Search (MRBFS) as an enhancement of the Gnutella's algorithm. In MRBFS, only a subset of neighbors are selected randomly for forwarding. They also proposed an intelligent search mechanism which stores the performance of the queries previously done for each neighbor. This memory storage is then used to direct the subsequent queries. Following the ideas of Kalogeraki et al., Menascé [155] proposed that only a subset of neighbors are randomly selected for forwarding. Yang and Garcia-Molina [153] proposed the Directed BFS (DBFS), which selects the first neighbor based on one of several heuristics and further uses BFS for forwarding the query. They also proposed the use of local indices for replicating resources to a certain radius of hops from a node. In Gnutella2 [156] a trial query is sent to the neighbors and, on the basis of obtained results, an estimate of how widely the actual query should be forwarded is calculated.

In the DFS strategies, selection of the neighbor chosen for the query forwarding is performed by means of heuristics. The main problem related to use of this strategy is the proper choice of this heuristic. A popular heuristic employed with this aim is the random walker which selects the neighbor randomly. The random walker terminates when a predefined number of hops have been travelled or when enough resources have been found. Lv et al. [152] studied the use of multiple random walkers which periodically check the query originator in order to verify if the query should still be forwarded further. Tsoumakos and Roussopoulos [157] proposed an Adaptive Probabilistic Search (APS). The APS makes use of feedback from previous queries in order to tune probabilities for further forwarding of random walkers. Crespo and Garcia-Molina [158] proposed the routing indices, which provide shortcuts for random walkers in locating resources. Sarshar et al. [159] proposed the Percolation Search Algorithm (PSA) for power-law networks. The idea is to replicate a copy of resources to a sufficient number of nodes and thus ensure that resource discovery algorithm locates at least one replica of the resource.

The main limitation of the previous studies, for both BFS and DFS strategies, is that all the approaches are restricted to only one search strategy. On the contrary, for the same P2P network, in some conditions it is preferable to employ both BFS and DFS strategies. In order to obtain a flexible search strategy, which intelligently takes into account the working conditions of the P2P network, Vapa et al. [160] proposed a neural network based approach (NeuroSearch). This strategy combines multiple heuristics as inputs of a neural network in order to classify among all its neighbors those which will receive the query, thus it does not fix a priori the search strategy (breadth-first or depth-first) to be employed. Depending on the working conditions of the P2P network, NeuroSearch can alternate between both search strategies during a single query.

Since NeuroSearch is based on a neural network, it obviously follows that an initial training is needed. The resulting optimization problem is very challenging because neural networks have a large number of weights varying from minus to plus infinity. In addition, in order to obtain a robust search strategy it is required that training is performed in various working conditions of a P2P network. It is therefore required that many queries are executed, thus making the training problem computationally expensive and the optimization environment noisy.

8.2 NeuroSearch - Neural Network Based Query Forwarding

As highlighted above, NeuroSearch [160] is a neural network-based approach for solving the resource discovery problem. NeuroSearch combines different local information units together as an input to multi-layer perceptron (MLP) neural network [161]. Multi-layer perceptron is a non-linear function approximator, which is organized into different layers: an input layer, one or more hidden layers and an output layer. Adjacent layers are connected together with weights, these weights are the parameters of the function approximator to be determined by the learning process. Hidden and output layers contain neurons, which take a weighted sum of outputs from the previous layer and use a non-linear transfer function to produce output to the next layer. NeuroSearch uses two hidden layers, both having 10 neurons and two different transfer functions in hidden and output layers. The structure of this neural network has been selected on the basis of previous studies carried out by means of the P2PRealm simulation framework [162].

We characterize the query forwarding situation with a model consisting of 1)the previous forwarding node, 2)the currently forwarding node and 3)the receiver of the currently forwarding node. Upon receiving a query, the currently forwarding node selects the first of its neighbors and determines the inputs, related to that neighbor, of the neural network. Then, the neural network output is calculated. This output establishes whether or not the query will be forwarded to the neighbor. Next, all other neighbors including the previous forwarding node, are processed in a similar manner by means of the same neural network. If some of the neighbors were forwarded, then new query forwarding situations will occur until all forwarding nodes have decided not to forward query further.

Fig. 8.1 shows the functioning of a P2P network with neural network based forwarding.

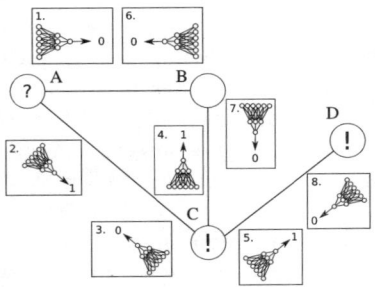

Fig. 8.1. Query Forwarding

The circles shown in the figure represent the peers of the P2P network. The arcs between the peers represent the Transmission Control Protocol (TCP) communication links between the peers. The rectangles represent a neural network evaluation for different neighbors. More specifically, node A denoted with a question mark begins a query. It attempts to forward the query to node B. The neural network in rectangle 1. outputs zero and therefore the query is not forwarded. Instead the second evaluation for node C, shown in rectangle 2, outputs one and the query is forwarded to node C. Then node C attempts to forward the query to neighbor nodes and the nodes B and D receives the query. In the last steps nodes B and D do not forward the query further and the query ends. The query enters nodes C and D denoted with an exclamation mark thereby locating two resource instances.

8.2.1 The Set of the Neural Network Inputs

The MLP uses constant, binary and discrete valued inputs as an information for making forwarding decisions. Each input I_j is a neuron and all 22 inputs I form the input layer.

The following input is constant:

(1) *Bias* takes value 1. Bias is needed in MLP neural networks to approximate functions with non-zero output in case of zero input.

The following inputs are binary:

(2) *Sent* scores 1 if the query has already been forwarded to the receiver. Otherwise it scores 0.

(3) *CurrentVisited* scores 1 if the query has already been received by the currently forwarding node, else it scores 0.

(4) *From* is a binary variable indicating whether a query was received from this receiver. *From* scores 1 if the current query was received from this receiver. Otherwise it scores 0.

(5) *RandomNeighbor* scores 1 for a randomly selected receiver and 0 for other receivers in the current node.

(6) *EnoughReplies* scores 1, if through the query path used by the current query an equal number or more resources have been found as were given in RepliesToGet input parameter (see below). Otherwise *EnoughReplies* scores 0.

The following inputs are discrete:

(7) *Hops* is the number of edges the query has travelled in a P2P network (see definition of *Hops* in section 8.1).

(8) *ToNeighbors* is the number of neighbors connected to the receiver.

(9) *CurrentNeighbors* is the number of neighbors connected to the currently forwarding node.

(10) *FromNeighbors* is the number of neighbors connected to the previous forwarding node.

(11) *InitiatorNeighbors* is the number of neighbors connected to the query initiator.

(12) *NeighborsOrder* is a number associated to each neighbor connected to the forwarding peer. The *NeighborsOrder* is assigned by ascent sorting and enumerating $(0, 1, 2...)$ the neighbors according to their degree. By degree of a peer node we mean the number of neighbors connected to it.

(13) *FromNeighborsOrder*, indicates the *NeighborsOrder* of the previous forwarding node.

(14) *RepliesNow* is the number of replies the query locates in its query path.

(15) *PacketsNow* is the number of packets the query produces in its query path.

(16) *RepliesToGet* is the number of resources that need to be located.

(17) *Forwarded* is the number of times the currently forwarding node has forwarded the query.

(18) *NotForwarded* is the number of times the current node did not forward the query.

(19) *DecisionsLeft* is the number of forwarding "decisions" the current node will still make for the current query message i.e. how many neighbors have not yet been evaluated for forwarding the query message.

(20) *SentCounter* is the number of times the current query has already been forwarded to the receiver.

(21) *CurrentVisitedCounter* is the number of times the query has already been received by the currently forwarding node.

(22) *BranchingResourcesMissing* estimates how many resources on average should still be located from the current query path. First the estimate is set

to the value of *RepliesToGet*. The estimate is updated each time the current
node has made all the forwarding decisions. If the current node contained the
queried resource, the value is decreased by one. The estimate is then updated
depending on whether the current value is positive or negative. In case of a positive value, the current value is divided with the number of neighbors receiving
the query. In case of a negative value, the current value is multiplied by the
number of neighbors, which will receive the query.

8.2.2 Input Scaling

To ensure that all inputs are in the range of $[0, 1]$ the discrete inputs need to be
scaled. The discrete inputs can be classified into three categories according to
their original range of variability.

(a) Inputs in the range of $[0, \infty]$ are *Hops, NeighborsOrder, FromNeighborsOrder, RepliesNow, PacketsNow, RepliesToGet, Forwarded, NotForwarded, DecisionsLeft, SentCounter* and *CurrentVisitedCounter* and they are scaled with the
function $s(I_j) = \frac{1}{I_j+1}$

(b)Inputs in the range of $[1, \infty]$ are *ToNeighbors, CurrentNeighbors, FromNeighbors* and *InitiatorNeighbors* and they are scaled with $s(I_j) = \frac{1}{I_j}$

(c)*BranchingResourcesMissing* is in the range of $[-\infty, \infty]$ and it is scaled with
the sigmoid function $s(I_{22}) = \frac{1}{1+e^{-I_{22}}}$

The scaled inputs I are then given to the neural network.

8.2.3 Calculation of the Neural Network Output

The neurons on the hidden layers contain the transfer function $t(a) = \frac{2}{1+e^{-2a}} - 1$
where a is the sum of the outcoming weighted outputs from the previous (input
or first hidden) layer and *Bias*.

The output layer neuron contains a transfer function

$$u(a) = \begin{cases} 0 \text{ if } a < 0 \\ 1 \text{ if } a \geq 0 \end{cases} \tag{8.1}$$

where a is the sum of the outcoming weighted outputs from the second hidden
layer. The output function is thus defined as follows:

$$O = f(I) = u \left(w_{3,1} + \sum_{l=2}^{L} w_{3,l} t \left(w_{2,1,l} + \sum_{k=2}^{K} w_{2,k,l} t \left(\sum_{j=1}^{J} w_{1,j,k} s_j (I_j) \right) \right) \right) \tag{8.2}$$

where J is the number of inputs, K is the number of neurons on the second layer,
L is the number of neurons on the third layer, $w_{1,j,k}$ is the weight from the jth
input to k^{th} neuron on the first hidden layer, $w_{2,k,l}$ is the weight from the kth
neuron on the first hidden layer to l^{th} neuron on the second hidden layer and
$w_{3,l}$ is the weight from the lth neuron on the second hidden layer to the output

neuron, $w_{2,1,l}$ is the bias weight associated to the second hidden layer and $w_{3,1}$ is the bias weight associated to the output layer.

Output O can take a boolean value indicating whether the query is forwarded to the neighbor node currently being taken into consideration. The neural network output is calculated separately for each neighbor node and after the calculations, the query is sent to neighbor nodes which had an output value 1.

8.3   The Optimization Problem

The neural network described above is supposed to handle the communication and data transfer between a couple of peers. As in all cases of the neural networks, its proper functioning is subject to correctly executed training. The training of a neural network consists of determination of the set of weight coefficients W. As shown in formula (8.2), the weights can be divided into three categories on the basis of the layer to which they belong to. There are 22 input neurons and 10 neurons on both the hidden layers. Since one input is constant ($Bias$) the total amount of weights is $22 * 9 + 10 * 9 + 10 = 298$. The weights can take values in the range $[-\infty, \infty]$.

8.3.1   Fitness Formulation

In order to estimate the quality of a candidate solution, the performance of the P2P network is analyzed with the aid of a simulator whose working principles are described in [162]. More specifically, the set of weights is given to the simulator and a certain number n of queries are performed. The total fitness over the n queries is given by:

$$F = \sum_{j=1}^{n} F_j(W) \tag{8.3}$$

where F_j is the fitness contribution from each query. It is important to remark that multiple queries are needed in order to ensure that the neural network is robust in different query conditions. In addition, for each query the amount of Available Resources (AR) instances is known. Thus, AR is a constant value given a priori.

For each query, the simulator returns two outputs:

(a) the number of query packets P used in the query
(b) the number of found resource instances R during the query

For details regarding the simulation see [162]. These outputs are combined in the following way in order to determine each F_j:

$$F_j = \begin{cases} 0 & \text{if } P > 300 \\ 1 - \frac{1}{P+1} & \text{if } P \le 300 \text{ AND } \quad R = 0 \\ 50 * R - P & \text{if } P \le 300 \text{ AND } 0 < R < \frac{AR}{2} \\ 50 * \frac{AR}{2} - P & \text{if } P \le 300 \text{ AND } \quad \frac{AR}{2} < R \end{cases} \tag{8.4}$$

In formula (8.4), the constant values 300 and 50 have been set according to the criterion explained in [160].

The first condition in (8.4) ensures that the neural network should eventually stop forwarding the queries. The second condition controls that if no resources are found then the neural network increases the number of query packets sent to the network. The third condition states that if the number of found resources is not enough then the neural network develops only by locating more resources. The fourth condition ensures that when half of the available resource instances are found from the network the fitness grows if the neural network uses fewer query packets. The fourth condition also upperbounds F_j to $50 * \frac{AR}{2} - \frac{AR}{2} = 49 * \frac{AR}{2}$, because it is imposed that when half of the resource instances are found P is at minimum $\frac{AR}{2}$.

Thus, the problem of discovering resources in the P2P network consists of the maximization of F in a 298-dimension continuous space:

$$\max\left(F\left(W\right)\right) \text{ in } [-\infty, \infty]^{298} \qquad (8.5)$$

Due to the necessity of ensuring robustness of the neural network in different queries, the fitness value varies with the chosen query. The querying peer and the queried resource need to be changed to ensure that the neural network is not just specialized for searching resources from one part of the network or one particular resource alone. Since n (in our case $n = 10$) queries are required and they are chosen at random, fitness F is noisy. This noise does not have any peculiarity and therefore it can hardly be approximated by a known distribution function. Let us indicate with PN the distribution of this noise and thus re-formulate the problem in equation (8.5)

$$\max\left(F\left(W\right) + Z\right) \text{ in } [-\infty, \infty]^{298}; Z \sim PN \qquad (8.6)$$

8.3.2 Features of the Decision Space and the Fitness Landscape

As highlighted above, the optimization problem is highly multivariate and is defined in a continuous domain. It obviously follows that the problem is quite challenging due to a high dimensionality. In addition, presence of the noise enhances the difficulty of the problem because it introduces some "false" optima into the landscape which disturb the functioning of any optimization algorithm [163,164].

Due to the structure of each F_j (see equation (8.4)), the fitness landscape contains discontinuities. In particular, it is relevant to observe that due to the first condition in (8.4) the fitness landscape contains some plateaus with a null value as well as some other areas which take non-null values and contain a variability. In order to give a rough description of the fitness landscape, the following test has been designed. 2 million candidate solutions have been pseudo-randomly sampled by means of a uniform distribution within the decision space. Fig. 8.2 and 8.3 show the histogram and distribution curve, respectively, related to this test. It should be noted that the y-axis has a logarithmic scale. Fig. 8.2 shows that about half the points take a null fitness value and Fig. 8.3 shows that

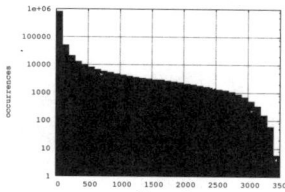
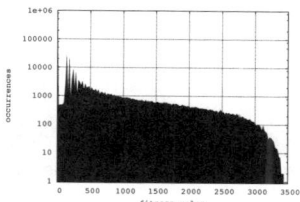

Fig. 8.2. Histogram of Fitness Values   **Fig. 8.3.** Distribution of Fitness Values

the distribution curve contains a very high, sharp peak in zero and other lower sharp peaks before 500. This obviously means the fitness landscape contains some plateau areas for low fitness values (up to 500) and a variational area for high fitness values. In other words the fitness landscape is locally flat and contains several areas having a small variation in fitness values [165, 166]. This feature of the fitness landscape makes the optimization problem very challenging since many optimization algorithms can easily stagnate or prematurely converge in a suboptimal plateau.

8.4 The Adaptive Global-Local Memetic Algorithm

In order to solve the problem in (8.5), an Adaptive Global-Local Memetic Algorithm (AGLMA) has been implemented.

8.4.1 Initialization

An initial sampling made up of S_{pop}^i individual has been executed pseudo-randomly with a uniform distribution function over the interval $[-0.2, 0.2]$. This choice can be briefly justified in the following way. The weights of the initial set of neural networks must be small and comparable among each other in order to avoid one or a few weights dominating with respect to the others as suggested in [167, 168].

8.4.2 Parent Selection and Variation Operators

All individuals of the population S_{pop} undergo recombination and each parent generates an offspring. The variation occurs as follows. For each candidate solution i is associated a self-adaptive vector h_i which represents a scale factor for the exploration. More specifically, at the first generation the self-adaptive vectors h_i are pseudo-randomly generated with uniform distribution within $[-0.2, 0.2]$.

At the subsequent generations each self-adaptive vector is updated according to [167, 168]:

$$h_i^{k+1}(j) = h_i^k(j) e^{(\tau N_j(0,1))} \text{ for } j = 1, 2...n, \tag{8.7}$$

where k is the index of generation, j is the index of variable ($n = 298$), $N_j(0,1)$ is a Gaussian random variable and $\tau = \frac{1}{\sqrt{2\sqrt{n}}} = 0.1659$. Each corresponding candidate solution W_i is then perturbed according to the following formula [167, 168]:

$$W_i^{k+1}(j) = W_i^k + h_i^{k+1}(j) N_j(0, 1) \quad \text{for} \quad j = 1, 2...n \tag{8.8}$$

It is interesting to observe that each component $h_i^k(j)$ of the self-adaptive vector at the kth generation can be seen as the standard deviation of a Gaussian perturbation.

8.4.3   Fitness Function

In order to take into account the noise, function F is calculated n_s times (where n_s stands for number of samples) and an *Explicit Averaging* technique is applied [164, 169]. More specifically each set of weights of a neural network (candidate solution) is evaluated by means of the following formula:

$$\hat{F} = F_{mean}^i - \frac{\sigma^i}{\sqrt{n_s}} \tag{8.9}$$

where F_{mean}^i and σ^i are respectively the mean value and standard deviation related to the n_s samples performed to the ith candidate solution.

The penalty term $\frac{\sigma^i}{\sqrt{n_s}}$ takes into account the distribution of the data and the number of performed samples [170]. Since the noise strictly depends on the solution under consideration, it follows that for some solutions the value of σ^i is relatively small (stable solutions) and so the penalization is small. On the other hand, other solutions could be unstable and score 0 during some samples and a high performance value during other samples. In these cases σ^i is quite large and the penalization must be significant.

8.4.4   Local Searchers

Two local search algorithms with different features in terms of search logic and pivot rule have been employed. These local searchers have the role of supporting the evolutionary framework, offering new search directions and exploiting the available genotypes [171, 172]. The **Simulated Annealing** (SA) metaheuristic [173], [174] has been chosen since it offers an exploratory perspective in the decision space which can choose a search direction leading to a basin of attraction different from where starting point W_0 is. The exploration is performed by using the same mutation scheme as was described in equations (8.7) and (8.8) for an initial self-adaptive vector h_0 pseudo-randomly sampled in $[-0.2, 0.2]$. The main reason for employing the SA in the AGLMA is that the evolutionary framework should be assisted in finding better solutions which improve the available genotype while at the same time exploring areas of the decision space not yet explored. It accepts with a certain probability solutions with worse performance in order to obtain a global enhancement in a more promising basin of attraction. In addition, the exploratory logic aims to overcome discontinuities of the fitness landscape and to "jump" into a plateau having better performance. For these reasons the SA has been employed as a "global" local searcher.

The application of the SA local searcher can be successful in most cases, in the early generations, and in the late generations as well. Moreover, due to its structure the SA can efficiently offer solutions in unexplored basins of attractions and, therefore, prevent an undesired premature convergence. The most delicate issue related to the SA is choice of parameters. The SA has two parameters which are the budget and the initial temperature $Temp^0$. The budget has been fixed at 600 fitness evaluations (in order to have a constant computational cost for the SA). The setting of the initial temperature $Temp^0$ is performed as explained in section 8.4.5. The temperature $Temp$ is reduced according to a hyperbolic law following the suggestions in [175].

The **Hooke-Jeeves Algorithm** (HJA) [176, 177] is a deterministic local searcher which has a steepest descent pivot rule. Briefly the implemented HJA consists of the following. An initial radius d_0 (in our implementation $d_0 = 0.5$) an initial candidate solution W_0 and a direction exploratory matrix are required. In this implementation a standard identity matrix I has been chosen due to the hypercubic features of the decision space. Let us indicate with $I(m,:)$ the mth row of the direction matrix with $m = 1, 2..n$ ($n = 298$).

The HJA consists of an exploratory move and a pattern move. Indicating with W_{cb} the current best candidate solution and with d the generic radius of the search, the HJA during the exploratory move samples the points $W_{cb}(m) + dI$ $(m,:)$ with $m = 1, 2..n$ and the points $W_{cb}(m) - dI(m,:)$ with $m = 1, 2..n$ only along those directions which turned out unsuccessful during the "+" move. Then, if a new current best is found W_{cb} is updated and the pattern move is executed. If a new current best is not found, d is halved and the exploration is repeated.

The HJA pattern move is an aggressive attempt of the algorithm which aims to exploit promising search directions. Rather than centering the following exploration at the most promising explored candidate solution (W_{cb}), the HJA tries to move further [178]. The algorithm centers the subsequent exploratory move at $W_{cb} \pm dI(m,:)$ ("+" or "-" on the basis of the best direction). If this second exploratory move does not outperform $\hat{F}(W_{cb})$ (the exploratory move fails), then an exploratory move with W_{cb} as the center is performed. The HJA stops either when $d < 0.01$ or when the budget condition of 1000 fitness evaluation is reached.

The HJA is supposed to efficiently exploit promising solutions enhancing their genotype in a meta-Lamarckian logic and thus assist the evolutionary framework in quickly climbing the basin of attractions. In this sense the HJA can be considered as a kind of "local" local searcher integrated in the AGLMA.

8.4.5   Adaptation

An adaptation has been implemented taking into account the features of this kind of fitness landscape in order to design a robust algorithm [179, 171]. At the end of each generation the following parameter is calculated:

$$\psi = 1 - \left| \frac{\hat{F}_{avg} - \hat{F}_{best}}{\hat{F}_{worst} - \hat{F}_{best}} \right| \tag{8.10}$$

where $\hat{F}_{worst}$, $\hat{F}_{best}$, and $\hat{F}_{avg}$ are the worst, best, and average of the fitness function values in the population, respectively.

As highlighted in [166], ψ is a fitness-based measurement of the fitness diversity which is well-suited for flat fitness landscapes. The employment of this parameter, taking into account the presence of plateaus in the fitness landscape. ψ, measures the population diversity in terms of fitness and is relative to the range of the fitness values $[\hat{F}_{best}, \hat{F}_{worst}]$ in the population. Thus, even when all fitness values are very similar, leading to $\hat{F}_{best}$ and $\hat{F}_{worst}$ being close to each other, ψ still gives a well scaled measure, since it uses the relative distance of $\hat{F}_{avg}$ from $\hat{F}_{best}$. The population has high diversity when $\psi \approx 1$ and low diversity when $\psi \approx 0$. A low diversity means that the population is converging (possibly in a suboptimal plateau). This parameter has been used in order to control coordination among the local searchers and a dynamic population size.

8.4.6   Coordination of the Local Searchers

ψ has been employed in order to execute an adaptive coordination of the local searchers so as to let them assist the evolutionary framework in the optimization process.

The SA is activated by the condition $\psi \in [0.1, 0.5]$. This adaptive rule is based on the observation that for values of $\psi > 0.5$, the fitness diversity is high and then the evolutionary framework needs to have a high exploitation of the available genotypes (see [180], [166] and [181]). In other words, under this condition the evolutionary framework does not require the assistance of a local searcher. On the other hand, if $\psi < 0.5$ the fitness diversity is decreasing and the application of the SA can introduce a new genotype in the population which can prevent a premature convergence. Basically, the SA has the potential to detect new promising solutions outside a suboptimal plateau into which the population could have fallen. In this sense, the SA has been employed as a local searcher with "global" exploratory features. The condition regarding the lower bound of usability of the SA ($\psi > 0.1$) is due to the consideration that if $\psi < 0.1$ convergence is approaching and the fitness value has already been drastically reduced.

Thus, the SA has the role of exploiting already existing good genotypes but nevertheless to explore other areas of the decision space. Due to its structure, the SA could lead new search directions but its application can lead to a solution which is worse than that which it started with. For this reason, in our implementation it is applied to the second best individual. The initial temperature $Temp^0$ has to be chosen for this local searcher. It is adaptively set to be $Temp^0 = \left| \hat{F}_{avg} - \hat{F}_{best} \right|$. This means that the algorithm does not accept worse solutions when the convergence has practically occurred.

The HJA is activated when $\psi < 0.2$ and is applied to the solution with best performance. The basic idea behind this adaptive rule is that the HJA has the role of quickly improving the best solution while staying in the same basin of attraction. In this light, the action of the HJA can be seen as purely "local". The condition $\psi < 0.2$ means that the HJA is employed when there are some chances that optimal convergence is approaching. An early application of this local searcher can be inefficient since a high exploitation of solutions having poor fitness values would not lead to significant improvements of the population.

It should be noted that in the range $\psi \in [0.1, 0.2]$ both the local searchers are applied to the best two individuals of the population. This range is very critical for the algorithm because the population is tending towards a convergence but still has not reached such a condition. In this case, there is a high risk of premature convergence due to the presence of plateaus and suboptimal basins of attraction or false minima introduced by noise. Thus, the two local searchers are supposed to "compete and cooperate" within the same generation, merging the "global" search power of the SA and the "local" search power of the HJA under supervision of the evolutionary framework.

An additional rule has been implemented. When the SA has succeeded in enhancing the starting solution, the algorithm attempts to further enhance it by the application of the HJA. This choice can be justified by the consideration that when the SA succeeds, it returns a solution having better performance with a genotype (usually) quite different from the starting one and, therefore, belonging to a region of the decision space which has not yet been exploited.

8.4.7 Dynamic Population Size in Survivor Selection

The adaptation controls the population size whose dynamic variation has two combined roles. The first is to massively explore the decision space and thus prevent a possible premature convergence (see [182], [180]), the second is to *Implicitly Average* in order to compensate for noise by means of the evaluations of similar individuals [169]. The population is resized at each generation according to the formula:

$$S_{pop} = S_{pop}^f + S_{pop}^v \cdot (1 - \psi), \qquad (8.11)$$

where S_{pop}^f and S_{pop}^v are the fixed minimum and maximum sizes of the variable population S_{pop}, respectively.

The coefficient ψ is then used to dynamically set the population size [183,184] in order to prevent a premature convergence and stagnation. According to the first role, when the population is highly diverse a small number of solutions need to be exploited. When $\psi \approx 0$ the population is converging and a larger population size is required to increase the exploration. The main idea is that if a population is in a suboptimal plateau an increase of the population size enhances the chances of detecting new promising areas of the decision space and thus prevent premature convergence. On the other hand, if the population is spread out in the decision space it is highly desirable that the most promising solution leads the search and that the algorithm exploits this promising search direction.

According to the second role, it is well-known that large population sizes are helpful in defeating the noise (*Implicitly Averaging*) [185,186]. Furthermore, recent studies [187,170] have noted that the noise jeopardizes proper functioning of the selection mechanisms, especially in cases of low fitness diversity since the noise introduces a disturbance in pair-wise comparison. Therefore, the AGLMA

Pseudo-Random Initial Sampling of the weights W and self-adaptive parameters h;
Fitness evaluation of the initial population by $\hat{F} = F^i_{mean} - \frac{\sigma^i}{\sqrt{n_s}}$;

Calculate $\psi = 1 - \left| \frac{\hat{F}_{avg} - \hat{F}_{best}}{\hat{F}_{worst} - \hat{F}_{best}} \right|$;
while budget conditions and $\psi > 0.01$
  for all the individuals i
    for all the variables j
      $h_i(j) = h_i(j) e^{\left(\tau N_j(0,1)\right)}$;
      $W_i(j) = W_i + h_i(j) N_j(0,1)$;
    end-for
  end-for
  Fitness evaluation of the population by $\hat{F} = F^i_{mean} - \frac{\sigma^i}{\sqrt{n_s}}$;
  Sort the population made up of parents and offsprings according to their fitness values;
  if $\psi \in [0.1, 0.5]$
    Execute the SA on the individual with the 2nd best performance;
    if $\psi < 0.2$
      Execute the HJA on the individual with the best performance;
    end-if
    if the SA succeeds
      Execute the HJA on the individual enhanced by the SA;
    end-if
  end-if
  Calculate $S_{pop} = S^f_{pop} + S^v_{pop} \cdot (1 - \psi)$;
  Select the S_{pop} best individuals to the subsequent generation;
  Calculate $\psi = 1 - \left| \frac{\hat{F}_{avg} - \hat{F}_{best}}{\hat{F}_{worst} - \hat{F}_{best}} \right|$;
end-while

Fig. 8.4. AGLMA pseudo-code

aims to employ a large population size in critical conditions (low diversity) and a small population size when a massive averaging is unnecessary.

After the calculation of S_{pop} in equation (8.11), the AGLMA selects for the subsequent generation, among parents and offspring, the S_{pop} candidate solutions having the best performance.

The algorithm stops when either a budget condition on the number of fitness evaluations is satisfied or ψ takes a value smaller than 0.01.

Fig. 8.4 shows the pseudo-code of the AGLMA.

8.5 Numerical Results

For the AGLMA 30 simulation experiments have been executed. Each experiment has been stopped after 1500000 fitness evaluations. At the end of each generation, the best fitness value has been saved. These values have been averaged over the 30 experiments available. The average over the 30 experiments defines the Average Best Fitness (ABF). Analogously, 30 experiments have been carried out with the Checkers Algorithm (CA) described in [167, 168] according to the implementation in [160], and the ACA which is the CA with the fitness as shown in (8.9) and the adaptive population size as shown in (8.11). In addition a standard real valued Genetic Algorithm (GA) has been run for the problem under study. The GA employs an arithmetic blend crossover and a Gaussian mutation. For the same P2P network, the BFS according to the implementation in

Gnutella and the random walker DFS proposed in [152] have been applied. Table 8.1 shows the parameter settings for the three algorithms and the optimization results. The final fitness $\hat{F}^b$ obtained by the most successful experiment (over the 30 sample runs), the related number of query packets P used in the query and the number of found resource instances R during the query are given. In addition the average best fitness at the end of the experiments $< \hat{F} >$, the final fitness of the least successful experiment $\hat{F}^w$ and the related standard deviation are shown. Since the BFS follows a deterministic logic, thus only one fitness value is shown. On the contrary, the DFS under study employs a stochastic structure and thus the same statistic analysis as that of GA, CA, ACA and AGLMA over 30 experiments has been carried out.

Numerical results in Table 8.1 show that the methods employing the neural network approach are more promising than the classical methods for P2P networks. Moreover, AGLMA and ACA outperform the CA and the AGLMA slightly outperformed the ACA in terms of final solution found. The GA performed significantly worse than the other optimization algorithms.

Fig. 8.5 shows a graphical representation of the solution in the most successful experiment (over the 30 carried out) returned by the proposed AGLMA. An index of the weights are shown on the x-axis and the corresponding weight values are shown on the y-axis (see the crosses in figure).

As shown in Fig. 8.5, according to AGLMA, we propose a neural network having a set of 298 weights, which take small values. More specifically, the proposed neural network contains 296 weight values between -1 and 1. On the contrary, two weights belonging to the first hidden layer take the values of around -1.5 and 1.5.

Table 8.1. Parameter setting and numerical results

PARAMETER	AGLMA	CA	ACA	GA	BFS	DFS
EVOLUTIONARY FRAMEWORK						
S_{pop}^i	30	30	30	30	–	–
S_{pop}	$\in [20, 40]$	30	$\in [20, 40]$	30	–	–
sample size n_s	10	–	10	–	–	–
SIMULATED ANNEALING						
initial temperature $Temp^0$	adaptive	–	–	–	–	–
temperature decrease	hyperbolic	–	–	–	–	–
maximum budget per run	600	–	–	–	–	–
HOOKE-JEEVES ALGORITHM						
exploratory radius	$\in [0.5, 0.01]$	–	–	–	–	–
maximum budget per run	1000	–	–	–	–	–
NUMERICAL RESULTS						
P	350	372	355	497	819	514
R	81	81	81	85	81	81
$\hat{F}^b$	3700	3678	3695	3366	3231	3536
$< \hat{F} >$	3654	3582	3647	2705	–	3363
$\hat{F}^w$	3506	3502	3504	0	–	3056
std	36.98	37.71	36.47	1068	–	107.9

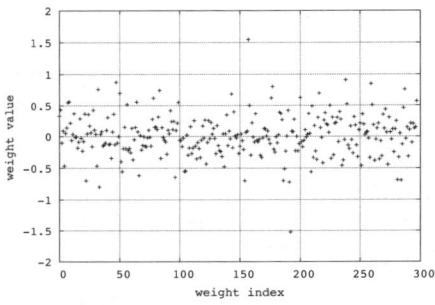

Fig. 8.5. Distribution of Neural Network Weights

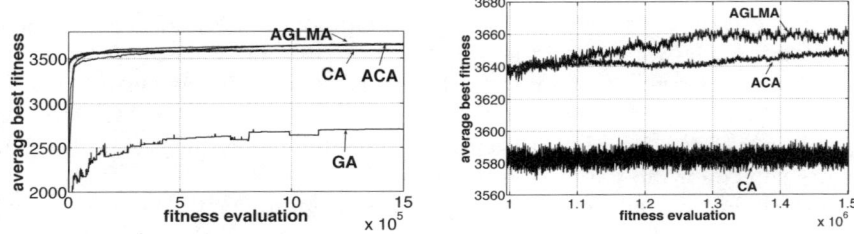

Fig. 8.6. Comparison of the algorithmic **Fig. 8.7.** Comparison of the algorithmic performance   performance (zoom)

Fig. 8.6 shows the comparison of the performance over the 1.5×10^6 fitness evaluations and Fig. 8.7 shows a zoom detail of the algorithmic performance.

Fig. 8.6 shows that the AGLMA has a slower convergence than the CA and the ACA but it reaches a final solution having better performance. It is also clear that the ACA has intermediate performance between the CA and AGLMA. The ACA trend, in early generations, has a rise quicker than the AGLMA but slower than the CA. On the other hand, in late generations, the ACA outperforms the CA but not the AGLMA. As shown in Fig. 8.6, the GA performed much worse than the CA structured algorithms (CA, ACA, AGLMA) also in terms of convergence speed.

It can be remarked that the ACA can be seen as an AGLMA which does not employ local searchers but only executes *Implicit* (dynamic population size) and *Explicit Averaging* (n_s re-samples and modified fitness). In other words, the ACA does not contain the memetic components but does contain the noise filtering components. Fig. 8.7 shows that the ACA and the AGLMA are much more robust to noise than the CA. In fact, as shown in Fig. 8.7, the trend of the CA performance contains a high amplitude (about 20) and frequency ripple around a mean value, while the ACA and AGLMA performance are roughly monotonic. The oscillatory trend of the CA performance is due to an incorrect estimation of candidate solutions. The quick initial rise of the CA performance is, according to our interpretation, also due to an overestimation of an unstable solution. On the contrary, the ACA and the AGLMA efficiently filter the noise and select only reliable solutions for the subsequent generations.

Regarding effectiveness of the local searchers, the comparison between the ACA and the AGLMA shows that the AGLMA slightly outperforms the ACA tending to converge to a solution having a better performance. Moreover it is shown that after 1.5×10^6 fitness evaluations, the trend of the AGLMA still continues to grow whilst the other trends seem to have reached a final value.

8.6 Conclusion

This chapter proposes an Adaptive Global Local Memetic Algorithm (AGLMA) for performing the training of a neural network, which is employed as computational intelligence logic in P2P resource discovery. The AGLMA employs averaging strategies for adaptively executing noise filtering and local searchers in order to handle the multivariate fitness landscape. These local searchers execute the global and local search of the decision space from different perspectives. The numerical results show that the application of the AGLMA leads to a satisfactory neural network training and thus to an efficient P2P network functioning. The comparison with two popular metaheuristics present in literature shows that the proposed approach seems to be promising in terms of final solution found and reliability in noise environment. Matching with another algorithm with intermediate features highlights the effectiveness of each algorithmic component integrated in the proposed algorithm.

The proposed neural network along with the learning strategy carried by the AGLMA allows the efficient location of resources with little query traffic. Thus, the user of the P2P network obtains plentiful amounts of information about resources without consuming a large portion of his own bandwidth for query traffic.

Acknowledgements

We wish to thank Teemu Keltanen and Andrea Caponio for their kind support in analyzing the data.

Part III

Constrained Problems

9

An Iterative Heuristic Algorithm for Tree Decomposition

Nysret Musliu

Institute for Information Systems, Vienna University of Technology, Vienna, Austria
musliu@dbai.tuwien.ac.at

Summary. Many instances of NP-hard problems can be solved efficiently if the treewidth of their corresponding graph is small. Finding the optimal tree decompositions is an NP-hard problem and different algorithms have been proposed in the literature for generation of tree decompositions of small width. In this chapter we present a new iterated local search algorithm to find good upper bounds for treewidth of an undirected graph. The iterated local search algorithm consists from the construction phase, and includes the mechanism for perturbation of solutions, and the mechanism for accepting of solutions for the next iteration. In the construction phase the solutions are generated by the heuristics which search for good elimination ordering of nodes of graph based on moving of only vertices that produce the largest clique in the elimination process. We proposed and evaluated different perturbation mechanisms and acceptance criteria. The proposed algorithm is tested on DIMACS instances for vertex colouring, and is compared with the existing approaches in the literature. The described algorithm has a good time performance and for several instances improves the best existing upper bounds for the treewidth.

Keywords: Graph Decomposition, Tree Decomposition, Constraint Satisfaction Problem, Heuristics, Iterated Local Search.

9.1 Introduction

The concept of tree decomposition is very important due to the fact that many instances of constraint satisfaction problems and in general NP-hard problems can be solved in polynomial time if their treewidth is bounded by a constant. The process of solving problems with bounded treewidth includes two phases. In the first phase the tree decomposition with small upper bound for treewidth is generated. The second phase includes solving a problem (based on the generated tree decomposition) with a particular algorithm such as for example dynamic programming. The efficiency of solving of problem based on its tree decomposition depends from the width of the tree decomposition. Thus it is of high interest to generate tree decompositions with small width.

Tree decomposition has been used for several applications, like combinatorial optimization problems, expert systems, computational biology etc. The use of tree decomposition for inference problems in probabilistic networks is shown

C. Cotta and J. van Hemert (Eds.): Recent Advances in Evol. Comp., SCI 153, pp. 133–150, 2008.
springerlink.com   © Springer-Verlag Berlin Heidelberg 2008

in [188]. Koster et al [189] propose the application of tree decompositions for frequency assignment problem. Tree decomposition has also been applied for problem of vertex cover on planar graphs [190]. Further, the solving of partial constraint satisfaction problems (e.g. MAX-SAT) with tree decomposition based method has been investigated in [191]. In computational biology tree decompositions has been used for protein structure prediction [192] etc.

In this paper we investigate the generation of tree decompositions of undirected graphs. The concept of tree decompositions has been first introduced by Robertson and Seymour [193]:

Definition 1. *(see [193], [194]) Let $G = (V, E)$ be a graph. A tree decomposition of G is a pair (T, χ), where $T = (I, F)$ is a tree with node set I and edge set F, and $\chi = \{\chi_i : i \in I\}$ is a family of subsets of V, one for each node of T, such that*

1. *$\bigcup_{i \in I} \chi_i = V$,*
2. *for every edge $(v, w) \in E$, there is an $i \in I$ with $v \in \chi_i$ and $w \in \chi_i$, and*
3. *for all $i, j, k \in I$, if j is on the path from i to k in T, then $\chi_i \cap \chi_k \subseteq \chi_j$.*

The width of a tree decomposition is $max_{i \in I} |\chi_i| - 1$. The treewidth of a graph G, denoted by $tw(G)$, is the minimum width over all possible tree decompositions of G.

Figure 9.1 shows a graph G (19 vertices) and a possible tree decomposition of G. The width of shown tree decomposition is 5.

For the given graph G the treewidth can be found from its triangulation. Further we will give basic definitions, explain how the triangulation of graph can be constructed, and give lemmas which give relation between the treewidth and the triangulated graph.

Two vertices u and v of graph $G(V, E)$ are neighbours, if they are connected with an edge $e \in E$. The neighbourhood of vertex v is defined as: $N(v) := \{w | w \in V, (v, w) \in E\}$. A set of vertices is clique if they are fully connected. An edge

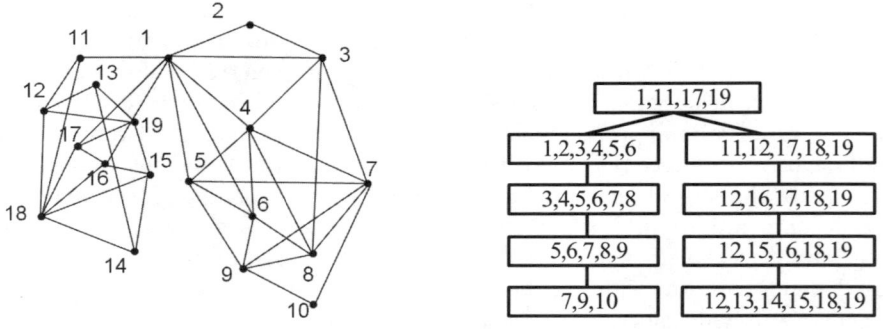

Fig. 9.1. A graph G (left) and a tree decomposition of G (right)

connecting two non-adjacent vertices in the cycle is called chord. The graph is triangulated if there exist a chord in every cycle of length larger than 3.

A vertex of a graph is simplicial if its neighbours form a clique. An ordering of
nodes $\sigma(1, 2, \ldots, n)$ of V is called a perfect elimination ordering for G if for any
$i \in \{1, 2, \ldots, n\}$, $\sigma(i)$ is a simplicial vertex in $G[\sigma(i), \ldots, \sigma(n)]$ [195]. In [196]
it is proved that the graph G is triangulated if and only if it has a perfect
elimination ordering. Given an elimination ordering of nodes the triangulation
H of graph G can be constructed as following. Initially $H = G$, then in the
process of elimination of vertices, the next vertex in order to be eliminated is
made simplicial vertex by adding of new edges to connect all its neighbours in
current G and H. The vertex is then eliminated from G. This process is repeated
for all vertices in the ordering.

The process of elimination of nodes from the given graph G is illustrated
in Figure 9.2. Suppose that we have given the following elimination ordering:
$10, 9, 8, \ldots$. The vertex 10 is first eliminated from G. When this vertex is eliminated no new edges are added in the graph G and H (graph H is not shown
in the figure), as all neighbours of node 10 are connected. Further from the re-
mained graph G the vertex 9 is eliminated. To connect all neighbours of vertex
9, two new edges are added in G and H (edges $(5, 8)$ and $(6, 7)$). The process
of elimination continues until the triangulation H is obtained. A more detailed
description of the algorithm for constructing a graph's triangulation for a given
elimination ordering is found in [194].

For generation of tree decomposition during the vertex elimination process, first
the nodes of tree decomposition are created. This is illustrated in Figure 9.2. When
vertex 10 is eliminated a new tree decomposition node is created. This node contains the vertex 10 and all other vertices which are connected with this vertex
in current graph G. Further the next tree node with vertices $\{5, 6, 7, 8, 9\}$ is created when the vertex 9 is eliminated. To the end of elimination process all tree

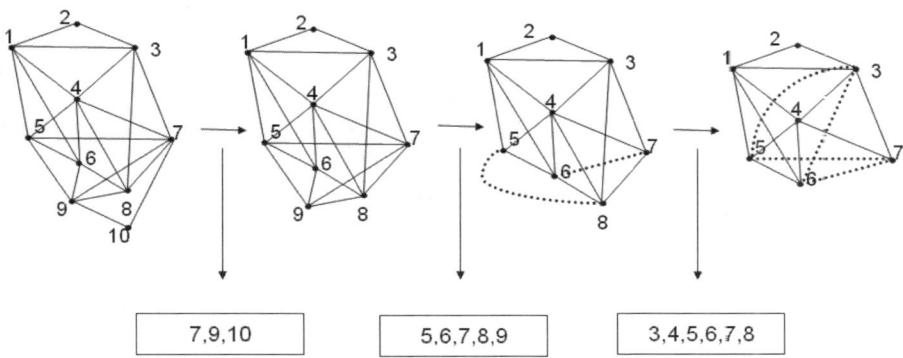

Fig. 9.2. Illustration of the elimination of nodes 10, 9, and 8, and generation of tree
decomposition nodes during the construction of triangulated graph

decomposition nodes will be created. The created tree nodes should be connected, such that the connectedness condition for the vertices is fulfilled. This is the third condition in the tree decomposition definition. To fulfil this condition the tree decomposition nodes are connected as following. The tree decomposition node with vertices $\{10, 9, 8\}$ that is created when vertex 10 is eliminated, is connected with the tree decomposition node which will be created when the next vertex in the ordering which appear in $\{10, 9, 8\}$ is eliminated. In this case the node $\{10, 9, 8\}$ should be connected with the node created when vertex 9 is eliminated, because this is the next vertex in the ordering that is contained in $\{10, 9, 8\}$. This rule is further applied for connection of other tree decomposition nodes, and from the graph a part of tree decomposition in Figure 9.1 will be constructed.

The treewidth of a triangulated graph can be calculated based on its cliques. For the given triangulated graph the treewidth is equal to its largest clique minus 1 [197]. Moreover, the largest clique of trinagulated graph can be calculated in polynomial time. The complexity of calculation of the largest clique for the triangulated graphs is $O(|V| + |E|)$ [197]. For every graph $G = (V, E)$, there exists a triangulation of G, $\overline{G} = (V, E \bigcup E_t)$, with $tw(\overline{G}) = tw(G)$. Thus, finding the treewidth of a graph G is equivalent to finding a triangulation $\overline{G}$ of G with minimum clique size (for more information see [194]).

9.1.1   Algorithms for Tree Decompositions

For the given graph and integer k, deciding whether the graph has a tree decomposition with a treewidth at most k is an NP-hard problem [198]. To solve this problem different complete and heuristic algorithms have been proposed in the literature. Examples of complete algorithms for tree decompositions are [199], [200], and [201]. Gogate and Dechter [200] reported good results for tree decompositions by using the branch and bound algorithm. They showed that their algorithm is superior compared to the algorithm proposed in [199]. The branch and bound algorithm proposed in [200] applies different pruning techniques, and provides anytime solutions, which are good upper bounds for tree decompositions. The algorithm proposed in [201] includes several other pruning and reduction rules and is used successfully for small graphs.

Heuristic techniques for generation of tree decompositions with small width are mainly based on searching for a good perfect elimination ordering of graph nodes. Several heuristics that run in polynomial time have been proposed for finding a good elimination ordering of nodes. These heuristics select the ordering of nodes based on different criteria, such as the degree of the nodes, the number of edges to be added to make the node simplicial etc.

Maximum Cardinality Search (MCS) proposed by Tarjan and Yannakakis [202], initially selects a random vertex of the graph to be the first vertex in the elimination ordering. The next vertex will be picked such that it has the highest connectivity with the vertices previously selected in the elimination ordering. The ties are broken randomly. MCS repeats this process iteratively until all vertices are selected.

The min-fill heuristic first picks the vertex which adds the smallest number of edges when eliminated (the ties are broken randomly). The selected vertex is made simplicial and it is eliminated from the graph. The next vertex in the ordering will be any vertex that adds the minimum number of edges when eliminated from the graph. This process is repeated iteratively until the whole elimination ordering is constructed.

Minimum degree heuristic picks first the vertex with the minimum degree. Further, the vertex that has the minimum number of unselected neighbours will be chosen as the next node in the elimination ordering. This process is repeated iteratively.

MCS, min-fill, and min-degree heuristics run in polynomial time and usually produce tree decomposition in a reasonable amount of time. According to [200] the min-fill heuristic performs better than MCS and min-degree heuristic. Although these heuristics give sometimes good upper bounds for tree decompositions, with more advanced techniques, usually better upper bounds can be found for most problems. Min-degree heuristic has been improved by Clautiaux et al [195] by adding a new criterion based on the lower bound of the treewidth for the graph obtained when the node is eliminated. For other types of heuristics based on the elimination ordering of nodes see [194].

Metaheuristic approaches have also been used for tree decomposition. Simulated annealing was used by Kjaerulff [203] for similar problem to tree decomposition. Application of genetic algorithm for tree decompositions is presented in [204]. The algorithm proposed by [204] with some changes in fitness function has been tested on different problems for tree decompositions in [205] A tabu search approach for generation of the tree decompositions has been proposed by Clautiaux et al [195]. The authors reported good results for DIMACS vertex colouring instances [206]. Their approach improved the previous results in literature for 53% of instances. Some of the results in [195] have been further improved by Gogate and Dechter [200]. The reader is referred to [207] for other approximation algorithms, and the information for lower bound algorithms.

In this chapter we propose new heuristic algorithms with the aim of improving existing upper bounds for tree decomposition and reducing the running time of algorithms for different problems. Two simple heuristics for searching of good elimination orderings of vertices are proposed. These local search heuristics are based on changing of positions of nodes in ordering, which cause the largest clique when eliminated. The proposed heuristics are exploited by a new iterated local search algorithm in the construction phase. The iterative local search algorithm applies iteratively the construction heuristic and additionally includes the perturbation mechanism and the solution acceptance criteria. These algorithms have been applied in 62 DIMACS instances for vertex colouring. For several problems we report new upper bounds for the treewidth, and for most of problems the tree decomposition is generated in a reasonable amount of time. Our results have been compared with the results reported in [194], [200], [195], and [205] which to our best knowledge report the best results known yet in literature considering the width of tree decompositions for these instances. For up to date information for the best upper and lower bounds for treewidth for different instances the reader is referred to TreewidthLIB:http://www.cs.uu.nl/ hansb/treewidthlib/.

9.2 An Iterative Local Search Algorithm

As described in the previous section, the generation of tree decomposition with small width can be done by finding an appropriate elimination ordering which produces a triangulated graph with smallest maximum clique size. In this section we present an algorithm which searches among the possible ordering of nodes to find small treewidth for the given graph. The algorithm contains a local search heuristic for constructing a good ordering, and the iterative process, during which the algorithm calls the local search techniques with the initial solution that is produced in previous iteration. The algorithm includes also a mechanism for acceptance of a candidate solution for the next iteration. Although the constructing phase is very important, choosing the appropriate perturbation in each iteration as well as the mechanism for acceptance of solution are also very important to obtain good results using an iterative local search algorithm. The proposed algorithm is presented in Algorithm 1.

The algorithm starts with an initial solution which takes an order of nodes as they appear in the input. Better initial solutions can also be constructed by using other heuristics which run in polynomial time, such as Maximum Cardinality Search, min-fill heuristic, etc. However, as the proposed method usually finds fast

---

Algorithm 9.1. Iterative heuristic algorithm - IHA

Generate initial solution $S1$

$BestSolution = S1$

while Termination Criteria is not fulfilled **do**
  $S2 = ConstructionPhase(S1)$

  if Solution $S2$ fulfils the acceptance criteria **then**
    $S1 = S2$
  else
    $S1 = BestSolution$
  end if

  Apply perturbation in solution $S1$

  Update $BestSolution$ if solution $S2$ has better (or equal) width than the current best solution

end while

RETURN $BestSolution$

---

a solution produced by these heuristics, our algorithm starts with an ordering of nodes given in the input.

After construction of the initial solution the iterative phase starts. In this phase iteratively the construction method is called, then the solution produced in the construction phase is tested if it fulfils the acceptance criteria, and the perturbation mechanism is applied. The construction phase includes the local search procedure which is used to improve the input solution. We propose two different local search techniques that can be used in the construction phase. These techniques are described in Section 9.2.1. The solution returned from the construction phase will be accepted for the next iteration if it fulfils the specific criteria determined by the solution acceptance mechanism. We experimented with different possibilities for the acceptance of the solution returned from the construction phase. These variants are described in Section 9.2.3. If the solution does not fulfil the acceptance criteria this solution will be discarded and the best current found solution is selected. In the selected solution the perturbation mechanism is applied. Different possibilities are used for perturbation. Types of perturbation used for experimentation are described in Section 9.2.2. The perturbed solution is given as an input solution in the next call of the construction phase. This process continues until the termination criterion is fulfilled. For the termination criteria we used a time limit.

9.2.1   Local Search Techniques

We propose two local search methods for generation of a good solution which will be used as an initial solution with some perturbation in the next call of the same local search algorithm. Both techniques are based on the idea of moving only those vertices in the ordering, which cause the largest clique during the elimination process. The motivation for using this method is the reduction of the number of solutions that should be evaluated. The first proposed technique (LS1) is presented in Algorithm 2.

The proposed algorithm applies a simple heuristic. In the current solution a vertex is chosen randomly among the vertices that produce the largest clique in the elimination process. Then the selected vertex is moved from its position. We experimented with two types of moves. In the first variant the vertex is inserted in a random position in the elimination ordering, while in the second variant the vertex is swapped with another vertex located in a randomly selected position, i.e. the two chosen vertices change their position in the elimination ordering. The swap move was shown to give better results. The heuristic will stop if the solution is not improved after a certain number of iterations. We experimented with different $MAXNotImprovments$. LS1 alone is a simple heuristic and usually can not produce good results for tree decomposition. However, by using this heuristic as a construction heuristic in the iterated local search algorithm (see Algorithm 1) good results for tree decomposition are obtained.

The second proposed heuristic (LS2) is presented in Algorithm 3. This technique is similar to algorithm LS1. However, this technique differs from LS1 considering the exploration of the neighbourhood. In LS2 in some of iterations the

---

Algorithm 9.2. Local Search Algorithm 1 - LS1 (InputSolution)

$BestLSSolution = InputSolution$
$NrNotImprovments = 0$

while $NrNotImprovments < MAXNotImprovments$ **do**
    In current solution ($InputSolution$) select a vertex in the elimination ordering
    which causes the largest clique when eliminated - ties are broken randomly if
    there are several vertices which cause the clique equal with the largest clique

    Swap this vertex with another vertex located in a randomly chosen position

    if the current solution is better than $BestLSSolution$ **then**
        $BestLSSolution = InputSolution$
        $NrNotImprovments = 0$
    else
        $NrNotImprovments + +$
    end if

end while

RETURN $BestLSSolution$

---

neighbourhood of solution consist from only one solution which is generated with swapping of vertex in the elimination ordering which causes the largest clique, with another vertex located in the randomly chosen position. The use of this neighbourhood in particular iteration will depend from the parameter p, which determines the probability of applying this neighbourhood in each iteration. We experimented with different values for parameter p. With probability $1-p$, other type of neighbourhood will be explored. The neighbourhood of current solution in this case consists from all solutions which can be obtained by swapping of a vertex in the elimination ordering which causes the largest clique, with the vertices in the elimination ordering, which are its neighbours. The best solution from the generated neighbourhood is selected for the next iteration in the LS2. Note that in this technique the number of solutions that have to be evaluated is much larger than in LS1. In particular in the first phase of search the node which causes the largest clique usually has many neighbours and thus the number of solution to be evaluated when the second type of neighbourhood is used is equal to the size of the largest clique produced during the elimination process.

9.2.2   Perturbation

During the perturbation phase the solution obtained by local search procedure is perturbed and the newly obtained solution is used as an initial solution for the new call of the local search technique. The main idea is to avoid the random restart. Instead of random restart the solution is perturbed with a bigger move(s)

Algorithm 9.3. Local Search Algorithm 2 - LS2 (InputSolution)

$BestLSSolution = InputSolution$
$NrNotImprovments = 0$

while $NrNotImprovments < MAXNotImprovments$ **do**
  With probability p:

  Select a vertex in the elimination ordering which causes the largest clique (ties are broken randomly)

  Swap this vertex with another vertex located in the randomly chosen position

  With probability $1 - p$:
  Select a vertex in the elimination ordering which causes the largest clique (ties are broken randomly)

  Generate neighbourhood of the solution by swapping the selected vertex with its neighbours, i.e. all solutions are generated by swapping the selected vertex with its neighbours

  Select the best solution from the generated neighbourhood

  if the current solution is better than $BestLSSolution$ **then**
    $BestLSSolution = CurrentSolution$
    $NrNotImprovments = 0$
  else
    $NrNotImprovments + +$
  end if

end while

RETURN $BestLSSolution$

---

as those applied in the local search technique. This enables some diversification that helps to escape from the local optimum, but avoids beginning from scratch (as in case of random restart), which is very time consuming. We propose three perturbation mechanisms:

- RandPert: N vertices are chosen randomly and they are moved into new random positions in the ordering.
- MaxCliquePer: All nodes that produce the maximal clique in the elimination ordering are inserted in a new randomly chosen positions in the ordering.
- DestroyPartPert: All nodes between two positions (selected randomly) in the ordering are inserted in the new randomly chosen positions in the ordering.

The perturbation RandPert just perturbs the solution with a larger random move and would be kind or random restart if N is very large. Keeping N smaller avoids restarting from completely new solution, and the perturbed solution differs not much from the previous solution. MaxCliquePer concentrates on moving of only vertices which produce maximal clique in the elimination ordering. The basic idea for this perturbation is to apply a technique similar to min-conflict heuristic, by moving only the vertices that cause large treewidth. DestroyPart-Pert is similar to RandPert, except that the selected nodes to be moved are located near each other in the elimination ordering.

Determining the number of nodes N that will be moved is complex and may be dependent on the problem. To avoid this problem we propose an adaptive perturbation mechanism that takes into consideration the feedback from the search process. The number of nodes N varies from 2 to some number y (determined experimentally), and the algorithm begins with small perturbation with $N = 2$. If during the iterative process (for a determined number of iterations) the local search technique produces solutions with same tree width for more than 20% of cases, the size of perturbation is increased by 1, otherwise the size of N will be decreased by 1. This enables an automatic change of perturbation size based on the repetition of solutions with the same width.

We applied each perturbation mechanism separately, and additionally considered combination of two perturbations. The mixed perturbation applies (in Algorithm 1) two perturbations: RandPert, and MaxCliquePer. The algorithm starts with RandPert, and switches alternatively between two perturbations if the solution is not improved for a determined number of iterations. Note that we experimented with different sizes of perturbation sizes for each type of perturbation. The experimental evaluation of different perturbations is presented in Section 9.3.

9.2.3   Acceptance of Solution in Iterated Algorithm

Different techniques can be applied for acceptance of the solution obtained by the local search technique. If the solution is accepted it will be perturbed and will serve as an initial solution for the next call of one of the local search techniques. We experimented with the following variants for acceptance of solution for the next iteration (see Algorithm 1):

- Solution returned from the construction phase is accepted only if it has a better width than the best current existing solution.
- Solution returned from the construction phase is always accepted.
- Solution is accepted if its treewidth is smaller than the treewidth of the best yet found solution plus x, where x is an integer.

The first variant for acceptance of solution is very restrictive. In this variant the solution from the construction phase will be accepted only if it improves the best existing solution. Otherwise, the best existing solution is perturbed and it is used as input solution for next call of the construction phase. In second variant of acceptance of solution, the iterated local search applies the perturbation in a solution returned from the construction phase, independently from the quality of produced solution. The third variant is between the first and the second variant, and in this case the solution which does not improve the best existing solution can be accepted for the next iteration, if its width is smaller than the best found width plus some bound.

9.3 Evaluation of Algorithms

The algorithms described in Section 9.2 are experimentally tested in DIMACS vertex colouring instances. Using our algorithm we experimented with two proposed local search techniques for construction phase, different perturbation, different acceptance criteria, swap move, and different termination criteria for the local search procedures. For algorithm LS2 we experimented with different probabilities for p. Considering the acceptance of solution in iterated local search we experimented with three variants described in Section 9.2.3. For the third variant we experimented with $x = 2$ and $x = 3$. We did experiments with three types of perturbations: RandPert, MaxCliquePer, and DestroyPartPer. Additionally, we experimented with combination of RandPert and MaxCliquePer. For each type of perturbation mechanism we experimented with different perturbation sizes.

In Table 9.1 results for different perturbations mechanisms for 20 DIMACS problems are presented. These problems are selected among the hardest problems in each class of DIMACS problems. The results for all problems and the comparison with the results in the literature are presented in the next section. The perturbation mechanisms shown in Table 9.1 are the following:

- (P1) RandPert with size of perturbation 3
- (P2) RandPert with the size of perturbation 8
- (P3) MaxCliquePer
- (P4) DestroyPartPer with perturbation size 5
- (P5) DestroyPartPer with perturbation size 10
- (P6) Mixed perturbation (RandPert+MaxCliquePer) with the size of perturbation 3
- (P7) Mixed perturbation with perturbation size 8
- (P8) Mixed perturbation with the adaptive perturbation (with size 2-11)

In the mixed perturbation are used both RandPert and MaxCliquePer perturbations. Initially RandPert is applied. Further the algorithm switches alternatively between two perturbations RandPert and MaxCliquePer, when IHA runs for 100 iterations without improvement of best existing solution.

The results in Table 9.1 presents the average width of tree decompositions over 5 runs for each example. Maximal run time of each run is 500 seconds, and the algorithm stops after 200 seconds of non improvement of the best solution. Based on the results give in the Table 9.1 we can conclude that considering the tree width the best results are obtained with the perturbation P5 and P8. Similar results are obtained with perturbations P1. Perturbation which includes only moving of nodes with largest cliques gives the worse results and in general for other perturbations, if the size of perturbation is large the results are worse.

Table 9.1. Comparision of different perturbation mechanisms

Instance	P1	P2	P3	P4	P5	P6	P7	P8
games120.col	33	33.4	32.8	33	33.2	**32.2**	32.4	32.4
queen14_14	145	146	146.6	143.6	144.6	**142.6**	145.6	143.2
queen15_15	168.4	168.4	168	168.4	169.2	167.2	167	**165.4**
queen16_16	**189.2**	193.4	194.2	191.4	193.2	191.4	191	190.8
inithx.i.3.col	**35**	**35**	**35**	**35**	**35**	**35**	**35**	**35**
miles500.col	24.2	25.2	**23.2**	24.6	25.4	24.2	23.8	24.2
myciel7.col	67.2	**66**	67.2	67.2	66.6	69	**66**	**66**
school1.col	189	195.2	226.4	189.8	196.2	193.4	199.4	**187.4**
school1_nsh	186.2	174	181.6	**165.6**	169.2	173	165.8	170.6
zeroin.i.3.col	32.8	32.8	**32.6**	32.8	33	**32.6**	33	**32.6**
le450_5a.col	**263.6**	280.6	301.4	271.4	290.2	272	278.8	279.4
le450_15b.col	**278.4**	284.6	287.4	279.6	290.6	**278.4**	282	279.6
le450_25a.col	240	245.4	253	244.4	251.8	**234.8**	240.8	240.2
le450_25b.col	237.4	244.6	248.2	241.6	253.2	**235.4**	243.2	235.8
le450_25c.col	**331.2**	339.8	353.6	336.8	339.8	340	336.4	334
le450_25d.col	337.4	346	351.6	341	343.6	**337**	339	337.2
DSJC125.1.col	61.6	62.2	**61.2**	**61.2**	63.2	63	62	61.8
DSJC125.5.col	108.2	108.4	**108**	108.4	108.4	108.2	108.2	**108**
DSJC250.1.col	**171.2**	172.6	176	172	176.2	171.6	171.6	171.6
DSJC250.5.col	230.6	231.2	231	**230.2**	231	230.6	231	**230.2**

The current best results presented in this paper are obtained with the iterative heuristic algorithm (IHA) and these parameters: LS1 algorithm (see Algorithm 2) is used in the construction phase and this algorithm stops if the solution does not improve for 10 iterations ($MAXNotImprovments = 10$). In the perturbation phase are used both RandPert and MaxCliquePer perturbations. Initially RandPert with $N = 2 - 11$ is applied. Further the algorithm switches alternatively between two perturbations RandPert and MaxCliquePer, when IHA runs for 100 iterations without improvement of a solution. For accepting of solution in IHA the third variant is used. The solution produced in construction phase is accepted if its width is smaller than the width of the best current solution plus 3.

9.3.1   Comparison with Results in Literature

In this section we report on computational results obtained with the current implementation of methods described in this paper. The results for 62 DIMACS vertex colouring instances are given. These instances have been used for testing of several methods for tree decompositions proposed in the literature (see [194], [195], and [200]). Our algorithms have been implemented in C++ and the current experiments were performed with a Intel Pentium 4 CPU 3GHz, 1GB RAM.

Table 9.2. Algorithms comparison regarding treewidth for DIMACS graph colouring instances

Instance	$|V|/|E|$	KBH	TabuS	BB	GA-b	GA-avg	IHA-b	IHA-avg	IHA-sd
anna	138 / 986	12	12	12	12	12	12	12	0
david	87 / 812	13	13	13	13	13	13	13	0
huck	74 / 602	10	10	10	10	10	10	10	0
homer	561 / 3258	31	31	31	31	31	31	31.2	0.42
jean	80 / 508	9	9	9	9	9	9	9	0
games120	120 / 638	37	33	-	32	32	32	32.2	0.42
queen5_5	25 / 160	18	18	18	18	18	18	18	0
queen6_6	36 / 290	26	25	25	26	26	25	25	0
queen7_7	49 / 476	35	35	35	35	35.2	35	35	0
queen8_8	64 / 728	46	46	46	45	46	45	45.3	0.48
queen9_9	81 / 1056	59	58	59	58	58.5	58	58.1	0.31
queen10_10	100 / 1470	73	72	72	72	72.4	72	72.3	0.67
queen11_11	121 / 1980	89	88	89	87	88.2	87	87.7	0.94
queen12_12	144 / 2596	106	104	110	104	105.7	**103**	104.4	1.17
queen13_13	169 / 3328	125	122	125	121	123.1	121	122.2	1.22
queen14_14	196 / 4186	145	141	143	141	144	**140**	142.6	1.64
queen15_15	225 / 5180	167	163	167	162	164.8	162	166.3	2.62
queen16_16	256 / 6320	191	186	205	186	188.5	186	188.2	1.47
fpsol2.i.1	269 / 11654	66	66	66	66	66	66	66	0
fpsol2.i.2	363 / 8691	31	31	31	32	32.6	31	31.1	0.31
fpsol2.i.3	363 / 8688	31	31	31	31	32.3	31	31.2	0.42
inithx.i.1	519 / 18707	56	56	56	56	56	56	56	0
inithx.i.2	558 / 13979	35	35	**31**	35	35	35	35	0
inithx.i.3	559 / 13969	35	35	**31**	35	35	35	35	0
miles1000	128 / 3216	49	49	49	50	50	49	49.2	0.42
miles1500	128 / 5198	77	77	77	77	77	77	77	0
miles250	125 / 387	9	9	9	10	10	9	9.3	0.48
miles500	128 / 1170	22	22	22	24	24.1	22	23.5	1.08
miles750	128 / 2113	37	36	37	37	37	36	36.9	0.56
mulsol.i.1	138 / 3925	50	50	50	50	50	50	50	0
mulsol.i.2	173 / 3885	32	32	32	32	32	32	32	0
mulsol.i.3	174 / 3916	32	32	32	32	32	32	32	0
mulsol.i.4	175 / 3946	32	32	32	32	32	32	32	0
mulsol.i.5	176 / 3973	31	31	31	31	31	31	31	0
myciel3	11 / 20	5	5	5	5	5	5	5	0
myciel4	23 / 71	11	10	10	10	10	10	10	0
myciel5	47 / 236	20	19	19	19	19	19	19	0
myciel6	95 / 755	35	35	35	35	35	35	35.4	0.84
myciel7	191 / 2360	74	66	**54**	66	66	66	67.2	1.68
school1	385 / 19095	244	188	-	185	192.5	**178**	190.5	17.1
school1_nsh	352 / 14612	192	162	-	157	163.1	**152**	156.4	3.59
zeroin.i.1	126 / 4100	50	50	-	50	50	50	50	0
zeroin.i.2	157 / 3541	33	32	-	32	32.7	32	32.7	0.48
zeroin.i.3	157 / 3540	33	32	-	32	32.9	32	32.6	0.51

Table 9.3. Algorithms comparison regarding treewidth for DIMACS graph colouring instances

Instance	$|V|/|E|$	KBH	TabuS	BB	GA-b	GA-avg	IHA-b	IHA-avg	IHA-sd
le450_5a	450 / 5714	310	256	307	**243**	248.3	244	250	3.39
le450_5b	450 / 5734	313	254	309	248	249.9	**246**	249.3	2.4
le450_5c	450 / 9803	340	272	315	**265**	267.1	266	273	5.77
le450_5d	450 / 9757	326	278	303	265	265.6	265	267.2	2.39
le450_15a	450 / 8168	296	272	-	265	268.7	**262**	267.9	3.31
le450_15b	450 / 8169	296	270	289	265	269	**258**	266.7	4.76
le450_15c	450 / 16680	376	359	372	351	352.8	**350**	355.4	3.71
le450_15d	450 / 16750	375	360	371	**353**	356.9	355	357.5	1.71
le450_25a	450 / 8260	255	234	255	225	228.2	**216**	222.6	4.35
le450_25b	450 / 8263	251	233	251	227	234.5	**219**	227.2	4.39
le450_25c	450 / 17343	355	327	349	**320**	327.1	322	327.4	2.63
le450_25d	450 / 17425	356	336	349	**327**	330.1	328	332.3	3.19
dsjc125.1	125 / 736	67	65	64	61	61.9	**60**	61.1	0.87
dsjc125.5	125 / 3891	110	109	109	109	109.2	**108**	108	0
dsjc125.9	125 / 6961	119	119	119	119	119	119	119	0
dsjc250.1	250 / 3218	179	173	176	169	169.7	**167**	168.6	0.84
dsjc250.5	250 / 15668	233	232	231	230	231.4	**229**	230.1	0.73
dsjc250.9	250 / 27897	243	243	243	243	243.1	243	243	0

We compare our results with the results reported in [194], [195], and [200]. Additionally we include the recent results obtained by Genetic Algorithm [205]. The results reported in [194] are obtained in Pentium 3, 800 Mhz processor. Results reported in [195] are obtained with Pentium 3, 1GHz processor, and results reported in [200] are obtained with Pentium-4, 2.4 Ghz, 2GB RAM machine. Genetic algorithm [205] has been evaluated in a Intel(R) Xeon(TM) 3.20 GHz, 4 GB RAM. To our best knowledge these papers present the best existing upper bounds for treewidth for these 62 instances.

In Tables 9.2 and 9.3 the results for the treewidth for DIMACS graph colouring instances are presented. First and second columns of the tables present the instances and the number of nodes and edges for each instance. In column KBH are shown the best results obtained by algorithms in [194]. The TabuS column presents the results reported in [195], while the column BB shows the results obtained with the branch and bound algorithm proposed in [200]. Columns GA-b and GA-avg represents results obtained with Genetic Algorithm [205]. Column GA-b presents the best width obtained in 10 runs, and the column GA-avg gives the average of treewidth over 10 runs. The last three columns present results obtained by our algorithm proposed in this paper with the settings which were given in the previous section. In our algorithm are executed 10 runs for each instance. In column IHA-b is given the best width obtained in 10 runs for each instance, the column IHA-avg gives the average of treewidth over 10 runs, and the column IHA-sd presents the standard deviation of treewidth over 10 runs.

In Tables 9.4 and 9.5 for each instance is given the time (in seconds) needed to produce the treewidth presented in Tables 9.2 and 9.3 for all algorithms. The time

Table 9.4. Algorithms comparison regarding time needed for generation of tree decompositions

Instance	KBH	TabuS	BB	GA	IHA-b(avg)	IHA-b(sd)	IHA-t(avg)
anna	1.24	2776.93	1.64	213	0.1	0	11
david	0.56	796.81	77.6538	154	0.1	0	11
huck	0.24	488.76	0.041	120	0.1	0	11
homer	556.82	157716.56	10800	1118	127	89	327 8
jean	0.29	513.76	0.05	120	0	0	11
games120	5.2	2372.71	-	462	145.8	76.45	346.8
queen5_5	0.04	100.36	5.409	33	0.1	0	11
queen6_6	0.16	225.55	81.32	51	0.1	0.3	11.1
queen7_7	0.51	322.4	543.3	92	0.1	0	11
queen8_8	1.49	617.57	10800	167	28.8	51.9	229.8
queen9_9	3.91	1527.13	10800	230	5.2	8.06	206.2
queen10_10	9.97	3532.78	10800	339	28.3	44.84	229.3
queen11_11	23.36	5395.74	10800	497	29.6	35.6	230.6
queen12_12	49.93	10345.14	10800	633	106.7	122.31	304.2
queen13_13	107.62	16769.58	10800	906	3266.12	2739.4	10001
queen14_14	215.36	29479.91	10800	1181	5282.2	3496.39	10001
queen15_15	416.25	47856.25	10800	1544	3029.51	2740.95	10001
queen16_16	773.09	73373.12	10800	2093	7764.57	1819.88	10001

results given in [200] present the time in which the best solutions are found. The results given in [195] present the time of the overall run of the algorithm in one instance (number of iterations is 20000 and the algorithm stops after 10000 non-improving solutions). The running time of GA [205] is presented in column GA. For our algorithm are given the average time in which the best solution is found (IHA-b(avg)), the standard deviation (IHA-b(sd)) for the times in which the best solutions are found, and the time of the overall run of algorithm (IHA-total (avg)) in each instance (avg indicates that the average over ten runs is taken). IHA algorithm stops for easy instances after 10 seconds of non improvement of solution, for middle instances after 200 seconds, and for harder instances after 10000 seconds of non improvement of solution. The maximal running time of algorithm for each instance is set to be 10000 seconds.

Based on the results given in Tables 9.2 and 9.3 we conclude that considering the best result over 10 runs, our algorithm gives better results for 35 instances compared to [194] for the upper bound of treewidth, whereas algorithms in [194] does not outperform our algorithm in any instance. Comparing KBH to IHA average over 10 runs, KBH gives better results for 7 instances, and IHA-avg for 35 instances. Compared to the algorithm proposed in [195] our approach gives better upper bounds for 25 instances, whereas algorithm in [195] gives no better upper bounds than our algorithm. Comparing TabuS to our average, TabuS give better results for 21 instances, whereas IHA-avg gives better results for 18 instances. Further, compared to branch and bound algorithm proposed in [200] our algorithm gives better upper bounds for treewidth for 24 instances, whereas the branch and bound algorithm gives better results compared to our

Table 9.5. Algorithms comparison regarding time needed for generation of tree decompositions

Instance	KBH	TabuS	BB	GA	IHA-b(avg)	IHA-b(sd)	IHA-t(avg)
fpsol2.i.1	319.34	63050.58	0.587076	1982	4.8	1.54	15.8
fpsol2.i.2	8068.88	78770.05	0.510367	1445	8.4	3.74	19.4
fpsol2.i.3	8131.78	79132.7	0.492061	1462	8.7	3.86	19.7
inithx.i.1	37455.1	101007.52	26.3043	3378	10.2	2.04	21.2
inithx.i.2	37437.2	121353.69	0.05661	2317	11.7	5.6	22.7
inithx.i.3	36566.8	119080.85	0.02734	2261	10.6	2.63	21.6
miles1000	14.39	5696.73	10800	559	54.2	56.64	255.2
miles1500	29.12	6290.44	6.759	457	0.7	0.48	11.7
miles250	10.62	1898.29	1.788	242	2.9	3.69	13.9
miles500	4.37	4659.31	1704.62	442	81	70.36	282
miles750	8.13	3585.68	10800	536	112.2	86.85	313.2
mulsol.i.1	240.24	3226.77	1.407	671	0.1	0	11
mulsol.i.2	508.71	12310.37	3.583	584	0.8	0.42	11.8
mulsol.i.3	527.89	9201.45	3.541	579	0.5	0.52	11.5
mulsol.i.4	535.72	8040.28	3.622	578	0.9	0.32	11.9
mulsol.i.5	549.55	13014.81	3.651	584	1.1	0.56	12.1
myciel3	0	72.5	0.059279	14	0.1	0	11
myciel4	0.02	84.31	0.205	34	0.1	0	11
myciel5	2	211.73	112.12	80	0.1	0	11
myciel6	29.83	1992.42	10800	232	0.4	0.96	11.4
myciel7	634.32	19924.58	10800	757	18.2	17.23	219.2
school1	41141.1	137966.73	-	4684	5157.13	2497.5	10001
school1_nsh	2059.52	180300.1	-	4239	5468.9	2642.63	10001
zeroin.i.1	17.78	2595.92	-	641	0.1	0.3	11.1
zeroin.i.2	448.74	4825.51	-	594	43	72.84	244
zeroin.i.3	437.06	8898.8	-	585	22	38.82	223
le450_5a	7836.99	130096.77	10800	6433	7110.3	1789.35	10001
le450_5b	7909.11	187405.33	10800	6732	5989.9	2449.72	10001
le450_5c	103637.17	182102.37	10800	5917	4934.8	3367.84	10001
le450_5d	96227.4	182275.69	10800	5402	4033.8	2347.88	10001
le450_15a	6887.15	117042.59	-	6876	6191	2504.55	10001
le450_15b	6886.84	197527.14	10800	6423	6385.7	2646.55	10001
le450_15c	122069	143451.73	10800	4997	4368.9	3102.73	10001
le450_15d	127602	117990.3	10800	4864	3441.8	2767.31	10001
le450_25a	4478.3	143963.41	10800	6025	7377.9	1615.93	10001
le450_25b	4869.97	184165.21	10800	6045	6905.8	2241.92	10001
le450_25c	10998.68	151719.58	10800	6189	5345.9	2886.93	10001
le450_25d	11376.02	189175.4	10800	6712	4118.9	3464.8	10001
dsjc125.1	171.54	1532.93	10800	501	334.95	392.15	10001
dsjc125.5	38.07	2509.97	10800	261	66.0	72.73	267.0
dsjc125.9	55.6	1623.44	260.879	110	0.1	0	11.0
dsjc250.1	5507.86	28606.12	10800	1878	4162.4	2386.89	10001
dsjc250.5	1111.66	14743.35	10800	648	753.5	925.69	10001
dsjc250.9	1414.58	30167.7	10800	238	0.5	0.69	11.3

algorithm for 3 instances. Comparing this algorithm to our average, this algorithm gives better results for 11 examples, whereas IHA-avg is better for 24 instances. Considering comparison of GA and IHA, for the best width over 10 runs, our algorithm gives better results for 20 problems, whereas GA gives better results for 5 problems. For the average width in 10 runs, IHA-avg is better than GA-avg in 29 examples, whereas GA-avg is better than IHA-avg in 12 examples. Overall our algorithm is very good compared to other algorithms considering the width, and it gives new upper bounds for 14 instances (cells of table in bold).

Considering the time, a direct comparison of algorithms cannot be done, as the algorithms are executed in computers with different processors and memory. However, as we can see based on the results in Tables 9.4 and 9.5 our algorithm gives good time performance and for some instances it decreases significantly the time needed for generation of tree decompositions. Based on our experiments the efficiency of our algorithm is due to applying of LS1 algorithm in the construction phase of IHA. In LS1 only one solution is evaluated during each iteration. When using LS2 the number of solutions to be evaluated during most of iterations is much larger.

9.4 Conclusions

In this chapter, we presented a new heuristic algorithm for finding upper bounds for treewidth. The proposed algorithm has a structure of iterated local search algorithm and it includes different perturbation mechanisms and different variants for acceptance of solution for the next iteration. For the construction phase two simple local search based heuristics are proposed. Although the proposed constructive heuristics are simple, the whole iterated local search algorithm that uses these heuristics in a construction phase gives good results for tree decomposition. In particular using of construction method which includes only moving of nodes that produce the largest clique in the elimination ordering has been shown to be more efficient.

The proposed algorithm has been applied in 62 DIMACS vertex colouring instances, and our results have been compared with the best existing upper bounds for treewidth of these instances. The results show that our algorithm achieves good results for the upper bound of treewidth for different size of instances. In particular the algorithm improves the best existing treewidth upper bounds for many instances, and it has a good time performance.

For the future work we are considering the hybridization of our algorithm with the genetic algorithm for generation of tree decomposition. Furthermore, the algorithms described in this chapter can be used for generation of generalized hypertree decomposition (GHD). Generalized hypertree decomposition is a concept than includes additional condition and it is applied directly into the hypergraph. Good tree decomposition usually produce good generalized hypertree decomposition and the methods used for tree decomposition can be extended to generate GHD.

Acknowledgements

This paper was supported by the Austrian Science Fund (FWF) project: *Nr. P17222-N04, Complementary Approaches to Constraint Satisfaction.*

10

Search Intensification in Metaheuristics for Solving the Automatic Frequency Problem in GSM*

Francisco Luna[1], Enrique Alba[1], Antonio J. Nebro[1], and Salvador Pedraza[2]

[1] Department of Computer Science, University of Málaga (Spain)
{flv,eat,antonio}@lcc.uma.es
[2] Optimi Corp., Edif. Inst. Universitarios, Málaga (Spain)
Salvador.Pedraza@optimi.com

Summary. Frequency assignment is a well-known problem in Operations Research for which different mathematical models exist depending on the application specific conditions. However, most of these models are far from considering actual technologies currently deployed in GSM networks (e.g. frequency hopping). These technologies allow the network capacity to be actually increased to some extent by avoiding the interferences provoked by channel reuse due to the limited available radio spectrum, thus improving the Quality of Service (QoS) for subscribers and an income for the operators as well. Therefore, the automatic generation of frequency plans in real GSM networks is of great importance for present GSM operators. This is known as the Automatic Frequency Planning (AFP) problem. In this work, we focus on solving this problem for a realistic-sized, real-world GSM network with several metaheuristics featuring enhanced intensification strategies, namely $(1, \lambda)$ Evolutionary Algorithms and Simulated Annealing. This research line has been investigated because these algorithms have proven to perform the best for this problem in the literature. By using the same basic specialized operators and the same computational effort, SA has shown to outperform EAs by computing frequency plans which provoke lower interferences.

Keywords: Real-World Frequency Planning, Evolutionary Algorithms, Simulated Annealing.

10.1 Introduction

The success of the *Global System for Mobile* communication (GSM) [208], a multi-service cellular radio system, lies in efficiently using the scarcely available radio spectrum. The available frequency band is slotted into channels (or frequencies) which have to be allocated to the elementary transceivers (TRXs) installed in the base stations of the network. This problem is known as the Automatic Frequency Planning (AFP) problem, the Frequency Assignment Problem (FAP), or the Channel Assignment Problem (CAP). Several different problem types are

* This work has been partially funded by the Spanish Ministry of Education and Science and FEDER under contract TIN2005-08818-C04-01 (the OPLINK project).

C. Cotta and J. van Hemert (Eds.): Recent Advances in Evol. Comp., SCI 153, pp. 151–166, 2008.
springerlink.com

subsumed under these general terms and many mathematical models have been proposed since the late sixties [209, 210]. This work is focussed on concepts and models which are relevant for current GSM frequency planning [211] and not on simplified models of the abstract problem. In GSM, a network operator has usually a small number of frequencies (typically few dozens) available to satisfy the demand of several thousands TRXs. A reuse of these frequencies is therefore unavoidable. However, frequency reuse is limited by interferences which could lead the quality of service (QoS) for subscribers to be reduced to unsatisfactory levels. Consequently, the automatic generation of frequency plans in real GSM networks is a very important task for present GSM operators not only in the initial deployment of the system, but also in subsequent expansions/modifications of the network, solving unpredicted interference reports, and/or handling anticipated scenarios (e.g. an expected increase in the traffic demand in some areas). Additionally, several interference reduction techniques (e.g. frequency hopping, discontinuous transmission, or dynamic power control) [211] have been proposed to enhance the capacity of a given network while using the same frequency spectrum. These techniques are currently in use in present GSM networks and they must be carefully considered in AFP problems.

In this work we have addressed not only a real-world GSM network instance with real data and realistic size (more than 2600 TRXs to be assigned just 18 frequencies), but we have also used a commercial tool which implements accurate models for all the system components (signal propagation, TRX, locations, etc.) and actually deployed GSM interference reduction technologies. This tool allows us to evaluate the quality of the tentative frequency plans manipulated by the metaheuristic algorithms applied. Both the data as well as the software for evaluating the frequency plans are provided by Optimi Corp.^{TM}. It should be noted here that standard benchmarks like the Philadelphia instances, CELAR, and COST 259 [210] do not consider such technologies and therefore most of the proposed optimization algorithms are rarely faced with a real GSM frequency planning problem as we are actually doing in this work.

The AFP problem is a generalization of the graph coloring problem, and thus it is NP-hard [212]. As a consequence, using exact algorithms to solve real-sized instances of AFP problems is not practical, and therefore other approaches are required. Many different methods have been proposed in the literature [209] and, among them, metaheuristic algorithms have proved to be particularly effective. Metaheuristics [213, 214] are stochastic algorithms that sacrifice the guarantee of finding optimal solutions for the sake of (hopefully) getting accurate (also optimal) ones in a reasonable time. This fact is even more important in commercial tools, in which the GSM operator cannot wait for very long times to obtain a frequency plan.

This paper further investigates on the research lines presented in [215]. In this previous work, the real-world AFP problem has been addressed with Evolutionary Algorithms (EAs) [216]. Specifically, a fast and accurate $(1, \lambda)$ EA has been used. The main reason to make this decision had to do not only with the fact that these EAs do not need to recombine individuals, but also with its population-based evolutionary behavior and the low cost per iteration. After evaluating several configurations of the $(1, \lambda)$ EA, Luna *et al.* have concluded in [215] that the settings with the enhanced intensification capabilities reported the best frequency plans for the GSM network instance faced. Bearing this in mind, we have continued here the study about intensification strategies in two ways with the aim of engineering an improved algorithm that allows this complex problem to be solved with a higher accuracy. On the one hand, we have stressed even more the intensification features of the $(1, \lambda)$ EA. By using the same search operators, we have paid attention to the λ value. It is clear that if the $(1, \lambda)$ EAs stop when computing the same number of function evaluations, smaller values for λ make these algorithms iterate for longer, what therefore means a larger exploitation of the search experience, that is, enhanced intensification capabilities. While values of $\lambda = 10$ and $\lambda = 20$ have been tried in [215], we have analyzed here two new values $\lambda = 1$ and $\lambda = 5$. On the other hand, we have evaluated a second metaheuristic with deeply marked search intensification features: Simulated Annealing (SA) [217]. SA has been widely used in the literature for solving classical frequency assignment problems [209] and it has been included in this study not only because it fits perfectly in the kind of metaheuristics with the enhanced intensification search features, but also because it allows us to reuse the highly specialized search operators developed in [215], namely, solution initialization and perturbation. Let us now summarize the contributions of this work:

- We have studied two new configurations of the $(1, \lambda)$ EA that enhance its intensification capabilities with the aim of improving the results obtained previously for the AFP instance solved.
- By using the same operators as the $(1, \lambda)$ EA, a SA algorithm has been considered. This is the first time that this kind of technique is applied to our real-world approach of the AFP problem. The results indicate that SA clearly outperforms all the $(1, \lambda)$ EAs in the resulting frequency plans computed, what shows the suitability of this algorithm for solving this problem: a new best solution for the given instance has been found.

The paper is structured as follows. In the next section, we provide the reader with details on the frequency planning in GSM networks. Section 10.3 describes the two algorithms applied along with the search operators used. The results of the experiments are presented and analyzed in Section 10.4. Finally, conclusions and future lines of research are discussed in the last section.

10.2 Frequency Planning in GSM Networks

This section is devoted to presenting details on the frequency planning task for a GSM network. We first provide the reader with a brief description of the GSM architecture. Next, we give the relevant concepts to the frequency planning problem that will be used along this paper.

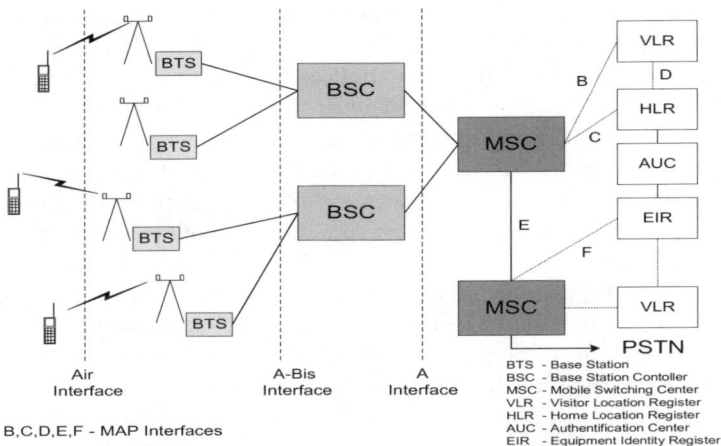

Fig. 10.1. Outline of the GSM network architecture

10.2.1   The GSM System

An outline of the GSM network architecture is shown in Fig. 10.1. As it can be seen, GSM networks are built out of many different components. The most relevant ones to frequency planning are the Base Transceiver Station (BTS) and the transceivers (TRXs). Essentially, a BTS is a set of TRXs. In GSM, one TRX is shared by up to eight users in TDMA (*Time Division Multiple Access*) mode. The main role of a TRX is to provide conversion between the digital traffic data on the network side and radio communication between the mobile terminal and the GSM network. The site at which a BTS is installed is usually organized in sectors: one to three sectors are typical. Each sector defines a cell.

The solid lines connecting components in Fig. 10.1 carry both traffic information (voice or data) as well as the "in-band" signaling information. The dashed lines are signaling lines. The information exchanged over these lines is necessary for supporting user mobility, network features, operation and maintenance, authentication, encryption, and many other functions necessary for the network's proper operation. Figure 10.1 shows the different network components and interfaces within a GSM network.

10.2.2   The Automatic Frequency Planning Problem

The frequency planning is the last step in the layout of a GSM network. Prior to tackling this problem, the network designer has to address some other issues: where to install the BTSs or how to set configuration parameters of the antennae (tilt, azimuth, etc.), among others [218]. Once the sites for the BTSs are selected and the sector layout is decided, the number of TRXs to be installed per sector has to be fixed. This number depends on the traffic demand that the corresponding sector has to support. The result of this process is a quantity of

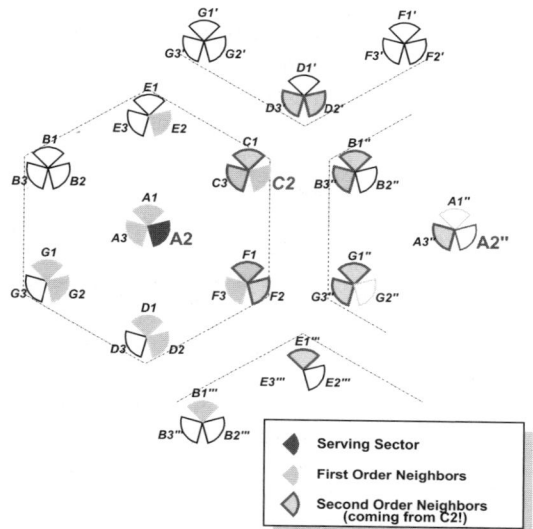

Fig. 10.2. An example of GSM network

TRXs per cell. A channel has to be allocated to every TRX and this is the main goal of the AFP [211]. Essentially, three kinds of allocation exist: Fixed Channel Allocation (FCA), Dynamic Channel Allocation (DCA), and Hybrid Channel Allocation. In FCA, the channels are permanently allocated to each TRX, while in DCA the channels are allocated dynamically upon request. Hybrid Channel Allocation (HCA) schemes combine FCA and DCA. Neither DCA nor HCA are supported in GSM, so we only consider FCA.

We now explain the most important parameters to be taken into account in GSM frequency planning. Let us consider the example network shown in Fig. 10.2, in which each site has three installed sectors (e.g. site A operates $A1$, $A2$, and $A3$). The first issue is the implicit topology which results from the previous steps in the network design. In this topology, each sector has an associated list of neighbors containing the possible handover candidates for the mobile residing in a specific cell. These neighbors are further distinguished into first order (those which can potentially provoke strong interference to the serving sector) and second order neighbors. In Fig. 10.2, $A2$ is the serving sector and the first order neighbors defined are $A1$, $A3$, $C2$, $D1$, $D2$, $E2$, $F3$, $G1$, $G2$, and $B1'''$, whereas, if we consider $C2$, second order neighbors of $A2$ are $F1$, $F2$, $C1$, $C3$, $D2'$, $D3'$, $A3''$, $B1''$, $B3''$, $G1''$, $G3''$, and $E1'''$.

As stated before, each sector in a site defines a cell; the number of TRXs installed in each cell depends on the traffic demand. A valid channel from the available spectrum has to be allocated to each TRX. Due to technical and regulatory restrictions, some channels in the spectrum may not be available in every cell. They are called locally blocked and they can be specified for each cell.

Each cell operates one Broadcast Control CHannel (BCCH), which broadcasts cell organization information. The TRX allocating the BCCH can also carry user data. When this channel does not meet the traffic demand, some additional TRXs have to be installed to which new dedicated channels are assigned for traffic data. These are called Traffic CHannels (TCHs).

In GSM, significant interference may occur if the same or adjacent channels are used in neighboring cells. Correspondingly, they are named co-channel and adj-channel interference. Many different constraints are defined to avoid strong interference in the GSM network. These constraints are based on how close the channels assigned to a pair of TRXs may be. These are called separation constraints, and they seek to ensure the proper transmission and reception at each TRX and/or that the call handover between cells is supported. Several sources of constraint separation exists: co-site separation, when two or more TRXs are installed in the same site, or co-cell separation, when two TRXs serve the same cell (i.e., they are installed in the same sector).

This is intentionally an informal description of the AFP problem in GSM networks. It is out the scope of this work to propose a precise model of the problem, since we use a proprietary software which is aware of all these concepts, as well as the consideration of all the existing interference reduction techniques developed for efficiently using the scarce frequency spectrum available in GSM.

10.3   Algorithms for Solving the AFP Problem

Both EAs and SA have been widely used for solving many existing flavors of the frequency assignment problem [209, 210, 211]. The main drawback of using EAs has to do with the fact that crossover operators such as single point crossover do not perform well on this problem [219]. Indeed, it does not make sense to randomly exchange different, possibly non-related assignments of two given frequency plans. With respect to EAs, our approach here is to use an $(1, \lambda)$ EA, since the recombination operator is not required. The SA algorithm does not need to recombine solutions either and therefore it is a very suitable approach for facing this problem. In fact, recent publications have successfully applied SA to solve not only classical [209] but also an advanced models of the problem (e.g., the work of Björklund et al. [220], which tackles frequency hopping optimization).

Next, after giving some insights on the fitness function of the optimization problem and the solution encoding used, the two optimization algorithms are described along with the methods used for generating the initial solutions and perturbing them.

10.3.1   Fitness Function

As it was stated before, we have used a proprietary application provided by Optimi Corp.^{TM}, that allows us to estimate the performance of the tentative frequency plans generated by the evolutionary optimizer. Factors like Frame

Erasure Rate, Block Error Rate, RxQual, and BER are evaluated. This commercial tool combines all aspects of network configuration (BCCHs, TCHs, etc.) including interference reduction techniques (frequency hopping, discontinuous transmission, etc.) in a unique cost function, C, which measures the impact of proposed frequency plans on capacity, coverage, QoS objectives, and network expenditures. This function can be roughly defined as:

$$C = \sum_v \left(CostIM\left(v\right) \cdot E\left(v\right) + CostNeighbor\left(v\right) \right), \tag{10.1}$$

that is, for each sector v which is a potential victim of interference, the associated cost is composed of two terms, a signaling cost computed with the interference matrix ($CostIM\left(v\right)$) that is scaled by the traffic allocated to v, $E\left(v\right)$, and a cost coming from the current frequency assignment in the neighbors of v. Of course, the lower the total cost the better the frequency plan, i.e., this is a minimization problem.

10.3.2 Solution Encoding

The solution encoding determines both the search space and the subsequent set of search operators that can be applied during the exploration of this search space. Let T be the number of TRXs needed to meet the traffic demand of a given GSM network. Each TRX has to be assigned a channel. Let $F_i \subset \mathbb{N}$ be the set of available channels for the transceiver i, $i = 1, 2, 3, \ldots, T$. A solution p (a frequency plan) is encoded as a T-length integer array $p = [f_1, f_2, f_3, \ldots, f_T]$, $p \in F_1 \times F_2 \times \cdots \times F_T$, where $f_i \in F_i$ is the channel assigned to TRX i. The fitness function (see the previous section) is aware of adding problem specific information to each transceiver, i.e., whether it allocates either a BCCH channel or a TCH channel, whether it is a frequency hopping TRX, etc.

As an example, Fig. 10.3 displays the representation of a frequency plan p for the GSM network shown in Fig. 10.2. We have assumed that the traffic demand in the example network is fulfilled by one single TRX per sector (TRX $A1$, TRX $A2$, etc.).

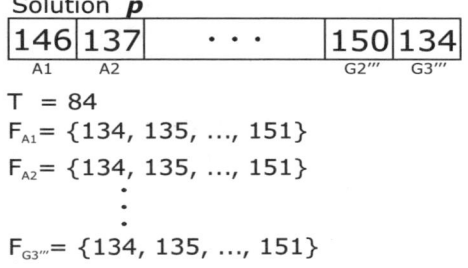

Fig. 10.3. Solution encoding example

10.3.3 $(1, \lambda)$ EA

This optimization technique firstly generates an initial solution. Next, the algorithm perturbs and evaluates this solution at each iteration, from which λ new ones are obtained. Then, the best solution taken from the newly generated λ individuals is moved to the next iteration. An outline of the algorithm is shown in Fig. 10.4. Other works using a similar algorithmic approach for the AFP problem can be found in [219, 221].

```
1:   s = new Solution();
2:   PAux = new Population(λ);
3:   init(s);
4:   evaluate(s);
5:   for iteration = 0 to NUMBER_OF_ITERATIONS do
6:     for i = 1 to λ do
7:       perturbation = perturb(s);
8:       evaluate(perturbation);
9:       PAux = addTo(PAux,perturbation);
10:    end for
11:    s = bestIndividual(PAux);
12:  end for
```

Fig. 10.4. Pseudocode of the $(1, \lambda)$ EA

As stated before, the seeding procedure for generating the initial solution and the perturbation operator are the core components defining the exploration capabilities of the $(1, \lambda)$ EA. The definition of these two procedures is detailed in Sections 10.3.5 and 10.3.6.

10.3.4   Simulated Annealing

The SA algorithm is a stochastic relaxation technique, which has its origins in statistical mechanics. SA can be seen as one way of trying to allow the basic dynamics of hill-climbing to be able to escape local optima of poor solution quality [217]. As the $(1, \lambda)$ EA, SA firstly generates an initial solution (line 3 in Fig. 10.5) and then it guides the original local search method in the following way. A new solution $s1$ is generated by perturbing the current one, $s0$ (line 8). Once evaluated, SA has to decide whether it keeps the current solution or it uses with the newly generated one. This decision is made in the accept() function that appears in line 10 of Fig. 10.5. This way, $s1$ is always accepted as the new current solution if $s1$ is better than $s0$, i.e., $v1 < v0$. To allow the search to escape a local optimum, moves that increase the objective function value are accepted with a probability $exp(-(v1 - v0)/T)$ if $s1$ is worse ($v1 > v0$), where T is a parameter called the "temperature". The value of T varies from a relatively large value to a small value close to zero. These values are controlled by a cooling

```
1:   t = 0;
2:   initialize(T);
3:   s0 = new Solution();
4:   v0 = evaluate(s0);
5:   repeat
6:     repeat
7:         t = t + 1;
8:         s1 = perturb(s0,T);
9:         v1 = evaluate(s0,T);
10:        if accept(v0,v1,T)
11:           s0 = s1;
12:           v0 = v1;
13:        end if
14:     until (t mod Markov_Chain_length) == 0
15:     T = update(T)
16: until 'loop stop criterion' satisfied
```

Fig. 10.5. Pseudocode of SA

schedule, which specifies the initial, and temperature values at each stage of the algorithm. We have used here a proportional cooling scheduling in which the temperature, T_i, at step i is computed as:

$$T_i = T_0 \left(\frac{T_0}{T_N} \right)^{\frac{i}{N}} \qquad (10.2)$$

where T_0 and T_N are the initial and final temperatures of the algorithm and N is the number of steps of the algorithm. An additional parameter of SA is the *Markov Chain Length*, that defines the number of consecutive steps at which the algorithm uses the same temperature in the acceptance criterion. The algorithm stops when a given temperature is reached or a preprogrammed number of iterations are computed.

As it can be seen, the operation and performance of this optimization technique strongly depends on the initial seeding and the perturbation used, since having a badly designed operators could guide the search towards low-quality regions of the search space.

10.3.5 Solution Initialization

Individuals are initialized site by site using a constructive method. For each site in the GSM network, a optimal frequency assignment is heuristically computed independently, without taking into account possible interferences from any other site. A simple greedy heuristic [209] is the method used (see Fig. 10.6 for its pseudocode). Given a site s, all its TRXs installed are randomly ranked (line 3). Then, random frequencies are assigned to the TRXs so that neither co-channel nor adjacent-channel interferences are provoked (lines 5 and 6). We want to note

```
1:   trxs = frequencies = ∅;
2:   trxs = TRXsFromSite(s);
3:   random_shuffle(trxs);
4:   for t in trxs do
5:      f = chooseInterferenceFreeFrequency(t,frequencies);
6:      assign(t,f);
7:      frequencies = insert(frequencies,t);
8:   end while
```

Fig. 10.6. Pseudocode of a greedy heuristic

that the available radio spectrum is large enough to generate optimal frequency assignments within a site most times, i.e., we are applying the greedy heuristic independently only to the TRXs within a site. This way we avoid the most important source of strong interference through the network: those involving TRXs installed in the same site or sector.

Finally, the individual undergoes a heuristic which is based on the *Dsatur with Costs* heuristic [211]. Here, all the TRXs of the network are ranked so the hardest ones to deal with are assigned first. The measure used for "hardest to deal with" consists of accurate information given by the simulator about the importance of the TRXs (interference provoked, capacity overload, etc.), rather than using the generalization of saturation and space degrees of *Dsatur with Costs*. Next, each TRX is assigned the frequency presently incurring the least additional interference. The main goal of this phase is to reach "good" solutions in very short times. This is usually a requirement within commercial applications, the context of this work. Using this greedy phase is also the main reason to use the $(1, \lambda)$ strategy. Indeed, the greedy step leads the search towards a local minimum and it makes the non-elitist $(1, \lambda)$ strategy suitable for the search because it avoids getting stuck in this region of the search space.

10.3.6 Perturbation Operator

In both the $(1, \lambda)$ EA and SA, the perturbation operator largely determines the search capabilities of the algorithm. The mechanism used is based on modifying the channels allocated to a number of transceivers. The operator first has to select the set of TRXs to be modified and, next, it chooses the new channels which will be allocated. The two proposed methods are as follows:

1. TRX Selection: At each operation, one single site is perturbed. The way of selecting the site is to choose first a TRX t and then the site considered is the one at which t is installed. The strategy used is Binary Tournament, which uses the same information from the simulator as the last greedy operation in the initialization method (see Section 10.3.5). Given two randomly chosen TRXs, this strategy returns the "hardest to deal with", i.e., the one which is preferred to be updated first. With this configuration, the perturbation

```
1:   trxs = TRXsFromSite(s);
2:   applySimpleGreedyHeuristic(s);
3:   trxs = rank(trxs);
4:   for t in trxs do
5:     f = chooseMinimumInterferenceFrequency(t,neighbors(s));
6:     assign(t,f);
7:   end for
```

Fig. 10.7. Pseudocode of frequency selection strategy

mainly promotes intensification. In [215], a random strategy was also proposed and evaluated, but it reported lower quality results and it has been no longer considered here.

2. Frequency Selection: Let s be the site chosen in the previous step. Firstly, s is assigned a interference-free frequency planning with the same strategy used in the initialization method. We have therefore avoided the strongest intra-site interferences. The next step aims at refining this frequency plan by reducing the interferences with the neighboring sites. The strategy proceeds iterating through all the TRXs installed in s. Again, these TRXs are ranked in decreasing order with the accurate information coming from the simulator. Finally, for each TRX t, if a frequency f, different from the currently assigned one, allows both to keep the intra-site interference-free assignment and to reduce the interference from the neighbors, then t is assigned f; otherwise, it does nothing. All the available frequencies for t are examined. A pseudocode of the method is included in Fig. 10.7. Note that this procedure guarantees that no interference will occur among the TRXs installed in a site.

10.4 Experiments

In this section we turn to present the experiments conducted to evaluate the new configurations of the $(1, \lambda)$ EA and the SA algorithm. We firstly give some details of the GSM network instance used. The experiments with different configurations of the evolutionary optimizer are presented and analyzed afterwards.

10.4.1 GSM Instance Used

The GSM network used here has 711 sectors with 2,612 TRXs installed. That is, the length of the individuals in the EA is 2,612. Each TRX has 18 available channels and therefore the search space to be explored has a size of $18^{2,612}$. Figure 10.8 displays the network topology. Each triangle represents a sectorized antenna in which operate several TRXs. As it can be seen, the instance presents clustered plus non-clustered zones where no classical hexagonal cell shapes exist (typically used in simplified models of the problem). Additional topological information indicates that, on average, each TRX has 25.08 first order neighbors

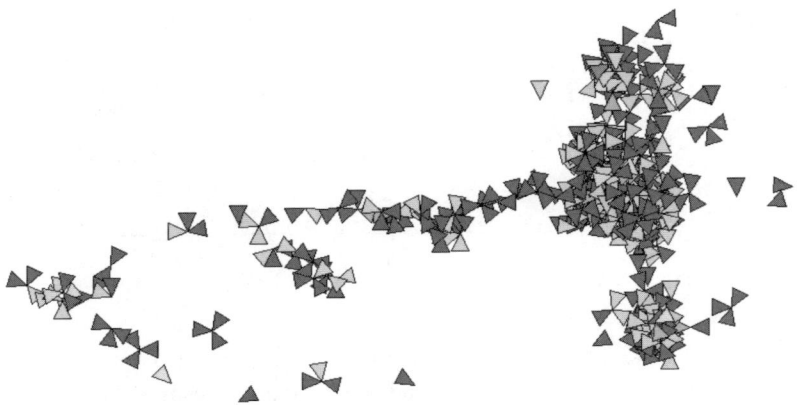

Fig. 10.8. Topology of the GSM instance used

and 96.60 second order neighbors, thus showing the high complexity of this AFP instance, in which the available spectrum is much smaller than the average number of neighbors. Indeed, only 18 channels can be allocated to TRXs with 25.08 potential first order neighbors. We also want to remark that this real network operates with advanced interference reduction technologies and it employs accurate interference information which has been actually measured at a cell-to-cell level (neither predictions nor distance-driven estimations are used).

10.4.2   Results

We have conducted experiments with two new configurations of the $(1, \lambda)$ EA, $\lambda = 5$ and $\lambda = 1$, and SA. The perturbation operator shared by both EAs and SA has the strongest intensification strategy which uses Binary Tournament as TRX selection scheme (the configurations with $\lambda_2 = 0$ in [215]). After some preliminary experimentation, the SA algorithm has been configured with the following values: $T_0 = 10$, $T_N = 0.1$, *Markov Chain Length* = 30 steps, and $N = 100,000$ steps. All the algorithms stop when 100,000 function evaluations have been computed.

We have made 30 independent runs of each experiment and the results obtained are shown in Table 10.1. Since we are dealing with stochastic algorithms and we want to provide the results with confidence, a precise statistical analysis has been performed throughout this work [222, 223]. After comparing the normality and homoscedasticity of the distributions, we have used the nonparametric Kruskal-Wallis test to compare their medians because some of them follow a gaussian distribution and some others do not. We always consider in this work a confidence level of 95% (i.e., significance level of 5% or p-value under 0.05) in the statistical tests, which means that the differences are unlikely to have occurred by chance with a probability of 95%. In this case, the test has been successful and the last row has been therefore marked with a "+" symbol.

Table 10.1. AFP costs of the $(1, \lambda)$ EAs and SA.

Algorithm	AFP Costs	
	$\tilde{x}$	IQR
(1,1) EA	10,462	2,760
(1,5) EA	4,947	78
(1,10) EA	4,680	148
(1,20) EA	4,787	151
SA	4,327	95
Statistical test	+	

Table 10.2. Post-hoc testing

(1,5)EA	+			
(1,10)EA	+	+		
(1,20)EA	+	+	−	
SA	+	+	+	+
	(1,1) EA	(1,5) EA	(1,10) EA	(1,20) EA

Looking for homogeneity in the presentation of the results, Table 10.1 respectively includes the median, $\tilde{x}$, and interquartile range, IQR, as measures of location (or central tendency) and statistical dispersion, because, on the one hand, some samples follow a gaussian distribution while others do not, and, on the other hand, mean and median are theoretically the same for gaussian distributions. Indeed, whereas the median is a feasible measure of central tendency in both gaussian and non-gaussian distributions, using the mean only makes sense for gaussian ones. The same holds for the standard deviation and the interquartile range. The best result has a gray colored background.

To further analyze the results statistically, we have then included a post-hoc testing phase in Table 10.2 which allows for a multiple comparison of samples. We have used the `multcompare` function provided by Matlab ©. This function automatically chooses the most appropriate type of critical value to be used in the multiple comparison, which ranges from the more conservative HSD or *Tukey-Kramer* method to less conservative *Scheffe's S* procedure [224]. The same confidence level has been kept for this testing phase ($\alpha = 0.05$). The "+"symbols in the table point out that most pairwise comparisons among the algorithms are significant.

Let us start analyzing the results of the two new $(1, \lambda)$ EAs using $\lambda = 1$ and $\lambda = 5$. Since the algorithms compute the same number of function evaluations (100,000 evals), (1,1) EA and (1,5) EA iterate 100,000 and 20,000 times, respectively. As stated before, the goal of these new experiments is to check whether intensifying the search leads this algorithm to computer more accurate frequency plans for our GSM network. The reported results are clear: the (1,10) EA is the most accurate algorithm out of the four EAs for the solved instance. Statistical confidence exists with respect to the (1,1) EA and the (1,5) EA (see the "+" symbols in Table 10.2).

On the one hand, the AFP costs get worse (increase) as the value of λ decreases, that is, the new balance between intensification and diversification reached by the two new configurations makes the algorithm to get stuck in frequency plans provoking stronger interference in the GSM network. We can explain the lower performance of these two new configs because, even though they are allowed to iterate for longer thus deeply exploiting the search experience, the number of solutions manipulated at each step of the algorithm is smaller and therefore the chance of generating and choosing a good solution for the next iteration is reduced. (Note that the current solution is always replace by one of the λ offsprings, even if they all have a higher AFP cost, i.e., a non elitist scheme.) This fact makes, since the beginning of the exploration, the search to be guided towards regions of the search space with low quality solutions. This is specially problematic in the (1,1) EA, in which the current solution is always replaced by the newly generated one at each step. To better illustrate this fact, Fig. 10.9 displays a graphical tracking of the AFP cost (fitness) of the current solution for the (1,1) EA, (1,5) EA, and (1,10) EA during the computation (the (1,20) EA behaves similarly to the $\lambda = 10$ config and its inclusion in this figure shows nothing relevant). Note the logarithmic scale in the y axis. It can be seen that the search of the (1,1) EA is almost random and its best solution is indeed the initial one which results from the greedy phase. This behavior is drastically reduced in the (1,5) EA. In general, the perturbation operator is able to generate an acceptable solution in five trials, which allows the search to progress properly. Finally, Fig. 10.9 clearly shows that the AFP cost in the (1,10) EA evolves smoothly, enabling the algorithm to always exploring high quality areas of the search space.

On the other hand, the AFP costs are also worse (higher) when $\lambda = 20$. This configuration iterates for half the number of steps of (1,10) EA (5,000 instead of 10,000) so as to compute the same number of function evaluation. As already explained in [215], iterating for longer, and thus enhancing the intensification capabilities, allows the algorithm to reach higher quality frequency plans (provoking lower interference). Therefore, we can conclude that the (1,10) EA achieves the best balance between intensification/diversification in the studied EAs for solving the given AFP problem instance.

As to the EAs, we also want to show that, however, the $\lambda = 5$ and $\lambda = 20$ settings have computed very competitive solutions compared to the (1,10) EA (4,947 and 4,787 against 4,680 on average) while keeping a very small IQR value at the same time. These IQR values in these three EAs indicate that they all are very robust, reaching very high quality frequency plans in all the independent runs. It should be noted that robustness is a highly desirable feature for commercial tools because it guarantee, to some extent, that the solver will always provide the user with a final solution (frequency plan) which is accurate enough.

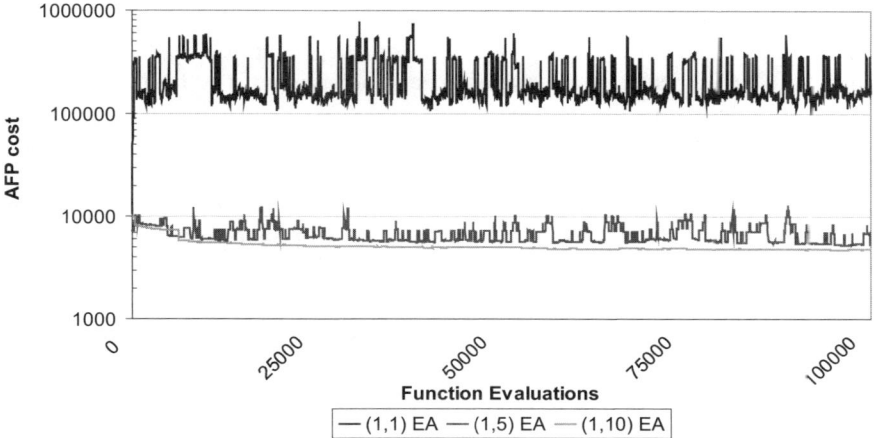

Fig. 10.9. Fitness of the current individual at each iteration of the three $(1, \lambda)$ EA

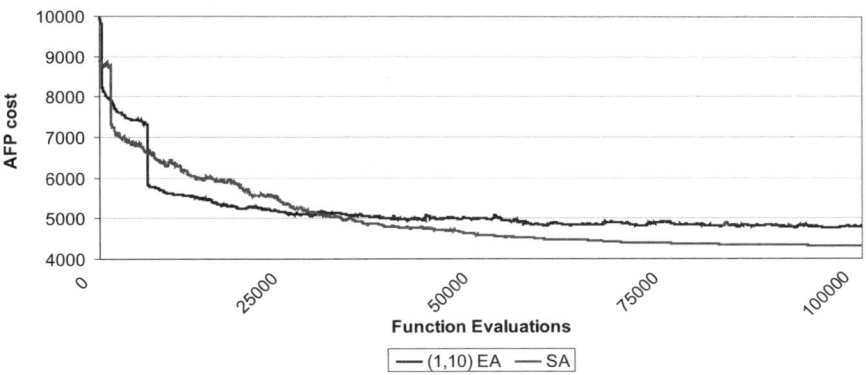

Fig. 10.10. AFP cost of the current individual at each iteration of the $(1, 10)$ EA and SA

The results of the SA algorithm are included in the fifth row of Table 10.1. With the configuration described at the beginning of this section, SA runs for 100,000 iterations (one evaluation per iteration). The AFP cost obtained by SA are clearly better (smaller) than those computed in any of the studied $(1, \lambda)$ EAs. This cost value (4,327) is the best solution found so far for this instance of the AFP problem. The cost reduction reached is 9.61%, 7.54%, 12.53%, and 58.64%, with respect to EAs with the $\lambda = 20$, $\lambda = 10$, $\lambda = 5$, and $\lambda = 1$ configurations, respectively. Additionally, the last row of Table 10.1 and the fourth row of Table 10.2 indicate that all the previous claims are supported with statistical confidence (at 95% level). Since both the solution initialization method and the perturbation operator are shared by the two kind of algorithms, we consider that the search engine of SA (acceptance criterion of worse solutions, cooling schedule, etc.) is the responsible for the reduction of the AFP costs. Figure 10.10 graphically shows the different behavior of the best EA ($\lambda = 10$) and the SA algorithm during one single run. (No logarithmic scale has been used here.) As expected, the (1,10) EA reaches better solutions than SA in the beginning because, at this phase, the probability of accepting a bad solution is higher in SA. However, while SA continues improving the solution until the end because fewer worse solutions are accepted, the EA finds difficulties at this stage due to its non-elitist scheme (the best solutions can be lost).

10.5   Conclusions and Future Work

This paper analyzes the utilization of $(1, \lambda)$ EAs and SA to solve the AFP problem in a real-world GSM network composed of 2,612 transceivers. Instead of using a mathematical formulation of this optimization problem, we have used a commercial application which allows the target frequency plannings to be accurately evaluated in a real scenario where current technologies are in use (e.g. frequency hopping, discontinuous transmission, dynamic power control, etc.).

We have used advanced operators for initializing the individuals and for generating the offspring. Then, different configurations for the evolutionary algorithms have been proposed and evaluated with $\lambda = 5$ and $\lambda = 1$, in order to check whether intensifying the search be means of reducing the value of λ (which consequently means iterating for longer and thus deeply exploiting the search experience). They have been compared to the configurations proposed in [215], with $\lambda = 10$ and $\lambda = 20$. The new experiments have shown that the configuration with $\lambda = 10$ reached the best (lowest) AFP costs among the EAs, i.e., it has computed the frequency plans with the lower interference in the network. This points out that, for the instance solved and with the initialization and perturbation operators used, decreasing the value of λ and iterating for longer make the $(1, \lambda)$ EA to reach worse (smaller) AFP costs. The config with $\lambda = 1$ performs specially badly because, at each step of the algorithm, the current solution is always replaced by the newly generated one. On the other hand, the (1,10) EA also improves upon the config with $\lambda = 20$, so therefore it can be stated that the (1,10) EA has reached the best balance between intensification/diversification among the EAs for the solved AFP problem instance. The results of SA clearly outperform those obtained by any of the EAs used. Since they share the same operators (initialization and perturbation), we can conclude that the search engine of this technique is very well suited for solving this problem.

As future work, we plan to develop new search operators and new metaheuristic algorithms to solve this problem. Their evaluation with the current instance and other real-world GSM networks is also an ongoing research line.

11

Contraction-Based Heuristics to Improve the Efficiency of Algorithms Solving the Graph Colouring Problem

István Juhos[1] and Jano van Hemert[2]

[1] Department of Computer Algorithms and Artificial Intelligence,
University of Szeged, Hungary
paper@juhos.info
[2] National e-Science Institute, School of Informatics, University of Edinburgh,
United Kingdom
j.vanhemert@ed.ac.uk

Summary. Graph vertex colouring can be defined in such a way where colour assignments are substituted by vertex contractions. We present various hypergraph representations for the graph colouring problem all based on the approach where vertices are merged into groups. In this paper, we explain this approach and identify three reasons that make it useful. First, generally, this approach provides a potential decrease in computational complexity. Second, it provides a uniform and compact way in which algorithms, be it of a complete or a heuristic nature, can be defined and progress toward a colouring. Last, it opens the way to novel applications that extract information useful to guide algorithms during their search. These approaches can be useful in the design of an algorithm.

Keywords: Graph Contraction, Graph Colouring, Graph Representation, Heuristics, Graph Homomorphisms, Evolutionary Computation.

11.1 Introduction

The *graph colouring problem* (GCP) is an important problem from the class of non-deterministic polynomial problems. It has many applications in the real world such as scheduling, register allocation in compilers, frequency assignment and pattern matching. To allow these applications to handle larger problems, it is important fast algorithms are developed. Especially as high-performance computing is becoming more readily available, it is worthwhile to develop algorithms that can make use of parallelism.

A graph $G = (V, E)$ consists of a set of vertices V and a set of edges $E \subseteq V \times V$ defines a relation over the set of vertices. We let $n = |V|$, $m = |E|$, $d(v)$ is the degree of a vertex $v \in V$ and A is the adjacency matrix of G. A colouring c of a graph is a map of colours to vertices ($c : V \to C$) where C is the set of colours used. There can be several such mappings, which can be denoted if necessary (e.g. $c_1, c_2, \dots$). A colouring c is a k-colouring iff $|C| = k$. It is a proper or valid colouring if for every $(v, w) \in E : c(v) \neq c(w)$. In general, when we refer to a

C. Cotta and J. van Hemert (Eds.): Recent Advances in Evol. Comp., SCI 153, pp. 167–184, 2008.
springerlink.com © Springer-Verlag Berlin Heidelberg 2008

colouring we mean a valid colouring, unless the context indicates otherwise. The chromatic number $\chi(G)$ of a graph G is the smallest k for which there exists a k-colouring of G. A colouring algorithm makes colouring steps, i.e., it progressively chooses an uncoloured vertex and then assigns it a colour. Let $t \in \{1, \ldots, n\}$ be the number of steps made.

The graph colouring approaches discussed here will construct a colouring for a graph by progressively contracting the graph. This condensed graph is then coloured. The colouring of intermediate graphs in examples serves only to help the reader follow the process. In reality, contractions themselves define the whole colouring process.

The operations that allow these contractions are naturally parallel. Moreover, they allow us to extract heuristic information to better guide the search. Two types of graph contractions exist in the literature. The first is the vertex-contraction; when unconnected or connected vertices are identified as one vertex, some original edges can either be kept or removed. The name 'contraction' can be seen as a merge or a form of coalescing ([225, 226, 227, 228, 229, 230, 231, 232, 233, 234, 235]) as well. The latter is commonly used in the domain of register allocation problems, which can be modeled as a graph colouring problem. In the colouring model of the register allocation, vertices are coalesced where this is safe, in order to eliminate move operations between distinct variables (registers). The second is the edge-contraction. This approach is similar to the idea described in case of vertices whenever two edges are merged together, vertices can be merged.

The purpose of the merging can either be simplification or the combination of several simple graphs into one larger graph. Both edge and vertex contraction techniques are valuable in proof by induction on the number of vertices of edges in a graph, where we can assume that a property holds for all contractions of a graph and use this to show it for the larger graph.

Usually, colouring algorithms use vertex merging for graph simplification and graph combination. Simplification is done by merging two or more unconnected vertices to get fewer vertices before or during colouring. In [225], [226] and [227] preprocessing of graphs is performed before colouring, where two vertices in a graph are merged to one if they are of the same colour in all colourings. This is analogous to studies of the development of a backbone or spine in the satisfiability problem [236, 237]. Here, the application of merging refers to removing one of two unconnected vertices. In fact, we could remove also edges that belong to the removed vertex. The only reason to perform these merges is to remove unnecessary or unimportant vertices from the graph in order to make it simpler. Those vertices that fulfil some specific condition, will be removed from the data structure, which describes the graph. This process will result in the loss of information.

The second approach is to consider two graphs, which have some colouring properties. For example the property could be that they are not k-colourable. Then, the two graphs are joined by merging vertices from both graphs to create a more complex graph, where the aim is that the original properties are inherited. In both cases the identified vertices get the same colour.

Register allocation can be modelled as a graph colouring problem, but it is modelled in other ways too. Coalescing is a terminology frequently used instead of merging, where two register are coalesced. If the problem is represented by graph colouring, coalescing is the same as merging unconnected vertices. Register coalescing techniques can be identified as vertex merging in a graph colouring problem [228, 229, 230, 231, 232].

Using the idea of graph contraction we will describe an efficient graph colouring model, where contractions are performed on unconnected vertices. We will define two well known and two novel heuristics based on this model. The model itself serves as a framework to describe graph colouring algorithms in a concise manner. Moreover, we show how the model allows an increase in the performance of certain algorithms by exploiting the data structures of the model through several heuristics.

11.2 Representing Solutions to Graph Colouring as Homomorphisms

We define merge operations to perform contraction of the original graph and subsequent contractions. A merge operation takes two unconnected vertices from a graph $G = (V, E)$ and produces a new graph $G' = (V', E')$ where these vertices become one hyper-vertex. If edges exist between another vertex and both the original vertices, then these become one hyper-edge. If $v_1, v_2 \in V$ are merged to $\{v_1, v_2\} \in V'$ and both $(v_1, u), (v_2, u) \in E$ then $(\{v_1, v_2\}, u) \in E'$ is called a hyper-edge. Examples of merge operations are shown in Figure 11.1. The merge operation is applied similarly to hyper-vertices.

By repeating merge operations we will end up with a complete graph. If during each merge operation we ensure only hyper-vertices and vertices are merged that have no edges between them, we then can assign all the vertices from the original graph that are merged into the same hyper-vertex one unique colour. The number of hyper-vertices in the final contracted graph corresponds to the number of colours in a valid colouring of the original graph. As Figure 11.1 shows, the order in which one attempts to merge vertices will determine the final colouring. Different colourings may use different number of colours. We will investigate a number of different strategies for making choices about which vertices to merge.

Graph colouring solvers make use of constraint checks during colouring. The adjacency checks to verify if assigned colourings are valid play a key role in the overall performance (see [238, 233, 234, 235]). The number of checks depends on the representation of the solution, the algorithm using the representation and the intrinsic difficulty of the problem. Graph colouring problem are known to exhibit an increase in difficulty in the so-called phase transition area [227]. In our experiments we will test several algorithms on problems in this area.

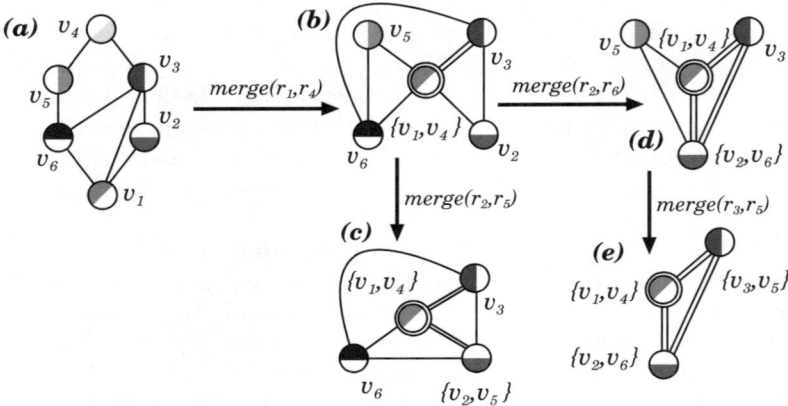

Fig. 11.1. Examples of the result of two different merge orders $P_1 = v_1, v_4, v_2, v_5, v_3, v_6$ and $P_2 = v_1, v_4, v_2, v_6, v_3, v_5$. The double-lined edges are hyper-edges and double-lined nodes are hyper-nodes. The P_1 order yields a 4-colouring (c), however with the P_2 order we get a 3-colouring (e).

Merge operations and the contracted graphs can reduce the number of constraint checks considerably [239]. There are two main possibilities to check if a colour can be assigned to a vertex. Either one examines the already coloured vertices for adjacency or one checks the neighbours of the vertex to validate the same colouring is not assigned to any of these. Using merge operations, we will have colour groups instead of sets of coloured vertices, thereby reducing the amount of checks required. In [235] a speed-up of about $\log n$ is derived both empirically and theoretically.

11.3   Vertex Colouring by Graph Homomorphisms

In this section we describe a valid vertex colouring of a graph G via several homomorphisms. Let ϕ be a homomorphism, we define ϕ in the following ways,

1. $\phi : V \to C$ defines an assignment from vertices to colours. It provides a valid colouring if $\phi(u) = \phi(v) \Rightarrow (u, v) \notin E$. Let the number of assigned colours be k, it defines the quality of the assignment when minimising the number of colours used (See Figure 11.2). We can represent the coloured graph by a composite structure $\langle G, C \rangle$. A homomorphism can be defined by atomic steps, where at each step one vertex is coloured. These steps will be defined in Section 11.3.1.
2. The second homomorphism groups vertices of the same colour together to form hyper-vertices, i.e., colour sets. The edges defined between normal and hyper-vertices (colour sets) provide an implicit representation of the colouring by the special graph formed this way. An example is shown in

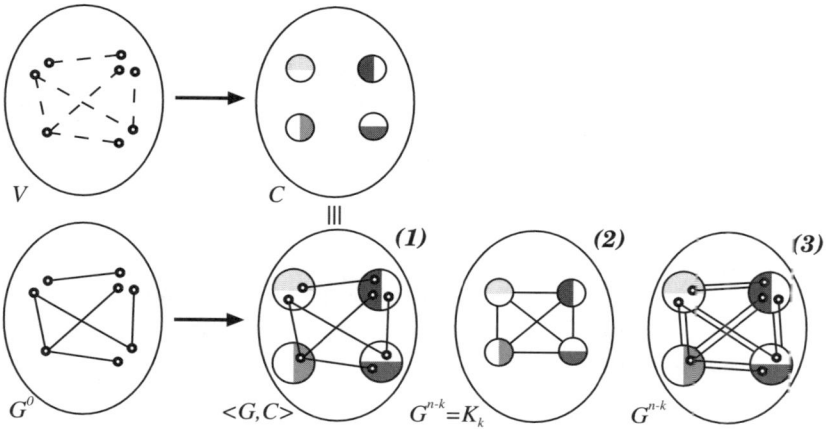

Fig. 11.2. Three homomorphisms from graphs to colourings

Figure 11.2. In this homomorphism, vertices grouped together form hyper-vertices as colour sets. The original vertices are associated with the colour set(s) where its adjacent vertices reside. When a normal vertex has more than one adjacent vertex in a colour set, then all the edges are folded into one hyper-edge. The cardinality of all the edges that were merged are assigned as a weight to the hyper-edge. This way we can keep track of the vertices in relation to the colours, and the cardinality allows us to measure the strength of such a relationship. We present a basic and a weighted representation of the contracted graph above, where vertices and edges are merged into one hyper-vertex or hyper-edge. The matrices that result from these merge operations are non-square, which is why the representation is called a table. Due to the basic and weighed edges, there are two types of representations, a *Binary* (BMT) and an *Integer* (IMT) type. The assigned A^{n-k} matrix dimension is $k \times n$. Columns represents normal vertices and rows refer to hyper-vertices, i.e., colour-sets. Zeros in a row define the members of the hyper-vertex. Non-zeros define edges between hyper-vertices and vertices.

3. The third homomorphism comes from the second by contracting normal nodes in color sets to one node. Let us define $\phi : G \rightarrow K$. It assigns a complete graph K to G. The number of vertices of K defines the quality of the colouring. The colouring is valid iff $\phi_V(u) = \phi_V(v) \Rightarrow (u,v) \notin E$. If we consider the original colour assignment then we merge all the vertices that have the same colour assigned. In other words, we can create the homomorphism by atomic sequential steps such as, colouring or merging one vertex at a time. Hence $n - k$ steps are needed to get K_k. An example is shown in Figure 11.2. In the previous homomorphism, the merge operations performed on G result in a complete graph on k vertices: K_k. Vertices assigned with the same colour are collapsed into a hyper-vertex, and all edges that connect to these vertices are collapsed into a hyper-edge. We provide a

representation of the resulting graph by two different matrices. The first matrix does not rely on the number of collapsed edges it is the basic adjacency matrix that represents the graph K_k, while the second does and it is the weighed adjacency matrix of the same graph. Here, the homomorphism is performed on the relevant matrices $\phi : A^0 \rightarrow A^{n-k}$. As the matrices that arise during the steps are square matrices, this model is called the *Merge Square* (MS) model. If we do not keep track of the cardinality of the edges we call this the *Binary Merge Square* model. Otherwise, we call this the *Integer Merge Square* model.

11.3.1   Merge Operations as Atomic Transformation Steps of the Homomorphisms

The different matrix representations require different kinds of merge operations. By combining two of these representations with two operations we will provide four colouring models. The representations are the Binary Merge Square or Integer Merge Square and the Binary Merge Table or Integer Merge Table, which are denoted by $A, \mathbb{A}, T, \mathbb{T}$ respectively. Merge Squares refers to the adjacency matrix of the merged graph. The Integer types assign weights to the (hyper-)edges according to the number of edges merged into hyper-edges. The Binary types collapse these edges simply into one common edge. The tables keep track of the vertices that were merged into a hyper-vertex and their cardinality. The graph depicted in Figure 11.3 and its adjacency matrix will be used in examples to explain the different models.

The zeroth element of the steps is the adjacency matrix of G in all cases:

$$A^0 = \mathbb{A}^0 = T^0 = \mathbb{T}^0 := A$$

We only deal with valid colourings, hence only non-adjacent vertices can be merged together. In case of Merge Squares representations columns and rows refer to the same (hyper-)vertex/colour-set. Thus, the condition of the merge depends on the relation between hyper-vertices; the coincidence of the appropriate row and column of the Merge Square must be zero. We can easily see that this condition is the same for Merge Tables (MT). Hence, we have to check

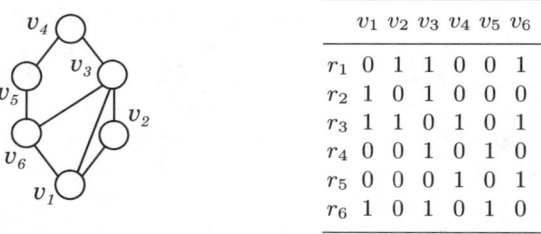

	v_1	v_2	v_3	v_4	v_5	v_6
r_1	0	1	1	0	0	1
r_2	1	0	1	0	0	0
r_3	1	1	0	1	0	1
r_4	0	0	1	0	1	0
r_5	0	0	0	1	0	1
r_6	1	0	1	0	1	0

Fig. 11.3. A graph G and its adjacency matrix: v-s refer to vertices and r-s refer to rows, i.e., colours

the adjacency between a normal vertex (which refers to an MT column) and a hyper-vertex/colour-set (which refers to an MT row). We can summarise the merge conditions by $a_{ij}^t = a_{ij}^t = t_{ij}^t = t_{ij}^t = 0$. Consequently $a_{ji}^t = a_{ji}^t = 0$ and thanks to the inherited graph property $a_{ii}^t = a_{jj}^t = a_{ii}^t = a_{jj}^t = 0$.

We define the following matrices:

$$P : \mathbf{e}_i \mathbf{e}_j' \qquad M : \mathbf{e}_j \mathbf{e}_j' \qquad W = P - M,$$

where P (Plus) will be used for addition (or bitwise OR operation) of the j-th row of a matrix to its i-th row. M (Minus) will support the subtraction of the j-th row from itself, thereby setting its components to zero. This could be done also by a bitwise exclusive or (XOR). W combines these operations together. Let $\mathbf{a}$ and $\mathbf{b}$ define the i-th and j-th row vector of a merge matrix in step t. We need only $n - k$ contraction steps to get a solution instead of n needed by the traditional colouring methods. As much hardware nowadays provide CPUs with vector operations, this opens the possibility to perform the atomic merge operations in one CPU operation, thereby increasing the efficiency.

We now define the four models formulated both as row/column operations and matrix manipulations. First the integer-based models and then the binary-based model, which do not track the number of edges folded into a hyper-edge.

Integer Merge Table (IMT) model

Row based formulation of the i-th and j-th row of $\mathbb{T}$ after merging j-th vertex into i-th: let $\mathbf{t}_i$ is the i-th row and $\mathbf{t}_{\cdot i}$ is the column vector.

$$\mathbf{t}_i^{t+1} = \mathbf{a} + \mathbf{b} \quad , \quad \mathbf{t}_j^{t+1} = \mathbf{0} \tag{11.1}$$

Matrix based formulation:

$$\mathbb{T}^{t+1} = \mathbb{T}^t + W\mathbb{T}^t = (I + W)\mathbb{T}^t \tag{11.2}$$

In the example below, rows r_1 and r_4 are merged, after which the row corresponding to colour c_4 is removed.

	v_1	v_2	v_3	v_4	v_5	v_6		v_1	v_2	v_3	v_4	v_5	v_6		v_1	v_2	v_3	v_4	v_5	v_6
r_1	0	1	1	0	0	1	$\{r_1, r_4\}$	0	1	2	0	1	1	$\{r_1, r_4\}$	0	1	2	0	1	1
r_2	1	0	1	0	0	0	r_2	1	0	1	0	0	0	r_2	1	0	1	0	0	0
r_3	1	1	0	1	0	1	r_3	1	1	0	1	0	1	r_3	1	1	0	1	0	1
r_4	0	0	1	0	1	0	r_4	0	0	0	0	0	0	r_5	0	0	0	1	0	1
r_5	0	0	0	1	0	1	r_5	0	0	0	1	0	1	r_6	1	0	1	0	1	0
r_6	1	0	1	0	1	0	r_6	1	0	1	0	1	0							

Integer Merge Square (IMS) model

Row/column based formulation: let $\mathbf{a}_i$ be the i-th row and $\mathbf{a}_{\cdot i}$ be the column vector and define $\mathbf{a}_j$ and $\mathbf{a}_{\cdot j}$ in the same way for the j-th row and column.

$$a_i^{t+1} = \mathbf{a} + \mathbf{b} \ , \quad a_j^{t+1} = \mathbf{0}' \tag{11.3}$$

$$a_{-i}^{t+1} = \mathbf{a}' + \mathbf{b}' \ , \quad a_{-j}^{t+1} = \mathbf{0} \tag{11.4}$$

Matrix based formulation:

$$\mathbb{A}^{t+1} = \mathbb{A}^t + W\mathbb{A}^t + \mathbb{A}^t W' \tag{11.5}$$

Since, $a_{ij}^t = 0$ and $a_{ji}^t = 0$, hence $W\mathbb{A}^t W' = 0$. Due to this fact we can rewrite (11.5) as

$$\mathbb{A}^{t+1} = (I + W)\mathbb{A}^t(I + W)' \tag{11.6}$$

In the example below, a merge square has caused both columns and rows to be merged. The result is an adjacency matrix of the merged graph with weights on the edges, which describe the number of edges that were merged.

	v_1	v_2	v_3	v_4	v_5	v_6		$\{v_1, v_4\}$	v_2	v_3	v_4	v_5	v_6		$\{v_1, v_4\}$	v_2	v_3	v_5	v_6
r_1	0	1	1	0	0	1	$\{r_1, r_4\}$	0	1	2	0	1	1	$\{r_1, r_4\}$	0	1	2	1	1
r_2	1	0	1	0	0	0	r_2	1	0	1	0	0	0	r_2	1	0	1	0	0
r_3	1	1	0	1	0	1	r_3	2	1	0	0	0	1	r_3	2	1	0	0	1
r_4	0	0	1	0	1	0	r_4	0	0	0	0	0	0	r_5	1	0	0	0	1
r_5	0	0	0	1	0	1	r_5	1	0	0	0	0	1	r_6	1	0	1	1	0
r_6	1	0	1	0	1	0	r_6	1	0	1	0	1	0						

Binary Merge Table Model (BMT) model

Row based formulations:

$$t_i^{t+1} = \quad \mathbf{a} \vee \mathbf{b} \ , \quad t_j^{t+1} = \mathbf{0} \tag{11.7}$$

$$t_i^{t+1} = \mathbb{t}_i^{t+1} - \mathbf{a} \bullet \mathbf{b} \ , \quad t_j^{t+1} = \mathbf{0} \tag{11.8}$$

$$\mathbf{a} \bullet \mathbf{b} = \quad diag(\mathbf{a}'\mathbf{b}) \quad = \sum_i (\mathbf{a}'\mathbf{b})e_i$$

Matrix based formulations:

$$T^{t+1} = T^t \vee PT^t - MT^t \tag{11.9}$$

$$T^{t+1} = \mathbb{T}^{t+1} - \sum_j (\mathbf{a}'\mathbf{b})E_{ji} \tag{11.10}$$

In the example below, row r_4 is merged with row r_1 to form $\{r_1, r_4\}$, after which r_4 is deleted.

	v_1 v_2 v_3 v_4 v_5 v_6		v_1 v_2 v_3 v_4 v_5 v_6		v_1 v_2 v_3 v_4 v_5 v_6
r_1	0 1 1 0 0 1	$\{r_1, r_4\}$	0 1 1 0 1 1	$\{r_1, r_4\}$	0 1 1 0 1 1
r_2	1 0 1 0 0 0	r_2	1 0 1 0 0 0	r_2	1 0 1 0 0 0
r_3	1 1 0 1 0 1	r_3	1 1 0 1 0 1	r_3	1 1 0 1 0 1
r_4	0 0 1 0 1 0	r_4	0 0 0 0 0 0	r_5	0 0 0 1 0 1
r_5	0 0 0 1 0 1	r_5	0 0 0 1 0 1	r_6	1 0 1 0 1 0
r_6	1 0 1 0 1 0	r_6	1 0 1 0 1 0		

Binary Merge Square (BMS) model

Row/column based formulations: let a_j be the j-th row and a_{-j} be the column vector.

$$a_i^{t+1} = \mathbf{a} \vee \mathbf{b} \quad , \qquad a_j^{t+1} = \mathbf{0}' \tag{11.11}$$
$$a_{-i}^{t+1} = (A_i^{t+1})' \quad , \qquad a_{-j}^{t+1} = \mathbf{0} \tag{11.12}$$

Matrix based formulations:

$$A^{t+1} = A^t \vee (PA^t + A^t P') - (MA^t + A^t M') \tag{11.13}$$
$$A^{t+1} = A^t \vee (PA^t P') - (MA^t M') \tag{11.14}$$

The example below shows a binary merge collapse, which does not perform any accounting of merged structures.

	v_1 v_2 v_3 v_4 v_5 v_6		$\{v_1, v_4\}$ v_2 v_3 v_4 v_5 v_6		$\{v_1, v_4\}$ v_2 v_3 v_5 v_6
r_1	0 1 1 0 0 1	$\{r_1, r_4\}$	0 1 1 0 1 1	$\{r_1, r_4\}$	0 1 1 1 1
r_2	1 0 1 0 0 0	r_2	1 0 1 0 0 0	r_2	1 0 1 0 0
r_3	1 1 0 1 0 1	r_3	1 1 0 0 0 1	r_3	1 1 0 0 1
r_4	0 0 1 0 1 0	r_4	0 0 0 0 0 0	r_5	1 0 0 0 1
r_5	0 0 0 1 0 1	r_5	1 0 0 0 0 1	r_6	1 0 1 1 0
r_6	1 0 1 0 1 0	r_6	1 0 1 0 1 0		

11.3.2   Information Derived from the Data Structures

First order structures are the cells of the representation matrices. They define the neighbourhood relation of the hyper-vertices for the binary and weighted relation for the integer models.

Secondary order structures or *co-structures* are the summary of the first order structures; i.e., the rows and columns in the representation matrices respectively. They form four vectors. Since, sequential colouring algorithms take steps, and the coloured and uncoloured part of the graphs are changing step-by-step

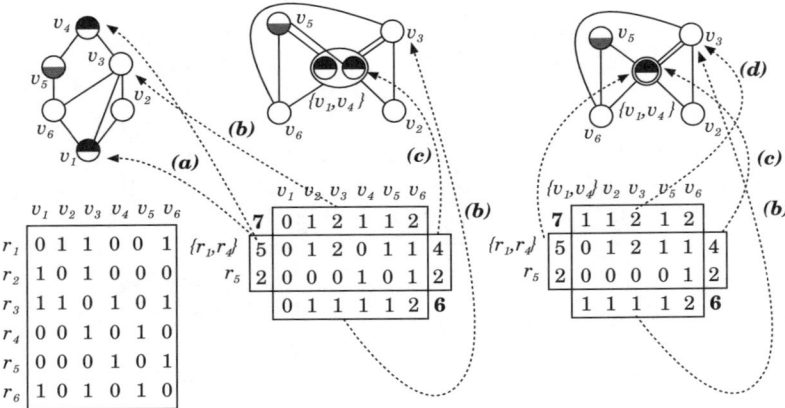

Fig. 11.4. The original graph, its induced sub-IMT and then its induced sub-IMS when colouring is in progress. (a) gives the sum of the degree of the nodes, (b) gives the number adjacent vertices already assigned a colour, (c) gives the degree of the hyper-vertex, and (d) gives the number of coloured hyper-edges.

it is worth to define these structures separately to the coloured/merged and uncoloured/non-merged sub-graphs. To identify these partial sums we use ^{col} and ^{unc} superscripts. We can obtain the sum of the rows and columns of binary merge matrices from their integer pairs by counting their non-zero elements. Hence, in Figure 11.4 only the left side of the sub-matrices sum the columns and the right side sum the non-zero elements to get the relevant summary in the binary matrix. This is the same for the columns, where the top vector is the sum of the rows and the bottom is the number of non-zeros. The second order structures are denoted by τ, using $_{t,b,l,r}$ indices as subscript to refer to the top, bottom, left and right vector.

Third order structures are formed by summarising the secondary order structures. These can be divided into two parts similar to the second order structures according to the coloured and uncoloured sub-graphs. These structures are denoted by ζ. In this study, they will be used in the fitness function of the evolutionary algorithm. The top-left sums the top vector (or the left vector) and the bottom-right sums the bottom vector (or the right vector). These are shown in bold font in Figure 11.4.

11.4   Heuristics Based on Merge Model Structures

before we define the heuristics, we first introduce two abstract sequential algorithms. These are appropriate for defining existing or novel GCP solvers in a concise manner.

Definition 1 (Sequential contraction algorithm 1 (SCA1))

1. *Choose a non-merged/uncoloured vertex v*
2. *Choose a non-neighbour hyper-vertex/colour-set r to v*
3. *Merge v to r*
4. *If there exists a non-merged vertex then continue with Step 1*

Definition 2 (Sequential contraction algorithm 2 (SCA2))

1. *Choose a hyper-vertex/colour-set r*
2. *Choose a non-neighbor non-merged/uncoloured vertex v to r*
3. *Merge v to r*
4. *If there exists a non-merged vertex then continue with Step 1*

The proceedings of the DIMACS Challenge II [240] states a core problem of the algorithm design for graph colouring: it is crucial to find an initially good solution, because each algorithm needs exhaustive searching to find an optimal solution, hence the problem becomes $\mathcal{NP}$-hard to solve. There are several polynomial algorithms (see [241, 242, 243, 244]), which provide a guarantee for the approximation of the colouring for a given number of colours. For example, ERDŐS HEURISTICS [242] guarantees at most $\mathcal{O}(n/log_\chi(n))$ number of colours in the worst case. Several algorithms have improved upon this bound, but many of them use the main technique introduced by Erdős. We will define this heuristics in our model. Another reference method is the well known heuristic of Brèlaz; the DSATUR algorithm, which works very well on graphs that have a small chromatic number. Two additional novel heuristics are introduced: the DOTPROD and COS heuristics.

DSatur

DSATUR is a SCA1 type algorithm. Where the choice of colouring the next uncoloured vertex v is determined by colour saturation degree. As colour saturation is calculated by observing the colours of neighbouring vertices, it requires $\mathcal{O}(n^2)$ constraint checks. However, if we merge the already coloured vertices, e.g., using MTs, we have the possibility to obtain this information by observing at most the number of hyper-vertices for v. Hence, $\mathcal{O}(nk_t)$ constraint checks are required, where k_t is the number of colours used in step t. The bottom co-structure of the relevant sub-IMT provide the saturation degree exactly which gives $\mathcal{O}(n)$ computational effort to find the largest one. Hence, IMT is an appropriate structure for the definition. Here, the choice for hyper-vertex/colour-set is done in a greedy manner.

Definition 3 (Non-merged/Uncoloured vertex choice of the DSatur$_{\text{imt}}$)

1. *Find those non-merged/uncoloured vertices which have the highest saturated degree: $S = \arg\max_u \tau_b^{col}(u) : u \in V^{unc}$*
2. *Choose those vertices from S that have the highest non-merged/uncoloured-degree: $N = \arg\max_v \tau_b^{unc}(v)$*
3. *Choose the first vertex from the set N*

DotProd

This heuristic is based on a novel technique, which is shown to be useful in [239]. Two vertices of a contracted graph are compared by consulting the corresponding BMS rows. The dot product of these rows gives a valuable measurement for the common edges in the graph. These values are the same in binary MT and MS but can differ between the binary and integer variations. Application of the DOTPROD heuristics to the BMS representation provides the *Recursive Largest First* (RLF) algorithm. Unfortunately the name RLF is somewhat misleading, since the largest first itself does not define exactly where largest first is relating to. The meaning of it differs throughout the literature. Here, we introduce a DOTPROD heuristic that is combined with the BMS representation and the SCA2 type algorithm, which results in the RLF of Dutton and Birgham [245]. We explore only thess combinations, but other combinations are possible, making the DotProd a more general technique.

Definition 4 (Non-merged/Uncoloured vertex choice of the DotProd$_{bms}$)

1. *Find those non-merged/uncoloured vertices which have the largest number of common neighbours with the first hyper-vertex (colour set): $S = \arg\max_u \langle u, r \rangle : u \in V^{unc}$*
2. *Choose the first vertex from the set S*

Cos

The COS heuristics is the second novel heuristics introduced here. It is derived from DOTPROD by normalisation of the dot product. As opposed to DOTPROD, COS takes in consideration the number of non-common neighbours as well. In the following definition we provide an algorithm of type SCA2 that uses the COS heuristics to choose the next vector:

Definition 5 (Non-merged/Uncoloured vertex choice of the Cos$_{bms}$)

1. *Find the non-merged (i.e., uncoloured) vertices that have the largest number of common neighbours with the first hyper-vertex (colour set), and that have the least number of constraints: $S = \arg\max_u \frac{\langle u, r \rangle}{\|u\|\|r\|} \equiv \arg\max_u \frac{\langle u, r \rangle}{\|u\|} \equiv \arg\max_u \frac{\langle u, r \rangle}{\tau_r(u)} : u \in V^{unc}$*
2. *Choose the first vertex from the set S*

Pál Erdős $O(n/\log n)$ number of colours guaranteed heuristic

The approach [242, page 245] is as follows. Take the first colour and assign it to the vertex v that has the minimum degree. Vertex v and its neighbours are removed from the graph. Continue the algorithm in the remaining sub-graph in the same fashion until the sub-graph becomes empty, then take the next colour and use the algorithm for the non-coloured vertices and so on until each vertex is assigned a colour.

All of representations are suitable as a basis for this heuristic. It uses SCA2, where the choice for the next target r-th (hyper-)vertex/colour-set for merging and colouring is greedy.

Definition 6 (Non-merged/Uncoloured vertex choice of the Erdős$_{bmt}$)

1. *Choose an uncoloured vertex with minimum degree.* $S = \arg \min_u \tau_b^{unc}(u)$
2. *Choose the first vertex from the set S*

The ERDŐS and DSATUR heuristics make use of the secondary order structures, whereas the other two, DOTPROD and COS, make use of the first order information. The ERDŐS heuristic uses similar assumptions as DSATUR but in the opposite direction, which becomes clear in the results from the experiments.

11.5   An Evolutionary Algorithm Based on Merge Models

We apply an evolutionary algorithm (EA) to guide the heuristics. It uses the BMT representation for colouring and the SCA1 contraction scheme. The genotype is a permutation of the vertices, i.e., the rows of the BMT. The phenotype is a final BMT, where no rows can be merged further. Vertex selection is defined by an initial random ordering of all vertices. The order is then evolved by the EA toward an optimum directed by both the fitness function and the genetic operators of the EA. The strategy for merging of selected vertices is guided by one of the DOTPROD and COS heuristics.

An intuitive way of measuring the quality of an individual (permutation) p in the population is by counting the number of rows remaining in the final BMT. This equals to the number of colours $k(p)$ used in the colouring of the graph, which needs to be minimised. When we know the optimal colouring[1] is χ then we may normalise the fitness function to $g(p) = k(p) - \chi$. This function gives a rather low diversity of fitnesses of the individuals in a population because it cannot distinguish between two individuals that use an equal number of colours. This problem is called the fitness granularity problem. We modify the fitness function introduced in [234] to allow the use of first and second order structures introduced in Section 11.3.2.

The fitness relies on the heuristic that one generally wants to avoid highly constraint vertices and rows in order to have a higher chance of successful merges at a later stage, commonly called a succeed-first strategy. It works as follows. After the last merge the final BMT defines the groups of vertices with the same colour. There are $k(p) - \chi$ over-coloured vertices, i.e., merged rows. Generally, we use the indices of the over-coloured vertices to calculate the number of vertices that need to be minimized (see $g(p)$ above). But these vertices are not necessarily responsible for the over-coloured graph. Therefore, we choose to count the hyper-vertices that violates the least constraints in the final hyper-graph. To cope better with the fitness granularity problem we should modify the $g(p)$ according to the constraints of the over-coloured vertices discussed previously.

---

[1] In our experiments χ is defined in advance.

The fitness function used in the EA is then defined as follows. Let $\zeta^{unc}(p)$ denote the number of constraints, i.e., non-zero elements, in the rows of the final BMT that belong to the over-coloured vertices, i.e., the sum of the smallest $k(p) - \chi$ values of the right co-structure of the uncoloured vertices. This is the uncoloured portion of the (right-bottom) third order structure. The fitness function becomes $f(p) = g(p)\zeta^{unc}(p)$. Here, the cardinality of the problem is known, and used as a termination criterium ($f(p) = 0$) to determine the efficiency of the algorithm. For the case where we do not know the cardinality of the problem, this approach can be used by leaving out the normalisation step.

Procedure EA$_{dot}$ and EA$_{cos}$

1. *population = generate initial permutations randomly*
2. *repeat*
   evaluate each permutation p:
   merge $p_j - th$ unmerged vertex v into hyper-vertex r by DOTPROD *or* COS
   calculate $f(p) = (k(p) - \chi)\zeta^{unc}(p)$
   population$_{xover}$ = xover(population, prob$_{xover}$)
   population$_{mut}$ = mutate(population$_{xover}$, prob$_{mut}$))
   population = select$_{2\text{-}tour}$(population $\cup$ population$_{xover}$ $\cup$ population$_{mut}$)
3. *until termination condition*

Fig. 11.5. The EA meta-heuristic uses directly the BMT structure and either the DOTPROD or COS merge strategy

   Figure 11.5 shows the outline of the evolutionary algorithm. It uses a generational model with 2-tournament selection and replacement, where it employs elitism of size one. This setting is used in all experiments. The initial population is created with 100 random individuals. Two variation operators are used to provide offspring. First, the 2-point order-based crossover (OX2) [246, in Section C3.3.3.1] is applied. Second, a simple swap mutation operator, which selects at random two different items in the permutation and then swaps. The probability of using OX2 is set to 0.4 and the probability for using the simple swap mutation is set to 0.6. These parameter settings are taken from previous experiments [234].

   The Erdős heuristic guarantees its performance. We omit this heuristics from the EA, as we would not be able to guarantee this property once embedded in the EA. A baseline version of the EA called EA_{noheur} serves a basis of the comparison. and the DSatur with backtracking is used as well, as it is commonly used as a reference method. Moreover, as this algorithm performs an exhaustive search it is useful to find the optimal solutions to some of the problem instances; some of the instances are too difficult for it to solve and we have to terminate it prematurely.

11.6   Experiments and Results

The test suites are generated using the well known graph k-colouring generator of Culberson [247]. It consists of k-colourable graphs with 200 vertices, where

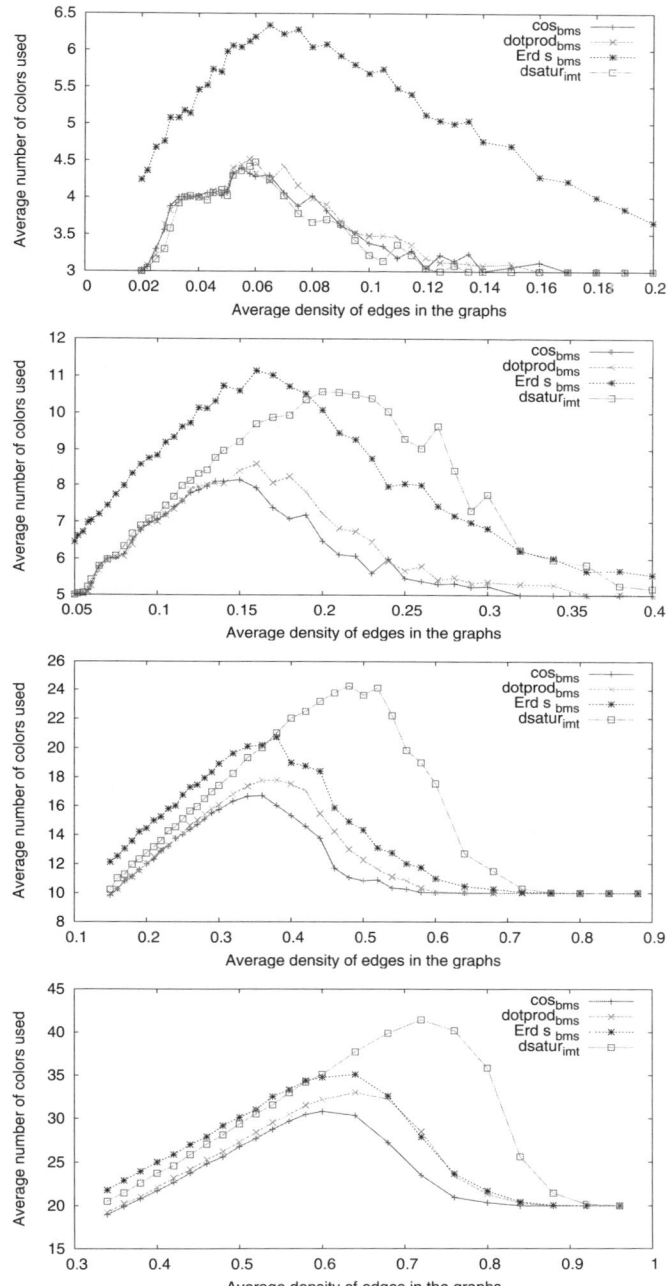

Fig. 11.6. Average number of colours used in sequential colouring heuristics. χ is 3, 5, 10, and 20, respectively in the sequence of figures.

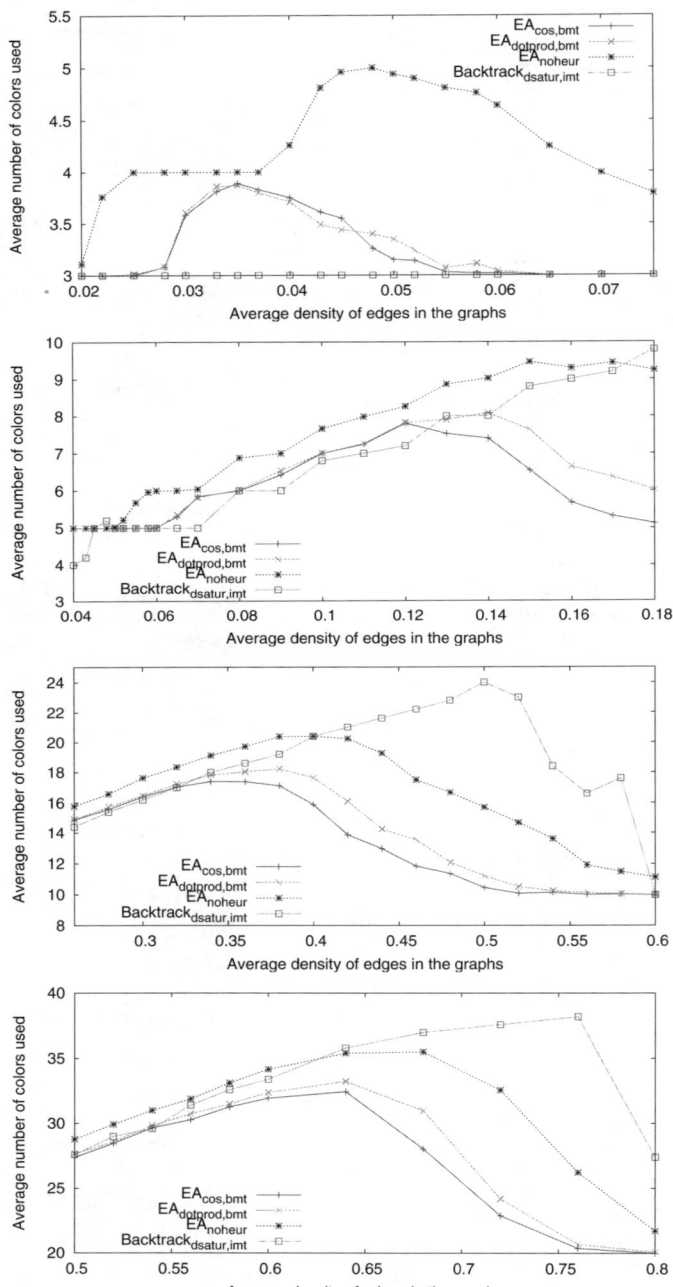

Fig. 11.7. Averaged number of colours used in the EA experiments. χ is 3, 5, 10, and 20, respectively in the sequence of the figures.

k is set to 3, 5, 10 and 20. For $k = 20$, ten vertices will form a colour set, therefore we do not use any larger chromatic number. The edge density of the graphs is varied in a region called the phase transition. This is where hard to solve problem instances are generally found, which is observed in the results as a typical easy-hard-easy pattern. The graphs are all generated to be equi-partite, which means that a solution should use each colour approximately as much as any other. The suite consists of groups where each group is a k-colouring with 20 unique instances.

In the first experiment, we compare the heuristics from Section 11.4 when embedded in DSatur. Figure 11.6 shows the results for each of the heuristics. Cos heuristics performed clearly better than the others except for the 3-colouring where DSATUR performs equally well. The DOTPROD is ranked second. While DSATUR performs well on sparse graphs having small chromatic number, ERDŐS heuristics performs well on graphs that require more colours, especially on dense graphs, i.e., that have a high average density of edges. Interesting is the location of the phase transitions. Figure 11.6 shows that it depends not only on the edge density of the graphs but also on the applied algorithm, especially the graph density where DSATUR exhibits its worst performance moves away from the others with increasing k. DSATUR and ERDŐS heuristics apply only second order information as opposed to the other two algorithms where first order information is used. The ERDŐS heuristic uses the secondary order structures in the opposite direction to DSATUR and our results show how that affects the performance as the effectiveness flips for different ends of the chromatic number of graphs.

While sequential algorithms make one run to get their result, EA experiments are performed several times due to the random nature of the algorithm. Therefore a reduced number of instances is selected from the previous experiments. One set consists of five instances, except for 3 and 5 colourable instances where the set contains ten instances because the diversity of the results of the solvers is low for small chromatic numbers. On each instance we perform ten independent runs and calculate averages over the total number of runs. 3-colouring needs more instances to get more confident results.

Figure 11.7 shows the results for the EA with two different heuristics. Also shown are results for the reference EA without heuristics and the DSATUR exhaustive algorithm. Similar to the previous experiments, the COS heuristics performs well, especially for larger k, and the DOTPROD is a close second. DSATUR is the strongest algorithm on 3-colourable graphs, always finding the optimum number of colours.

11.7 Conclusions

In this paper, we introduced four kinds of Merge Models for representing graph colouring problems. It forms a good basis for developing efficient graph colouring algorithms because of its three beneficial properties: a significant reduction in constraint checks, access to heuristics that help guide the search, and a compact description of algorithms. We showed how existing algorithms can be described in terms of the Merge Model concisely.

By incorporating the models in an exhaustive algorithm, DSATUR, and in a meta-heuristic, an evolutionary algorithm based on a permutation, we showed how the effectiveness of these algorithms can be improved.

Acknowledgements

This work is supported by the High Performance Computing Group of the University of Szeged. The second author is supported through the e-Science Data Information & Knowledge Transformation project funded by the Scottish Funding Council.

Part IV

Scheduling

12

Different Codifications and Metaheuristic Algorithms for the Resource Renting Problem with Minimum and Maximum Time Lags

Francisco Ballestín

Department of Statistics and OR, Public University of Navarra, Pamplona, Spain
francisco.ballestin@unavarra.es

Summary. Most work in project scheduling problems has been done with so-called regular objective functions, where a solution S never dominates another solution S' if no activity is scheduled earlier in S than in S'. This has provided with techniques and metaheuristic frameworks to create heuristic algorithms for these kinds of problems, something which does not exist in the nonregular case. One of these problems is the Resource Renting Problem, which models the renting of resources. In this paper we compare the quality of three algorithms for this problem, one adapted from the regular case, another specially designed for the problem and a third that might be adapted for other problems. We will also show that two of these algorithms are capable of outperforming a branch-and-bound procedure that exists for the problem.

Keywords: Project Scheduling, Temporal Constraints, Nonregular Objective Function, Metaheuristic Algorithms.

12.1 Introduction

Project scheduling consists in finding a time interval for the processing of activities whose start times are restricted by the start times of other activities. In many practical and academic uses of project scheduling each activity consumes some resources that are expensive and/or constrained. There are many different objective functions in this field, but they can be clearly distinguished between regular and nonregular. A regular measure of performance is a non-decreasing function of the activity completion times (in the case of a minimisation problem). The most important (regular) objective function is the makespan or project length, but there are others like the minimisation of the mean flowtime or the mean weighted tardiness. These performance measures - especially the makespan - have received much attention, which has led to many heuristic algorithms. Some of them can be easily adapted to other project scheduling problems with regular objective functions, "regular problems", sometimes even if we add additional restrictions. We refer to the surveys [359], [354] and [361].

In the last years scheduling problems with nonregular measures of performance have gained increasing attention (cf. [374]). Some popular nonregular measures of performance in the literature are the maximisation of the net present

C. Cotta and J. van Hemert (Eds.): Recent Advances in Evol. Comp., SCI 153, pp. 187–202, 2008.
springerlink.com

value of the project ([375]), the minimisation of the weighted earliness-tardiness penalty costs of the activities in a project ([350]), the Resource Levelling Problem (RLP, [352]) and the Resource Investment Problem (RIP), which models the purchase of resources ([369] and [370]). In the RIP, the usage of resources is associated to a cost of making this resource available, independently of the time. Hence, making resource units available means buying them. For many real-life projects, however, the use of resources is associated with time-dependent costs, e.g. for heavy machinery or manpower in civil engineering. That is why the resource renting problem has been proposed (see [371]) where, besides time-independent fixed renting costs, time-dependent variable renting costs are given for the resources. Hence, the RRP models the renting of resources. For resource acquisition via buying and renting see also [347]. In project scheduling problems with nonregular objective functions, "nonregular problems", there is apparently no common heuristic framework that can be adapted to new problems, and a specific algorithm should be theoretically developed each time. An exception to this are the methods developed in [368], some of which will be used in this paper. However, algorithms can be adapted from the regular case just by introducing several changes like the schedule generation scheme. Although this can be easily done, the problem is the quality offered by this approach. Besides, some procedures can be developed that can be used for several problems without making too many adjustments. The goals of this paper are threefold: first of all, we want to study the quality difference between an adaptation of an algorithm for a regular problem and a specially designed algorithm for the RRP. Secondly, the codification and crossover of that latter algorithm, already schematised in [349], deserve further attention and examples. Finally, we will also propose a heuristic which does not need much problem knowledge and that can be adapted to other nonregular problems. The next section contains the formal definition of the optimisation problem. In section 3 the three metaheuristic algorithms for the RRP are described. Finally, computational experience with the proposed procedures is presented in Section 4.

12.2 Preliminaries

12.2.1 Model of the Problem

In this section we follow [371] and [368]. We want to carry out a project which consists of $n+2$ activities numbered $0, 1, \ldots, n+1$, with dummy activities 0 and $n + 1$ representing the beginning and termination of the project, respectively. We denote V the set of all activities of the project, which coincides with the node set of a corresponding activity-on-node project network. Let $p_j \in Z^+$ be the duration (or processing time) and $S_j \in Z^+$ be the start time of activity j where $S_0 = 0$. Then S_{n+1} represents the project duration (or makespan) which is restricted by the constraint $S_{n+1} \leq \overline{d}$, being $\overline{d} \in Z^+$ a prescribed maximum project duration (deadline). In the project network, a minimum time lag $d_{ij}^{min} \in Z^+$ between the start of two different activities i and j, i.e., $S_i - S_j \geq d_{ij}^{min}$, is modelled as an arc $\langle i,j \rangle$ in the project network with weight $\delta_{ij} = d_{ij}^{min}$. A maximum time lag $d_{ij}^{max} \in Z+$ between the start of activities i and j means $S_j - S_i \leq d_{ij}^{max}$, and is modelled as an arc $\langle j,i \rangle$ with weight $\delta_{ij} = -d_{ij}^{max}$. The arc set of the project network is denoted by E. We assume that a set R of renewable resources is required for carrying out the project activities. Each activity j requires $r_{jk} \in Z^+$ ($j \in V, k \in R$) units of resource k during each period of its duration, where the dummy activities do not need any resources. The usage of resources incurs fixed and variable costs. Fixed renting costs c^f arise when bringing a unit of resource to service and often represents a delivery or transportation cost for resource unit rented. Variable renting costs c^v refer to one unit of resource and one unit of time for which the resource unit is rented. Accordingly, the provision of one unit of resource k for a time interval of t time units length leads to fixed costs of c_k^f and to variable costs of tc_k^v. We assume that $c_k^f \in Z^+ \backslash \{0\}$ and $c_k^v \in Z^+ \backslash \{0\}$ for all resources $k \in R$. Given a schedule S, let $A(S,t) := \{i \in V | S_i \leq t < S_i + p_i\}$ be the set of activities in progress at time t and let $r_k(S,t) := \sum_{i \in A(S,t)} r_{ik}$ be the amount of resource k required at time t. Without loss of generality we assume the condition of integrality for points in time where the capacities of the resources $k \in R$ can be increased or decreased. Given a schedule S, we have to decide on how many units of resource $k \in R$ are to be rented at each point in time $t \in [0, \overline{d}]$. That is, we should choose a renting policy $\varphi(S,.) := (\varphi_k(S,.))_{k \in R}$, where $\varphi_k(S,t)$ (or φ_{kt} for short) is the amount of resource k rented at time $t \in [0, \overline{d}]$. We can restrict ourselves to step functions $\varphi_k(S,.)$ with finitely many jump discontinuities and continuous from the right. Given renting policy $\varphi_k(S,.)$, $c_k^v \int_0^{\overline{d}} \varphi_k(S,t)dt$ represents the total variable renting cost for resource k and planning horizon $\overline{d}$. Let J_k be the finite set of jump discontinuities of function $\varphi_k(S,.)$ on interval $[0, \overline{c}]$ and min be the smallest of those jump points. For $t \in J_k \backslash \tau_{min}$, let $t := max\{\tau \in J_k | \tau < t\}$ be the largest jump point of function $\varphi_k(S,.)$ less than t and for $t \in J_k$, let

$$\Delta^+ \varphi_{kt} := \begin{cases} [\varphi_k(S,t) - \varphi_k(S,\tau_t)]^+, & \text{if } t > \tau_{min} \\ \varphi_k(S,\tau_{min}), & \text{otherwise} \end{cases} \quad (12.1)$$

be the increase in the amount or resource k rented at time t. Then the total fixed renting cost for resource k equals $c_k^f \sum_{t \in J_k} \Delta^+ \varphi_{kt}$.

A feasible renting policy $\varphi(S,.)$ must fulfil $\varphi_k(S,t) \geq r_k(S,t)$ for all $k \in R$ and $t \in [0, \overline{d}]$. Given schedule S, we will look for the optimal renting policy $\varphi(S,.)$, the feasible renting policy that minimises the total renting cost $\sum_{k \in R} \left[c_k^v \int_0^{\overline{d}} \varphi_k(S,t)dt + c_k^f \sum_{t \in J_k} \Delta^+ \varphi_{kt} \right]$.

The objective function f of the resource renting problem represents the total renting cost belonging to an optimal renting policy for schedule S and reads as follows

$$f(S) := \sum_{k \in R} min_{\varphi(S,\cdot) \geq r_k(S,\cdot)} \left[c_k^v \int_0^{\overline{d}} \varphi_k(S,t)dt + c_k^f \sum_{t \in J_k} \Delta^+ \varphi_{kt} \right] \quad (12.2)$$

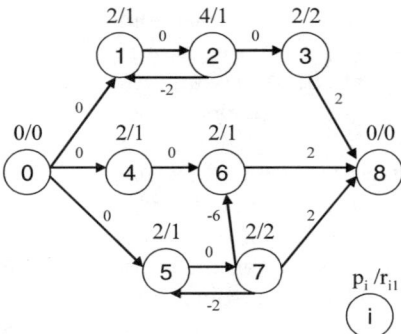

Fig. 12.1. Example problem

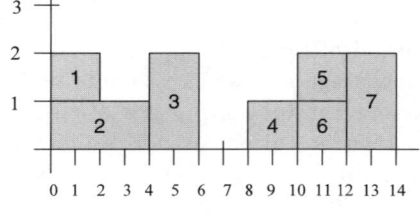

Fig. 12.2. S, a feasible solution for the problem

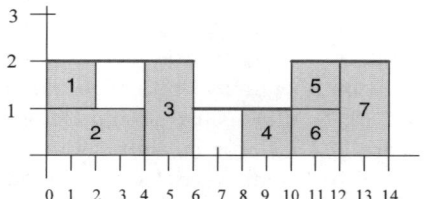

Fig. 12.3. A feasible renting policy for S

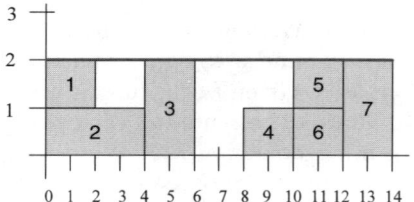

Fig. 12.4. Another feasible renting policy for S

Let $\varphi_k^*(S,.)$ be an optimal renting policy for schedule S and $k \in R$. The resource renting problem subject to temporal constraints RRP/max consists of finding a schedule S which satisfies and minimises objective function f. This problem is denoted $PS\infty|temp,\overline{d}|\Sigma\Sigma c_k^v\varphi_{kt}+c_k^f\Delta^+\varphi_{kt}$ (cf. [368]) and is NP-hard as an extension of the RIP /max (cf. [371]). It is not straightforward to calculate the optimal renting for a schedule, but polinomial procedures are available in [371] and [368]. We are going to comment some of the definitions taking as an example the project depicted in Figure 12.1. We will fix a deadline $\overline{d}$ of 14. A feasible solution for this project is shown in Figure 12.2, where both minimum and maximum time lags are observed and the deadline is not exceeded. As an example, note that the start time of activity 7 is 12, not more than 2 units bigger (exactly two units bigger) than the start time of activity 5. This is in concordance with the maximum time lag between activities 5 and 7 of 2 units. Let us compare several renting policies for this solution. As always, a feasible renting policy is $\varphi_1(t) = r(S,t)\forall t$ and other two feasible renting policies can be seen in Figure 12.3 (φ_2) and Figure 12.4 (φ_3). Table 12.2 shows the objective function of the three policies when different fixed and variable renting cost are given. The different scenarios considered are shown in Table 12.1. As an example, the fixed renting cost of 2 is calculated as follows. At time t = 0 we demand 2

Table 12.1. Scenarios considered

Scenario	Sc1	Sc2	Sc3
c^f	1	3	7
c^v	1	1	1
c^v/c^f	1	1/3	1/7

Table 12.2. Objective function for S for the different considered scenarios

Costs due to proc. cost			Rent cost	Total cost (obj. func.)		
Sc1	Sc2	Sc3	any Sc	Sc1	Sc2	Sc3
φ_1 $5 \cdot 1 = 5$	$5 \cdot 3 = 15$	$5 \cdot 7 = 35$	$20 \cdot 1 = 20$	25	35	55
φ_2 $3 \cdot 1 = 3$	$3 \cdot 3 = 9$	$3 \cdot 7 = 21$	$24 \cdot 1 = 24$	27	33	45
φ_3 $2 \cdot 1 = 2$	$2 \cdot 3 = 6$	$2 \cdot 7 = 14$	$28 \cdot 1 = 28$	30	34	42

units of the single resource. At time t = 6 we release one unit of the resource, which we need (and demand) at time t = 10. This means that we demand 3 units of resource at different times, which means $3 \cdot c^f$ units of fixed renting cost in the different scenarios. Regarding the variable renting cost of this renting policy, in interval $[0, 6[$ two units of the resource are available. Note that in interval $[2, 4[$ we only need one unit of the resource, but we would have to demand another unit of resource at time t = 4. However, this would be a different renting policy. In interval $[6, 10[$ one unit of the resource is available and in $[10, 14]$ two units of the resource are available. All in all, we have $(6-0) \cdot 2 + (10-6) \cdot 1 + (14-10) \cdot 2 = 24$ units available, which means $24 \cdot c^r$ units of variable renting cost and a total of $3 \cdot c^f + 24 \cdot c^r$ units of cost. When c^v/c^f is 1 or more (Sc1), it is more efficient to have few resources available at each time, because it is inexpensive to procure new resources when needed. In this case, solutions where the renting policy is very similar to the resource profile (in our case φ_1) are the best. When c^v/c^f approximates 0 (Sc3), it is very expensive to obtain new resources. Therefore it is better to have many of them available at each time, even if we do not use them all at each interval. The best solutions are then similar to $\varphi(t) = max(r(S,t)) \forall t$, i.e. in our case φ_3 is the best. Note that the limit, RRP with $c^v/c^f = 0$ is the same as RIP. In not so extreme scenarios the optimal policy does not look like φ_1 or φ_3. In Sc2, the best renting policy is φ_2.

12.2.2 Candidates for Optimal Solution

As shown in [368], it is enough to search in the space of quasistable schedules in order to find an optimal solution for the RRP. This means that, when we schedule an activity i in a partial schedule, we only have to look at integral points t in set $[ES_i, LS_i] \bigcap \{ES_i, LS_i, \{S_j + p_j, j \in C\}, \{S_j - p_i, j \in C\}\}$. $ES_i (LS_i)$ denotes the earliest (latest) start time where act i can be scheduled and C the set of activities already scheduled. Throughout the paper we will only take into account these points in time t when we schedule an activity, but we will not mention it again.

12.3   Metaheuristic Algorithms for the RRP

In this section we describe the metaheuristic algorithms we have applied to the RRP. The first ones, adapGA and specGA, share the same algorithmic structure but use a totally different codification and crossover. On the one hand, the codification and crossover of adapGA, an algorithm adapted from a Genetic Algorithm (GA) for the regular case, are the ones used in that algorithm and have proved their usefulness in those optimisation problems. On the other hand, these elements are completely new in specGA, the GA specially designed for the RRP. Both GA's can be regarded as memetic algorithms ([365] and [366]), since they include an improvement method after the recombination. Finally, a third metaheuristic algorithm IG is schematised. This algorithm does not need a deep knowledge of the problem and could be adapted to other nonregular optimisation problems.

12.3.1   Algorithm Adapted from a Regular Problem (adapGA)

One of the goals of this paper is to calculate the quality difference between performing the minimal adaptations from a specially designed algorithm and an algorithm for a regular problem. In order to achieve this we will use exactly the same framework for both algorithms and the only differences between them will be the codification and crossover operator. As a framework we have used a GA plus an Improvement Procedure (IP), that has been adapted (with more or fewer changes) to several problems with success. The original algorithm without IP appears in [357]. Some of the problems where it has been adapted are the multi-mode RCPSP ([358]), the RCPSP with preemption ([353]), the RCPSP with due dates ([351]) or the stochastic RCPSP ([348]). For an introduction into GAs, a heuristic meta strategy to solve hard optimisation problems that follow the basic principles of biological evolution, we refer to [143]. The codification used by adapGA will be a simplification of the activity list, a widely used codification in regular problems. The simplification is a permutation, which can be used due to the fact that there are no resource restrictions in our problem, and an activity list is a perturbation of the activities that fulfils the positive time lags between activities. The crossover used by adapGA will be the two-point crossover for permutations (cf. [357]). It will be followed by a perturbation similar to the one used in that algorithm. The outline of the algorithm for adapGA and specGA - without including the perturbation for adapGA - is as follows. After calculating the initial population POP with the multi pass algorithm MP plus IP applied to each solution, the procedure adds to POP in each iteration a new population of the same size with children generated by the crossover and improved by IP. The best individuals of this joint population constitute the population for the next iteration. The different adaptations of the GA for the RRP are explained in the next subsections.

Schedule Generation Scheme (SGS) and codification

In the book of [368], some priority rules were described for different types of objective functions. These priority rules consists basically in choosing an order for the activities - a permutation of the activities - and scheduling the activities in this order locally optimal. I.e., at the best time t with regards to the objective function in the feasible time window. We use the same untie rule as in [367]: we choose the smallest minimizer t when we schedule activity i if $i < \lceil V|/2 \rceil$, else we choose the greatest minimizer t. This method of creating one or several solutions can be transformed into a codification and a SGS for many nonregular objective functions: a perturbation of the activities and the SGS that schedules the activities in that order locally optimal.

Improvement Procedure

Regarding the IP, it is more difficult to devise general procedures useful for many nonregular objective functions. One possibility for the not sum-separable objective functions is the following. After using SGS mentioned in section 3.1.1 the activities are not longer locally optimal scheduled. A simple but effective IP (cf. [352]) consists in iteratively choosing a not already chosen activity, unscheduling it and then scheduling it locally optimal.

Procedure to calculate the initial population

With the priority rules described in 3.1.1 and the IP described in the foregoing section it is straightforward to create an initial population for an algorithm; one only needs to use some sort of sampling procedure or multi pass algorithm MP (cf. [360]). In our case, we use the regret-based random sampling method (cf. Drexl, 1991) with priority rules MST-GRR, with MST = minimum slack time $(LS_h - ES_h)$ and GRR = greatest resource requirements first. The second rule is used as an untie rule.

12.3.2 Specially Designed Algorithm (specGA)

As we have already mentioned, specGA also follows the structure presented in the foregoing section. However, it uses a different codification and crossover, described in the next section. This algorithm was schematised in [349].

Codification of a solution

We use $n + 2$ sets to codify a solution S, one for the whole schedule and another for each activity. The set for the schedule is framework(S), with framework(S) = $\{i \in V/S_i = d_{0i}$ or $S_i = -d_{n+1,i}\}$, where d_{0i} denotes the length of the longest path between activities 0 and i and $d_{i,n+1}$ is the length of the longest path between activities $n + 1$ and i if we introduce the arc $< n + 1, 0 >$ in the network with weight $\delta_{n+1,0} = -\overline{d}$. Besides, for each activity i we calculate the set of activities that finish exactly when i begins. We call this set before$(i, S) = j \in V/S_j + d_j = S_i$. It is straightforward to calculate these sets while building a schedule. In the schedule S of Figure 12.2, framework(S) = 1, 2, 7, because $d_{01} = d_{02} = 0 = S_1 = S_2$ and $-d_{n+1,7} = 12 = S_7$. Besides, before$(7, S)$ = $\{5, 6\}$ or before$(4, S)$ = $\emptyset$. This definition of codification is fully workable in combination with the crossover described in the next section. However, it is an incomplete codification in the sense that there might exist activities which do not belong to framework(S) or to any before(j, S) and for which before(i, S) = $\emptyset$. We are not going to dwell on this theme because the way to schedule activities - locally optimal - tends to schedule activities after or before other activities and so we will not often find such situations. At any rate, if we wanted to work with a complete codification we could use the concept of weighted spanning tree. See [368] for the definition and algorithms based on this concept.

Crossover

The crossover used in specGA tries to transfer good qualities from the parents to the children. The objective function somehow penalises the positive jumps in the resource utilisation; a good schedule will therefore have few points where resources are newly rented. So, the flow of resources is important, much more which activities are scheduled right after which others than where exactly an activity is scheduled. Our crossover relies on trying to schedule one after the other activities that are scheduled one after the other in the mother M and/or the father F. The pseudo-code from the operator is given in Scheme 1. Specifically, in order to obtain the daughter D from M and F, we first schedule the activities in framework(M) in the same interval as in M. Afterwards we select at each iteration an unscheduled activity and schedule it, until we end with a complete solution. The son can be obtained by changing the roles of the mother and the father. To further describe an iteration in the construction of the daughter, we need sets $C, \overline{C}$, Eleg1 and Eleg2. C and $\overline{C}$ contain the scheduled and unscheduled activities, respectively. Eleg1 and Eleg2 are subsets of $\overline{C}$ that are recalculated at each iteration. Suppose that an unscheduled activity i is scheduled right after (right before) a scheduled activity j in M and in F (a condition imposed on both parents). If i can be scheduled at a certain iteration right after (right before) j in D, then i belongs to Eleg1 at that iteration. If we select i to be scheduled at that iteration, we will mimic the behaviour of i in M and F and schedule it right after (right before) j. Note that this does not imply that the start time of i is the same in the different solutions. Eleg2 works in a similar way to Eleg1, the only difference is that we relax the condition and demand it just on one of the parents. At each iteration we randomly select an activity from Eleg1 or from Eleg2 if Eleg1 is empty and schedule it. We scrutinised the sets in this order because we prefer that the daughter inherits structures present in the two solutions at the same time. If both sets Eleg1 and Eleg2 are empty, we choose an activity i randomly and we schedule it at the best point in time with respect to the objective function.

Scheme 1. Pseudo-code to obtain the daughter D by recombining the mother M and the father F.

1. $i \in framework(M)$ do $S_i^D = S_i^M$, $\overline{C} = V \backslash framework(M)$, $C = \emptyset$.
2. **While** $\overline{C} = \emptyset$
   2.1 *Calculate Eleg1.* **If** *Eleg1 $\neq \emptyset$, select randomly an activity $i \in$ Eleg1.*
   2.2 **Else** *calculate Eleg2.* **If** *Eleg2 $\neq \emptyset$, select randomly an activity $i \in$ Eleg2.*
   2.3 **Else** *select randomly an activity $i \in \overline{C}$.*
   2.4 **If** $\exists j/i \in before(j, M) \cup before(j, F)$ *with* $S_j^D - d_i \in [ES_i, LS_i]$, *then*
     $t^* := S_j^D - d_i$. **Else if** $\exists j/j \in before(i, M) \cup before(i, F)$ *with* $S_j^D + d_j \in$
     $[ES_i, LS_i]$, **then**, $t^* := S_j^D + d_j$. **Else** *choose the best t^* available.*
   2.5 *Schedule i at t^*, $\overline{C} = \overline{C} \backslash \{i\}$, $C = C \cup \{i\}$. Update ES_i and $LS_i \forall i \in \overline{C}$.*
3. *Return solution D.*

$Eleg1 = \{i \in \overline{C} : \exists j \in C/i \in before(j, M) \cap before(j, F)$ with $S_j^D - d_i \in [ES_i, LS_i]$ or $\exists j \in C/j \in before(i, M) \cap before(i, F)$ with $S_j^D + d_j \in [ES_i, LS_i]\}$.

$Eleg2 = \{i \in \overline{C} : \exists j \in C/i \in before(j, M) \cup before(j, F)$ with $S_j^D - d_i \in [ES_i, LS_i]$ or $\exists j \in C/j \in before(i, M) \cup before(i, F)$ with $S_j^D + d_j \in [ES_i, LS_i]\}$.

Let us see an example of the crossover operator, using the mother M that appears as S in Figure 12.3 (with φ_2 as the renting policy) and the feasible father F appearing in Figure 12.5 (together with its optimal renting policy). If we consider $c^v = 1$ and $c^f = 3$, the objective functions of M and F are 33 and $3 \cdot (1 + 1 + 1 + 3) + 1 \cdot (1 \cdot 4 + 2 \cdot 4 + 3 \cdot 2 + 3 \cdot 2) = 42$, respectively. In the first step of

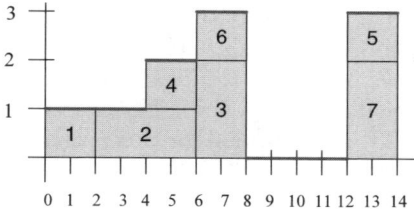

Fig. 12.5. A feasible solution for the problem F

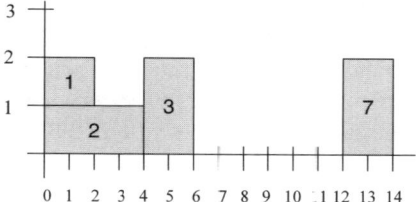

Fig. 12.6. First and second step in the construction of D

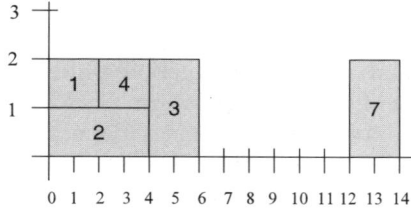

Fig. 12.7. Third step in the construction of D

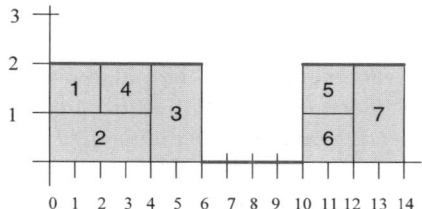

Fig. 12.8. Solution D = final outcome of the crossover operator Solution

the crossover, we schedule the activities in framework$(S) = \{1, 2, 7\}$ in the same time intervals as in M. The crossover relies on looking, at each moment, which activities are scheduled in M and/or in F right before or right after the activities already scheduled. For example, activity 5 is scheduled right before activity 7 in M but not in F and activity 6 is scheduled right after activity 2 in F, but not in M. This means that both activities belong to Eleg2 (activity 5 because $5 \in$ before$(7, M) \subseteq$ before$(7, M) \cup$ before$(7, F)$ and activity 6 because $2 \in$ before$(6, F) \subseteq$ before$(6, M) \cup$ before$(6, F)$). However, activity 3 is scheduled right after activity 2 both in M and in F, which means activity $2 \in$ before$(3, M) \cap$ before$(3, F)$ and thus activity $2 \in$ Eleg1. In fact, this is the only activity in Eleg1, so it is chosen to be scheduled in step 2 of the algorithm. Note that this activity does not begin at the same time in M and F. However, the "chain" 2-3 does occur in M and in F, and this is what we want the daughter to inherit. So, we schedule activity 3 after 2, obtaining the partial schedule depicted in Figure 12.6. To schedule the next activity we need to recalculate Eleg1 and Eleg2, which are Eleg1$= \emptyset$ and Eleg2 $= \{4, 5, 6\}$. We must select one activity randomly from Eleg2, let us say we select 4. Activity 4 is in Eleg2 because $4 \in$ before$(3, F)$. This means we have to schedule it before 3, obtaining the partial schedule of Figure 12.7. Note that now activity 6 becomes a candidate to be an element of Eleg1, because activity $4 \in$ before$(6, M) \cap$ before$(6, F)$. However, the maximum time lag between activities 6 and 7 of 6 units makes that $LS_6 = 6$, because activity 7 begins at time 12. If we scheduled activity 6 right after 4 in the partial schedule of Figure 12.7 it would begin at time 4, which would not be in concordance with this time lag. Saying it in another way, activity 6 does not belong to Eleg1 because $S_4^D + d_4 \notin [ES_6, LS_6]$, which is a part of the definition of activities in Eleg1 and Eleg2. So, we have again to resort to Eleg2 $= \{5, 6\}$. Both activities are scheduled before activity 7 in M, and this is plausible in the partial solution, so they are scheduled before 7, obtaining the final solution D depicted in Figure 12.8 (together with its optimal renting policy). The objective function of D is $3 \cdot (2+2) + 1 \cdot (2 \cdot 6 + 2 \cdot 4) = 12 + 20 = 32$, optimal for this problem.

12.3.3   Improved Algorithm (finalGA)

The algorithm presented in the previous section can be further improved by means of a diversification. The algorithm adapGA will be compared with specGA, but we will compare finalGA, the algorithm developed in this section, with a Branch and Bound (B&B) that exits in the literature. When certain conditions hold, finalGA build a new population by applying the function Different to each solution of the old population, followed by IP. The function Different calculates a new solution from each individual of the old population, randomly selecting and unscheduling n/2 activities. Afterwards it schedules them randomly one by one. Finally, IP is applied to the new solution. The conditions for the application of the diversification are that all the individuals share the same objective function or that itmax iterations without improvement of the worst individual have passed. The outline of the algorithm finalGA reads as follows:

Scheme 2. FinalGA

1. $POP = MP + IP(nPop); nit = 0.$
2. **For** $i = 1, \ldots,$ *niter*
   2.1 **If** $(Best = Worst$ or $nit = itmax)$
      2.1.1 **For every** $S \in POP$:
         2.1.1.1 $S' = Different(S).$
         2.1.1.2 $S' = IP(S').$
      2.1.2 $POP' = POP' \cap \{S'\}.$
      2.1.3 $POP = POP'$
   2.2 **Else** apply an iteration of specGA. **If** the worst solution is improved: nit $= 0$; **else** nit $= nit + 1.$
3. Return the best solution found in the algorithm.

We have fixed itmax $= 5$ in some preliminary tests. The algorithm finalGA uses two multi-pass procedures in the construction of the population. Apart from using MP (see section 3.1), it also uses MP2, where the schedule of an activity is biased.

12.3.4   An Iterated Greedy Approach for the RRP (IG)

In this section we are going to present a heuristic algorithm that obtains reasonable good quality solutions without using any special designed codification or crossover. The algorithm is a population-based Iterated Greedy IG ([363]). A basic IG begins with an initial solution and, iteratively, calculates a new solution from the solution model and decides whether to replace the solution model with the new solution. Each new solution is calculated by eliminating some elements of the solution model and reintegrating them, where it is assumed that a solution S can be represented as a sequence of solution components. Iterated greedy algorithms have been applied with success, for example, to the set covering problem ([364]), in the field of scheduling ([372]) and project scheduling ([352]). In this paper we are going to adapt the IG appearing in this last paper. Here is the outline of the IG for the RRP.

Scheme 3. IG

1. Generate an initial population P of size s_p.
2. Repeat until stopping condition is met
   2.1 Select a schedule S from P.
   2.2 Draw a uniformly distributed random number $u \in [0, 1[$.
   2.3 **If** $u < 0.5$ **then** apply operator M_A to S and obtain S'.
   2.4 **Else** apply operator M_B to S and obtain S'.
   2.5 Put $S' := IP(S')$.
   2.6 Select worst schedule S'' in population P.
   2.7 **If** $f(S') < f(S'')$ **then** replace S'' by S' in P.
3. Return the best solution found in the algorithm.

Since the applied IG is population-based, it can be also classified as an evolutionary algorithm without recombination.

Basic method of building a schedule

We use the same method to build schedules as in adapGA: to select an order for the activities and to schedule them in this order at the time t in its time windows that minimises the additional cost in the objective function. The essential difference with adapGA is that we now work with a preliminary fixed partial schedule and only schedule a subset of the activities. Obviously, the scheduled activities restrain the time window of the not yet scheduled activities. Finally, we must mention that there is no need to calculate the complete optimal renting policy each time. We have developed a procedure to update the optimal renting policy of a partial schedule when an activity is scheduled in an interval [t, t'].

Initial population

The initial population is similar to the one in the GAs, because the rule to choose the activity to be scheduled next is MST-GRR. However, a different method is used in order to introduce randomness in the construction of schedules. We also introduce some random bias in the activity selection, but employing the method introduced by [373] under the name of a β-biased random sampling. In each iteration, we follow deterministically the rule MST-GRD with probability β Otherwise, we select among the rest of unscheduled activities with the regret-based random sampling method. On average, $100(1 - \beta)$ is the percentage of iterations in which we deviate from the MST-GRD rule of Algorithm 2. The computational experiments discussed in Section 4 have been performed with a β value of 0.9, chosen after initial tests on a training set.

Operator M_A

This operator is exactly the same as in the IG for the RLP. We choose two random integers t', t'' with $0 = t' < t'' = S_{n+1}$. The activities to be unscheduled are the ones not beginning in $[t', t''[$. These activities are ordered according to their MST and subsequently scheduled. With a certain probability a the activity i to be scheduled is not scheduled at the best time t with respect to the increase in the objective function but at a random time t among the possible start times. After some preliminary experiments we have chosen $a = 0.2$ for our implementation.

Operator M_B

The idea behind this operator is to unschedule "critical" activities, those activities that most contribute to the objective function. In the RLP, rescheduling an activity i in a peak of the schedule is more beneficial than rescheduling it if is scheduled in a valley. In the RRP we want to penalise those activities that are scheduled in a peak just after a valley, because in that case we have a big amount of cost due to the fixed costs. Concretely, the penalisation of an activity i in schedule S is

$$
\begin{cases}
\sum_{k \in R} r_k\,(S, 0), & \text{if } S_i = 0 \\
\sum_{k \in R} \max\,(0, r_k\,(S, S_i) - r_k\,(S, S_i - 1)), & \text{if } S_i > 0
\end{cases}
\tag{12.3}
$$

The set U of activities to be unscheduled is constructed stepwise. Starting with $U = \emptyset$, in each iteration we select one real activity i from the set of remaining activities j according to a roulette-wheel strategy and add i to U. The slot sizes of activities j are chosen to be equal to the penalisations. After preliminary results we state that the inclusion of the second operator in the IG does not improve the average results of the algorithm. Therefore, the IG will only work with the first operator. We have included its definition since it was of use in the RLP and it can be useful for other problems.

12.4   Computational Results

In this section we will provide computational results for the different algorithms we have described in the foregoing paragraphs. First the metaheuristic algorithms will be compared to assert their ranking. Then we will compare the results of finalGA with those of a $B\&B$ algorithm of the literature.

12.4.1   Test Bed

The tests are based upon a test bed including 3 different sets, UBO10c, UBO20c and UBO50c, having each instance of UBOnc n activities. The instances have been generated using the problem generator ProGen/max by [362]. Several parameters such as the problem size n control the random construction of problem instances by ProGen/max. In addition, we fix a level for the deadline $\overline{d}$ of the project and for the cost quotient CQ, the ratio of variable renting costs and fixed renting costs, i.e. $c_k^v = CQc_k^f(k \in R)$. The test set includes 1800 problem instances for each combination of 10, 20 and 50 real activities. The settings for the cost quotient and the project deadline have been chosen to be $CQ \in \{0, 0.1, 0.2, 0.5, 1\}$, and $\overline{d} \in \{d_{0,n+1}, 1.1d_{0,n+1}, 1.25d_{0,n+1}, 1.5d_{0,n+1}, \}$. Note that the settings for the project deadline ensure that each generated RRP/max-instance possesses a feasible solution. Since there are already several instances for each combination of parameters each algorithm will only be run once. The sets UBO10c and UBO20c were used in [371] and we will compare our best algorithm with the $B\&B$ described in this paper in these sets. As far as we know, there are no other results for the RRP published in the literature. We will use UBO50c in order to compare the different heuristic algorithms we have developed. The comparison will be done by contrasting the average deviations with respect to a lower bound calculated by the *B&B*. All results refer to a Pentium personal computer with 1.4GHz clock pulse and 512MB RAM.

12.4.2   Comparison between the Metaheuristic Algorithms

In this section we compare the results of the different metaheuristic algorithms. We have imposed several time limits to compare the algorithms, concretely 1, 5, 10 and 25 seconds, allowing only a change in the population size in the algorithms. The graph in Figure 12.9 shows the average deviation with respect to the lower bound of the different algorithms and the different time limits. Using a Wilcoxon signed-rank test ([376]) we obtain that all differences between every pair of algorithms are significative for all time limits. This is true even in the case of specGA and finalGA and a time limit of 25 seconds. We can draw several conclusions based on the results. First of all, as somehow expected, the worst results come from adapGA. This fact eliminates the naive possibility of just adapting the algorithms from regular problems. We must also add that the poverty of the results would have been much greater had it not been for techniques such as the improvement procedure or the sampling procedure for the initial population. Secondly, the best results are obtained by the specially designed algorithm. However, the IG obtains a quality not far away from specGA and even from finalGA. The difference with these algorithms does not increase as the time limit increases, thus proving that a not so specialised algorithm can offer high quality solutions for short and long time limits for a problem with a nonregular objective function. Finally, the diversification introduced in specGA allows the algorithm to obtain better results, but the difference decreases as the time limit increases.

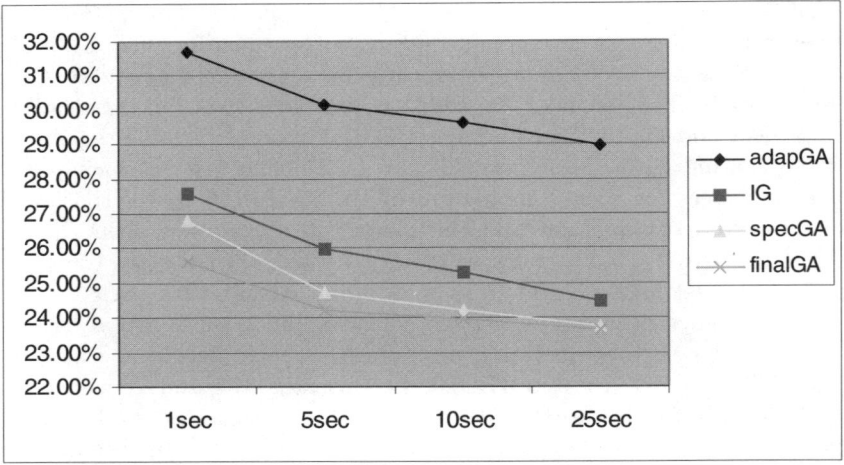

Fig. 12.9. Comparison among the different algorithms with different time limits

12.4.3   Comparison with $B\&B_N$

In this section we compare the results of finalGA with $B\&B_N$, the $B\&B$ from [371] and state the quality of the IG in small instances. We use UBO10c and UBO20c, presenting the results for the different combinations of the two most important parameters, CQ and $\overline{d}$. Table 12.3 contains the average deviation of finalGA with respect to the optimal solution found by the exact algorithm. $B\&B_N$ needed an average of 5.07 seconds to find them. With a time limit of 10 minutes, the $B\&B$ was not able to find all the optimal solutions in UBO20c. Table 12.4 shows the improvement average the metaheuristic obtains over the (truncated) $B\&B$. We have imposed a time limit of 0.5 seconds and 1 second on the metaheuristic algorithms in UBO10c and UBO20c respectively. The average CPU time of the $B\&B_N$ in UBO20 is 528.46 seconds. Note that all algorithms have been run with the same computer.

As we can see in the Tables, finalGA obtains near-optimal results in UBO10, being the global average over the 1800 instances 0.05%. Besides, it clearly outperforms the (truncated) $B\&B_N$ in UBO20. The reason is that $B\&B_N$ is not able to find all optimal solutions (only 14.78% of them) and in many of the rest of the cases the solution calculated by finalGA is of a higher quality. On average, the solutions of the metaheuristic are 10% better than those of the $B\&B$. If we analyse the effect of the parameters, the highest effect come from CQ. For lower CQ's the improvement is higher than 10%, whereas it is 2-5% for the larger CQ's . The effect of $\overline{d}/d_{0,n+1}$ is not so obvious, being the improvement slightly higher in cases 1.1 and 1.25, except when CQ = 0. Regarding the IG, its average deviation with respect to the optimal solutions in UBO10c is 0.25%, whereas the solutions obtained in UBO20c are only 0.71% worse than those obtained by finalGA.

Table 12.3. Averages of finalGA with respect to the optimal solutions in UBO10c

$\overline{d}/d_{0,n+1}$	0	0.1	0.2	0.5	1	TOTAL
1		0.00%	0.01%	0.00%	0.00%	0.00%	0.00%
1.1		0.01%	0.07%	0.04%	0.02%	0.01%	0.03%
1.25		0.03%	0.11%	0.16%	0.06%	0.03%	0.08%
1.5		0.14%	0.12%	0.11%	0.06%	0.01%	0.09%
Average	0.05%	0.08%	0.08%	0.04%	0.01%	0.05%

Table 12.4. Improvement average of finalGA with respect to $B\&B_N$ in UBO20c

$\overline{d}/d_{0,n+1}$	0	0.1	0.2	0.5	1	TOTAL
1		23.75%	13.98%	10.04%	5.23%	2.88%	11.18%
1.1		14.12%	17.42%	11.93%	6.52%	3.73%	10.70%
1.25		27.58%	15.43%	11.14%	6.32%	3.73%	12.84%
1.5		30.17%	12.30%	9.10%	5.17%	2.96%	11.94%
Average	23.90%	14.73%	10.56%	5.81%	3.32%	11.67%

12.5   Summary and Concluding Remarks

The RRP, a project scheduling problem with a resource-based objective function, enables us to model the renting of resources and is therefore interesting in practice. Its objective function is of the nonregular type and is therefore difficult to create a heuristic algorithm from scratch. In this paper we have used the RRP to compare three metaheuristic algorithms. The first one (adapGA) is one that uses a codification and crossover typically used in problems with regular objective function. We have observed that the direct use of these elements leads to bad results even using an improvement procedure and the "correct" Schedule Generation Scheme. The second algorithm (specGA) is a specially designed metaheuristic algorithm and the third algorithm (IG) is nearer the first than the second in terms of adaptation to the problem. The IG obtains quality solutions without having to devise a special codification or crossover. The best news is that the principles applied in this algorithm - fixing some activities and rescheduling the rest of them locally optimal - can be applied to several project scheduling problems with a nonregular objective function. Computational results show that both the specialised algorithm with a diversification (finalGA) as well as the IG obtain near-optimal solutions in UBO10c and already outperform the truncated B&B when the projects have 20 activities.

Acknowledgements

We thank Hartwig Nübel for providing us with the executable branch and bound file. This research was partially supported by the Ministerio de Educación y Ciencia under contract DPI2007-63100.

13

A Simple Optimised Search Heuristic for the Job Shop Scheduling Problem

Susana Fernandes[1] and Helena R. Lourenço[2]

[1] Universidade do Algarve, Faro, Portugal
   sfer@ualg.pt
[2] Univertitat Pompeu Fabra, Barcelona, Spain
   helena.ramalhinho@upf.edu

Summary. This paper presents a simple Optimised Search Heuristic for the Job Shop Scheduling problem that combines a GRASP heuristic with a branch-and-bound algorithm. The proposed method is compared with similar approaches and leads to better results in terms of solution quality and computing times.

Keywords: Job-Shop Scheduling, Hybrid Metaheuristic, Optimised Search Heuristics, GRASP, Exact Methods.

13.1   Introduction

The job shop scheduling problem has been known to the operations research community since the early 50's [392]. It is considered a particularly hard combinatorial optimisation problem of the NP-hard class [391] and it has numerous practical applications; which makes it an excellent test problem for the quality of new scheduling algorithms. These are main reasons for the vast bibliography on both exact and heuristic methods applied to this particular scheduling problem. The paper of Jain and Meeran, [392], includes an exhaustive survey not only of the evolution of the definition of the problem, but also of all the techniques applied to it.

Recently a new class of procedures that combine local search based (meta) heuristics and exact algorithms have been developed, we denominate them Optimised Search Heuristics (OSH) [389].

This paper presents a simple OSH procedure for the job shop scheduling problem that combines a GRASP heuristic with a branch-and-bound algorithm. In the next section, we introduce the job shop scheduling problem. In section 13.3, we present a short review of existent OSH methods applied to this problem and in section 13.4 we describe in detail the OSH method developed. In section 13.5, we present the computational results along with comparisons to other similar procedures applied to the job shop scheduling problem. Section 13.6 concludes this paper and discusses some ideas for future research.

C. Cotta and J. van Hemert (Eds.): Recent Advances in Evol. Comp., SCI 153, pp. 203–218, 2008.
springerlink.com   © Springer-Verlag Berlin Heidelberg 2008

13.2   The Job Shop Scheduling Problem

The Job Shop Scheduling Problem (JSSP) considers a set of jobs to be processed on a set of machines. Each job is defined by an ordered set of operations and each operation is assigned to a machine with a predefined constant processing time (preemption is not allowed). The order of the operations within the jobs and its correspondent machines are fixed a priori and independent from job to job. To solve the problem we need to find a sequence of operations on each machine respecting some constraints and optimising some objective function. It is assumed that two consecutive operations of the same job are assigned to different machines, each machine can only process one operation at a time and that different machines can not process the same job simultaneously. We will adopt the maximum of the completion time of all jobs - the makespan - as the objective function.

Formally let $O = \{0, 1, \ldots, o + 1\}$ be the set of operations with 0 and $o + 1$ dummy operations representing the start and end of all jobs, respectively. Let M be the set of machines, A the set of arcs between consecutive operations of each job and E_k the set of all possible pairs of operations processed by machine k, with $k \in M$. We define $p_i > 0$ as the constant processing time of operation i and t_i is the decision variable representing the start time of operation i. The following mathematical formulation for the job shop scheduling problem is widely used by researchers:

$$
(P) \quad
\begin{aligned}
&\min t_{o+1} \\
&\text{s.t. } t_j - t_i \geq p_i && (i,j) \in A && (1) \\
&\quad\ \, t_i \geq 0 && i \in O && (2) \\
&\quad\ \, t_j - t_i \geq p_i \vee t_i - t_j \geq p_j && (i,j) \in E_k,\ k \in M && (3)
\end{aligned}
$$

The constraints (1) state the precedence of operations within jobs and also that no two operations of the same job can be processed simultaneously (because $p_i > 0$). Expressions (3) are named "capacity constraints" and assure there are no overlaps of operations at the machines. A feasible solution for the problem is a schedule of operations respecting all these constraints.

The job shop scheduling problem is usually represented by a disjunctive graph [399] $G = (O, A, E)$. Where O is the node set, corresponding to the set of operations. A is the set of arcs between consecutive operations of the same job, and E is the set of edges between operations processed by the same machine. Each node i has weight p_i, with $p_0 = p_{o+1} = 0$. There is a subset of nodes O_k and a subset of edges E_k for each machine that together form the disjunctive clique $C_k = (O_k, E_k)$ of graph G. For every node j of $O \setminus \{0, o + 1\}$ there are unique nodes i and l such that arcs (i, j) and (j, l) are elements of A. Node i is called the job predecessor of node j - $jp(j)$ and l is the job successor of j - $js(j)$.

Finding a solution to the job shop scheduling problem means replacing every edge of the respective graph with a directed arc, constructing an acyclic directed graph $D_S = (O, A \cup S)$ where $S = \cup_k S_k$ corresponds to an acyclic union of sequences of operations for each machine k (this implies that a solution can be built sequencing one machine at a time). For any given solution, the operation processed immediately before operation i in the same machine is called the machine predecessor of i - $mp(i)$; analogously $ms(i)$ is the operation that immediately succeeds i at the same machine.

The optimal solution is the one represented by the graph D_S having the critical path from 0 to $o + 1$ with the smallest length.

13.3   Review of Optimised Search Heuristics

In the literature we can find a few works combining metaheuristics with exact algorithms applied to the job shop scheduling problem, designated as Optimised Search Heuristics (OSH) by Fernandes and Lourenço [389]. Different combinations of different procedures are present in the literature, and there are several applications of the OSH methods to different problems (see the web page of Fernandes and Lourenço)[1].

Chen, Talukdar and Sadeh [384] and Denzinger and Offermann [387] design parallel algorithms that use asynchronous agents information to build solutions; some of these agents are genetic algorithms, others are branch-and-bound algorithms.

Tamura, Hirahara, Hatono and Umano [405] design a genetic algorithm where the fitness of each individual, whose chromosomes represent each variable of the integer programming formulation, is the bound obtained solving lagrangian relaxations.

The works of Adams et al. [377], Applegate and Cook [379], Caseau and Laburthe [383] and Balas and Vazacopoulos [380] all use an exact algorithm to solve a sub problem within a local search heuristic for the job shop scheduling. Caseau and Laburthe [383] build a local search where the neighbourhood structure is defined by a subproblem that is exactly solved using constraint programming. Applegate and Cook [379] develop the shuffle heuristic. At each step of the local search the processing orders of the jobs on a small number of machines is fixed, and a branch-and-bound algorithm completes the schedule. The shifting bottleneck heuristic, due to Adams, Balas and Zawack [377], is an Iterated local Search with a construction heuristic that uses a branch-and-bound to solve the subproblems of one machine with release and due dates. Balas and Vazacopoulos [380] work with the shifting bottleneck heuristic and design a guided local search, over a tree search structure, that reconstructs partially destroyed solutions.

Lourenço [394] and Lourenço and Zwijnenburg [395] use branch-and-bound algorithms to strategically guide an iterated local search and a tabu search algorithm. The diversification of the search is achieved by applying a branch-and-bound method to solve a one-machine scheduling problem subproblem obtained from the incumbent solution.

In the work of Schaal, Fadil, Silti and Tolla [400] an interior point method generates initial solutions of the linear relaxation. A genetic algorithm finds integer solutions. A cut is generated based on the integer solutions found and the interior point method is applied again to diversify the search. This procedure is defined for the generalized job shop problem.

---

[1] http://www.econ.upf.edu/~ramalhin/OSHwebpage/index.html

The interesting work of Danna, Rothberg and Le Pape [385] "applies the spirit of metaheuristics" in an exact algorithm. Within each node of a branch-and-cut tree, the solution of the linear relaxation is used to define the neighbourhood of the current best feasible solution. The local search consists in solving the restricted MIP problem defined by the neighbourhood.

13.4   Optimised Search Heuristic - GRASP-B&B

We developed a simple Optimised Search Heuristic that combines a GRASP algorithm with a branch-and-bound method. Here the branch-and-bound is used within the GRASP to solve subproblems of one machine scheduling.

GRASP means "Greedy Randomised Adaptive Search Procedure", [388]. It is an iterative process where each iteration consists of two steps: a randomised building step of a greedy nature and a local search step. At the building phase, a feasible solution is constructed joining one element at a time. Each element is evaluated by a greedy function and incorporated (or not) in a restricted candidate list (RCL) according to its evaluation. The element to join the solution is chosen randomly from the RCL.

Each time a new element is added to the partial solution, if it has already more than one element, the algorithm proceeds with the local search step. The current solution is updated by the local optimum and this process of two steps is repeated until the solution is complete. Next, we describe the OSH method GRASP-B&B developed to solve the job shop scheduling problem. The main spirit of this heuristic is combining a GRASP method with a branch-and-bound to efficiently solve the JSSP.

13.4.1   The Building Step

In this section, we describe in detail the building step of the GRASP-B&B heuristic. We define the sequence of operations at each machine as the elements to join the solution, and the makespan $(max\,(t_i + p_i + q_i)\,, i \in O_k, k \in M)$ as the greedy function to evaluate them. In order to build the restricted candidate list (RCL), we find the optimal solution and optimal makespan, $f\,(x_k)$, for the one machine problems corresponding to all machines not yet scheduled. We identify the best $(\underline{f})$ and worst $(\overline{f})$ optimal makespans over all machines considered. A machine k is included in the RCL if $f\,(x_k) \geq \overline{f} - \alpha\,(\overline{f} - \underline{f})$, where $f\,(x_k)$ is the makespan of machine k and α is a uniform random number in $(0, 1)$. This semi-greedy randomised procedure is biased towards the machine with the higher makespan, the bottleneck machine, in the sense that machines with low values of makespan have less probability of being included in the restricted candidate list.

The building step requires a procedure to solve the one-machine scheduling problem. To solve this problem we use the branch-and-bound algorithm of Carlier [382]. The objective function of the algorithm is to minimize the completion time of all jobs. This one machine scheduling problem considers that to each job

---

Algorithm 13.1. Semigreedy(K)

$\alpha := Random\,(0, 1)$
$\overline{f} := max\,\{f\,(x_k)\,, k \in K\}$
$\underline{f} := min\,\{f\,(x_k)\,, k \in K\}$
$RCL := \{\}$
foreach $k \in K$
   if $f\,(x_k) \geq \overline{f} - \alpha\,\left(\overline{f} - \underline{f}\right)$
      $RCL := RCL \cup \{k\}$
   return $RandomChoice\,(RCL)$

---

j there are associated the following values that are obtained from the current so-lution: the processing time p_j, a release date r_j and an amount of time q_j that the job stays in the system after being processed. Considering the job shop problem and its disjunctive graph representation, the release date of each operation i - r_i is obtained as the longest path from the beginning to i, and its tail q_i as the longest path from i to the end, without the processing time of i.

The one-machine branch-and-bound procedure implemented works as follows: At each node of the branch-and-bound tree the upper bound is computed using the algorithm of Schrage [401]. This algorithm gives priority to higher values of the tails q_i when scheduling released jobs. We break ties by preferring larger processing times. The computation of the lower bound is based on the critical path with more jobs of the solution found by the algorithm of Schrage [401] and on a critical job, defined by some properties proved by Carlier [382]. The value of the solution with preemption is used to strengthen this lower bound. We introduce a slight modification, forcing the lower bound of a node never to be smaller than the one of its father in the tree. The algorithm of Carlier [382] uses some proven properties of the one machine scheduling problem to define the branching strategy, and also to reduce the number of inspected nodes of the branch-and-bound tree. When applying the algorithm to problems with 50 or more jobs, we observed that a lot of time was spent inspecting nodes of the tree, after having already found the optimal solution. So, to reduce the computational times, we introduced a condition restricting the number of nodes of the tree: the algorithm is stopped if there have been inspected more then n^3 nodes after the last reduction of the difference between the upper and lower bound of the tree (n is the number of jobs). We designated this procedure as Carlier_B&B(k), where k is the machine considered to be optimized, and output the optimal one-machine schedule and the respective optimal value.

The way the one-machine branch-and-bound procedure is used within the building step is described next. At the first iteration we consider the graph $D = (O, A)$ (without the edges connecting operations that share the same ma-chine) to compute release dates and tails. Incorporating a new machine in the solution means adding to the graph the arcs representing the sequence of oper-ations in that machine. In terms of the mathematical formulation, this means choosing one of the inequalities of the disjunctive constraints (3) correspondent to the machine. We then update the makespan of the partial solution and the release dates and tails of unscheduled operations, using the same procedure as the one used in the algorithm of Taillard [404]. We designate this procedure that computes the makespan of a partial solution x for the JSSP as $TAILLARD(x)$.

13.4.2   The Local Search Step

In order to build a simple local search procedure we need to design a neighbourhood structure (defined by moves between solutions), the way to inspect the neighbourhood of a given solution, and a procedure to evaluate the quality of each neighbour solution. It is said that a solution B is a neighbour of a solution A if we can achieve B by performing a neighbourhood defining move in A.

We use a neighbourhood structure very similar to the NB neighbourhood of Dell'Amico and Trubian [386] and the one of Balas and Vazacopoulos [380]. To describe the moves that define this neighbourhood we use the notion of blocks of critical operations. A block of critical operations is a maximal ordered set of consecutive operations of a critical path (in the disjunctive graph that represents the solution), sharing the same machine. Let $L(i, j)$ denote the length of the critical path from node i to node j. Borrowing the nomination of Balas and Vazacopoulos [380] we speak of forward and backward moves over forward and backward critical pairs of operations.

Two operations u and v form a forward critical pair (u, v) if:

1. they both belong to the same block;
2. v is the last operation of the block;
3. operation $js(v)$ also belongs to the same critical path;
4. the length of the critical path from v to $o + 1$ is not less than the length of the critical path from $js(u)$ to $o + 1$ ($L(v, o + 1) \geq L(js(u), o + 1)$).

Two operations u and v form a backward critical pair (u, v) if:

1. they both belong to the same block;
2. u is the first operation of the block;
3. operation $jp(u)$ also belongs to the same critical path;
4. the length of the critical path from 0 to u, including the processing time of u, is not less than the length of the critical path from 0 to $jp(v)$, including the processing time of $jp(v)$ ($L(0, u) + p_u \geq L(0, jp(v)) + p_{jp(v)}$).

Conditions 4 are included to guarantee that all moves lead to feasible solutions [380]. A forward move is executed by moving operation u to be processed immediately after operation v. A backward move is executed by moving operation v to be processed immediately before operation u.

The neighbourhood considered in the GRASP-B&B is slightly different from the one considered in Dell'Amico and Trubian [386] and Balas and Vazacopoulos [380] and it considers partial solutions obtained at each iteration of the GRASP-B&B heuristic. Therefore the local search is applied to a partial solution where a subset of all machines is scheduled. This neighbourhood is designated by $N(x, M \backslash K)$, where x is a partial solution, M is the set of all machines and K is the set of machines not yet scheduled in the building phase. When inspecting the neighbourhood $N(x, M\backslash K)$, we stop whenever we find a neighbour with a best evaluation value than the makespan of x. To evaluate the quality of a neighbour of a partial solution x, obtained by a move over a critical pair (u, v), we need only to compute the length of all the longest paths through the operations that were between u and v in the critical path of solution x. This evaluation is computed using the same procedure as the one used in the algorithm of Taillard [404], $TAILLARD(x)$. The local search phase consists in the two procedures described in Algorithm 13.2.

Algorithm 13.2. LocalSearch$(x, f(x), M\backslash K)$

LocalSearch$(x, f(x), M\backslash K)$

  $s := neighbour(x, f(x), M\backslash K)$
  while $s \neq x$
    $x := s$
    $s := neighbour(x, f(x), M\backslash K)$
  return s

Neighbour$(x, f(x), M\backslash K)$

  foreach $s \in N(x, M\backslash K)$
    $f(x) := evaluation(move(x \rightarrow s))$
    if $f(s) < f(x)$
      return s
  return x

Algorithm 13.3. GRASP-B&B$(runs)$

  $M = \{1, \ldots, m\}$
  for $r = 1$ to $runs$
    $x := \{\}$
    $K := M$
      while $K \neq \{\}$
        foreach $k \in K$
          $x_k := CARLIER\_B\&B(k)$
        $k^* := SEMIGREEDY(K)$
        $x := x \cup x_{k^*}$
        $f(x) := TAILLARD(x)$
        $K := K\backslash\{k^*\}$
        if $|K| < |M| - 1$
          $x := LOCALSEARCH(x, f(x), M\backslash K)$
    if x^* not initialised **or** $f(x) < f^*$
      $x^* := x$
      $f* := f(x)$
  return x^*

13.4.3   GRASP-B&B

In this section, we present the complete GRASP-B&B implemented, that considers the two phases previously described. Let *runs* be the total number of runs, M the set of machines of the instance and $f(x)$ the makespan of a solution x. The full GRASP-B&B method can be generally described by Algorithm 13.3.

This metaheuristic has only one parameter to be defined: the number of runs to perform (line (2)). The step of line (8) is the only one using randomness. When applied to an instance with m machines, in each run of the metaheuristic, the branch-and-bound algorithm is called $m \times (m + 1)/2$ times (line (7)); the local search is executed $m - 1$ times (line (13)); the procedure semigreedy (line (8)) and the algorithm of Taillard (line (10)) are executed m times.

13.5   Computational Results

We have tested the algorithm GRASP-B&B on the benchmark instances abz5-9 [377], ft6, ft10, ft20 [390], la01-40 [393], orb01-10 [379], swv01-20 [402], ta01-70 [403] and yn1-4 [406]. The algorithm has been run 100 times for each instance on a Pentium 4 CPU 2.80 GHz and coded in C.

We show the results of running the algorithm for each instance using the boxplots of RE_{UB}, the percentage of relative error to the best known upper bound (UB) calculated as follows:

$$RE_{UB}(x) = 100 \times \frac{f(x) - UB}{UB} \qquad (13.1)$$

The boxplots in figures 13.1, 13.2, 13.3, 13.4, 13.5 and 13.5 show that the quality achieved is more dependent on the ratio n/m than on the absolute numbers

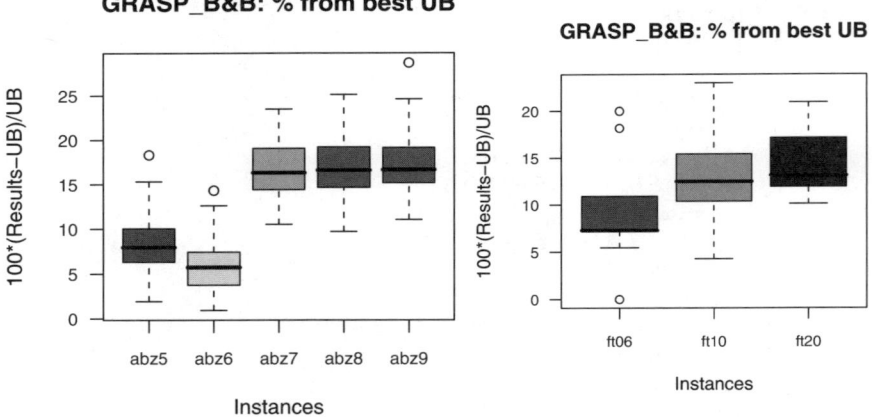

Fig. 13.1. GRASP-B&B: Boxplots of the RE_{UB} - abz and ft instances

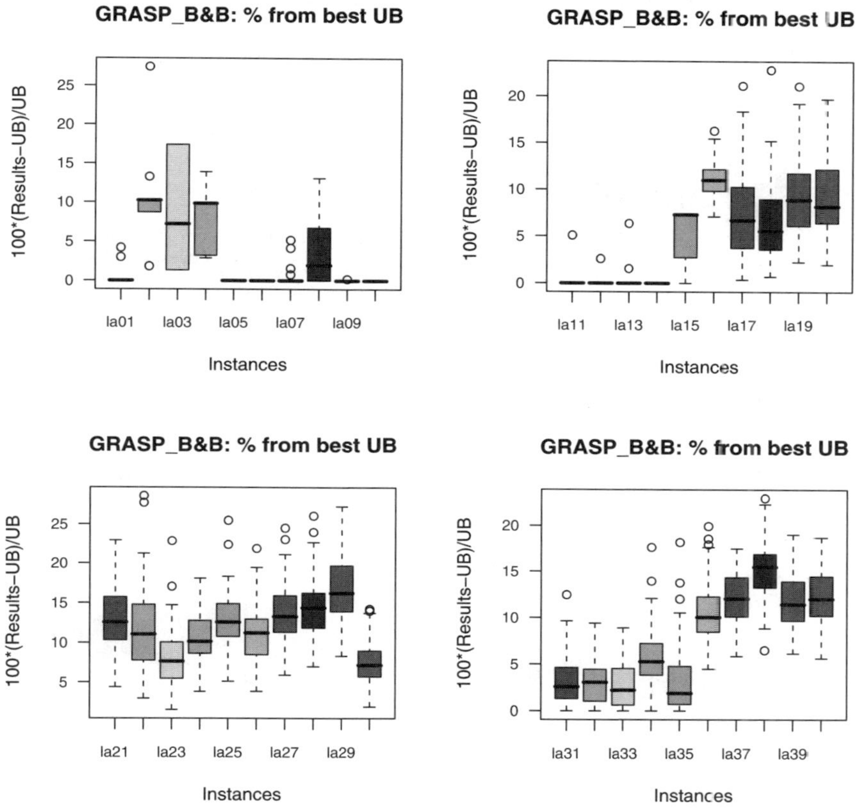

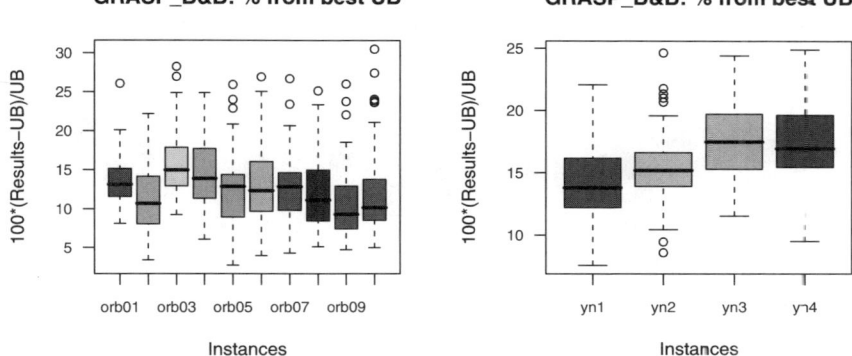

Fig. 13.2. GRASP-B&B: Boxplots of RE_{UB} - la instances

Fig. 13.3. GRASP-B&B: Boxplots of RE_{UB} - orb and yn instances

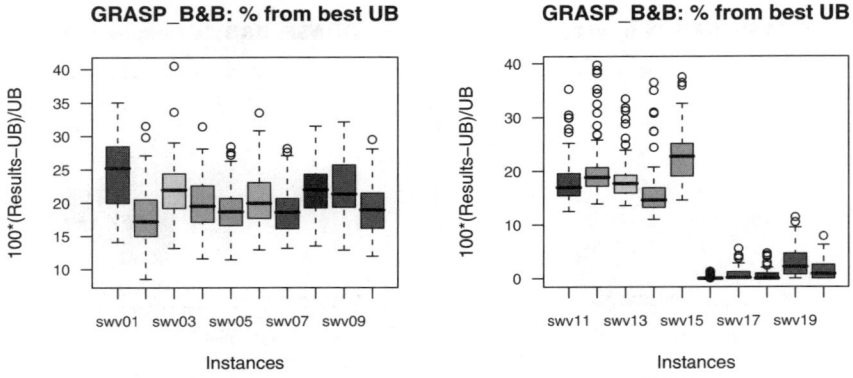

Fig. 13.4. GRASP-B&B: Boxplots of RE_{UB} - swv instances

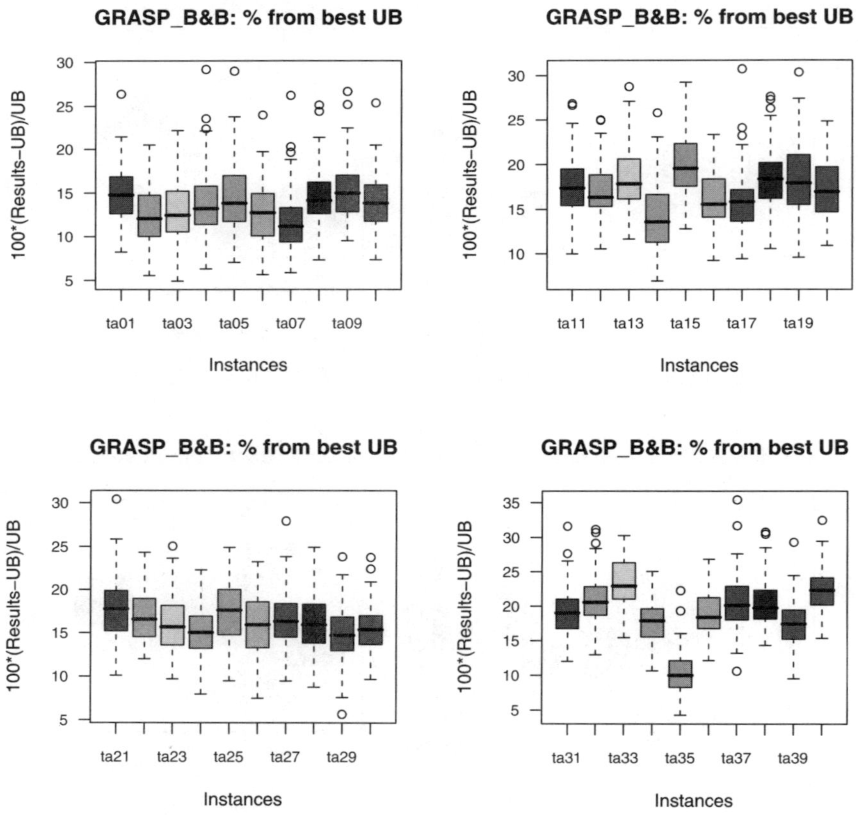

GRASP_B&B: % from best UB

GRASP_B&B: % from best UB

GRASP_B&B: % from best UB

GRASP_B&B: % from best UB

Fig. 13.5. GRASP-B&B: Boxplots of RE_{UB} - ta instances

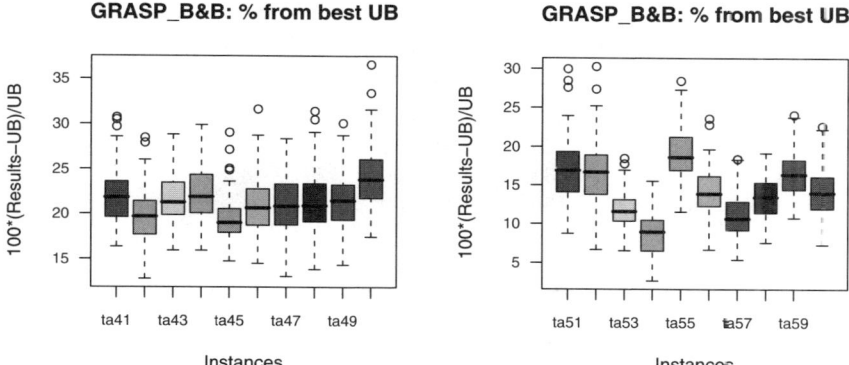

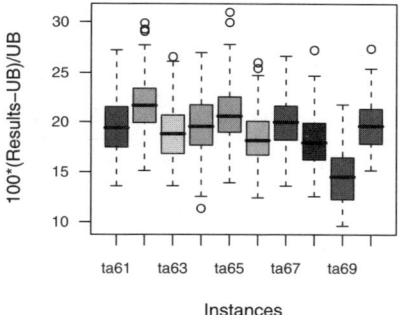

Fig. 13.5. (*continued*)

of jobs and machines. There is no big dispersion of the solution values achieved by the algorithm in the 100 runs executed, except maybe for instance la3. The number of times the algorithm achieves the best values reported is high enough, so these values are not considered outliers of the distribution of the results, except for instances ft06 and la38. On the other end, the worse values occur very seldom and are outliers for the majority of the instances. We gathered the values of the best known upper bounds from [392, 396, 397, 398].

13.5.1 Comparison to Other Procedures

GRASP-B&B OSH heuristic is a very simple GRASP algorithm with a construction phase very similar to the one of the shifting bottleneck. Therefore, we show comparative results to two other very similar methods: a simple GRASP heuristic

Table 13.1. GRASP-B&B vs GRASP: abz instances

instance	GRASP-B&B	ttime (s)	GRASP	time (s)
abz5	1258	0.7650	1238	6030
abz6	952	0.7660	947	62310
abz7	725	10.9070	667	349740
abz8	734	10.5160	729	365820
abz9	**754**	10.4690	758	343710

Table 13.2. GRASP-B&B vs GRASP: ft instances

instance	GRASP-B&B	ttime (s)	GRASP	time (s)
ft06	**55**	0.1400	55	70
ft10	970	1.0000	938	261290
ft20	1283	0.4690	1169	387430

Table 13.3. GRASP-B&B vs GRASP: orb instances

instance	GRASP-B&B	ttime (s)	GRASP	time (s)
orb01	1145	0.9850	1070	116290
orb02	918	0.9530	889	152380
orb03	1098	1.0150	1021	124310
orb04	1066	1.1250	1031	124310
orb05	911	0.8750	891	112280
orb06	1050	1.0460	1013	124310
orb07	414	1.0630	397	128320
orb08	945	1.0310	909	124310
orb09	978	0.9060	945	124310
orb10	991	0.8430	953	116290

of Binato, Hery, Loewenstern and Resende [381] and the Shifting Bottleneck heuristic by Adams, Balas and Zawack [377].

Comparison to GRASP of Binato, Hery, Loewenstern and Resende

The GRASP heuristic by Binato, Hery, Loewenstern and Resende [381] has a different building step in the construction phase, which consists in scheduling one operation at each step. In their computational results, they present the time in seconds per thousand iterations (an iteration is one building phase followed by a local search) and the thousands of iterations. For a comparison purpose we multiply these values to get the total computation time. For GRASP-B&B

Table 13.4. GRASP-B&B vs GRASP: la instances

instance	GRASP-B&B	ttime (s)	GRASP	time (s)
la01	**666**	0.1720	666	140
la02	667	0.1560	655	140
la03	605	0.2190	604	65130
la04	607	0.1710	590	130
la05	**593**	0.1100	593	130
la06	**926**	0.1710	926	240
la07	**890**	0.2030	890	250
la08	**863**	0.2970	863	240
la09	**951**	0.2810	951	290
la10	**958**	0.1410	958	250
la11	**1222**	0.2660	1222	410
la12	**1039**	0.2650	1039	390
la13	**1150**	0.3750	1150	430
la14	**1292**	0.2180	1292	390
la15	**1207**	0.9060	1207	410
la16	1012	0.7350	946	155310
la17	787	0.7660	784	60300
la18	854	0.7500	848	58290
la19	861	0.9690	842	31310
la20	920	0.8130	907	160320
la21	1092	2.0460	1091	325650
la22	**955**	1.7970	960	315630
la23	1049	1.8900	1032	65650
la24	**971**	1.8440	978	64640
la25	**1027**	1.7960	1028	64640
la26	**1265**	3.3750	1271	109080
la27	**1308**	3.5620	1320	110090
la28	1301	3.0000	1293	110090
la29	**1248**	3.2960	1293	112110
la30	1382	3.3280	1368	106050
la31	**1784**	7.0160	1784	231290
la32	**1850**	6.2350	1850	241390
la33	**1719**	7.9060	1719	241390
la34	**1721**	8.2810	1753	240380
la35	**1888**	5.6880	1888	222200
la36	1325	4.2650	1334	115360
la37	1479	4.7970	1457	115360
la38	1274	5.1090	1267	118720
la39	1309	4.4530	1290	115360
la40	1291	5.3910	1259	123200

Table 13.5. GRASP-B&B vs Shifting Bottleneck: abz instances

instance	GRASP-B&B	ttime (s)	Shifting Bottleneck	time (s)
abz5	**1258**	0.7650	1306	5.7
abz6	**952**	0.7660	962	12.67
abz7	**725**	10.9070	730	118.87
abz8	**734**	10.5160	774	125.02
abz9	**754**	10.4690	751	94.32

Table 13.6. GRASP-B&B vs Shifting Bottleneck: ft instances

instance	GRASP-B&B	ttime (s)	Shifting Bottleneck	time (s)
ft06	55	0.1400	55	1.5
ft10	**970**	1.0000	1015	10.1
ft20	**1283**	0.4690	1290	3.5

we present the total time of all runs (ttime), in seconds. As the tables 13.1, 13.2, 13.3 and 13.4 show, our algorithm is much faster. Whenever our GRASP achieves a solution not worse than theirs, we present the respective value in bold. This happens for 26 of the 58 instances whose results where compared.

Comparison to the Shifting Bottleneck Heuristic

The main difference of the Shifting Bottleneck procedure, [377], and GRASP-B&B is the random selection of the machine to be scheduled. In the Shifting Bottleneck the machine to be scheduled is always the bottleneck machine. The comparison between the shifting bottleneck procedure and the GRASP-B&B is also presented next, in tables 13.5, 13.6 and 13.7. Comparing the computation times of both procedures, the GRASP-B&B is slightly faster than the shifting bottleneck for smaller instances. Given the distinct computers used in the experiments we would say that this is not meaningful, but the difference does get accentuated as the dimensions grow.

Whenever GRASP-B&B achieves a solution better than the shifting bottleneck procedure, we present its value in bold. This happens in 29 of the 48 instances whose results where compared, and in 16 of the remaining 19 instances the best value found was the same.

13.6 Conclusions

In this work we present a very simple Optimised Search Heuristic, the GRASP-B&B to solve the Job-Shop Scheduling problem. This method is intended to be a starting point for a more elaborated metaheuristic, since it obtains reasonable solutions in very short running times. The main idea behind the GRASP-B&B

Table 13.7. GRASP-B&B vs Shifting Bottleneck: la instances

instance	GRASP-B&B	ttime (s)	Shifting Bottleneck	time (s)
la01	**666**	0.1720	666	1.26
la02	**667**	0.1560	720	1.69
la03	**605**	0.2190	623	2.46
la04	607	0.1710	597	2.79
la05	593	0.1100	593	0.52
la06	926	0.1710	926	1.28
la07	890	0.2030	890	1.51
la08	**863**	0.2970	868	2.41
la09	951	0.2810	951	0.85
la10	**958**	0.1410	959	0.81
la11	1222	0.2660	1222	2.03
la12	1039	0.2650	1039	0.87
la13	1150	0.3750	1150	1.23
la14	1292	0.2180	1292	0.94
la15	1207	0.9060	1207	3.09
la16	**1012**	0.7350	1021	6.48
la17	**787**	0.7660	796	4.58
la18	**854**	0.7500	891	10.2
la19	**861**	0.9690	875	7.4
la20	**920**	0.8130	924	10.2
la21	**1092**	2.0460	1172	21.9
la22	**955**	1.7970	1040	19.2
la23	**1049**	1.8900	1061	24.6
la24	**971**	1.8440	1000	25.5
la25	**1027**	1.7960	1048	27.9
la26	**1265**	3.3750	1304	48.5
la27	**1308**	3.5620	1325	45.5
la28	1301	3.0000	1256	28.5
la29	**1248**	3.2960	1294	48
la30	**1382**	3.3280	1403	37.8
la31	1784	7.0160	1784	38.3
la32	1850	6.2350	1850	29.1
la33	1719	7.9060	1719	25.6
la34	1721	8.2810	1721	27.6
la35	1888	5.6880	1888	21.3
la36	**1325**	4.2650	1351	46.9
la37	**1479**	4.7970	1485	6104
la38	**1274**	5.1090	1280	57.5
la39	**1309**	4.4530	1321	71.8
la40	**1291**	5.3910	1326	76.7

heuristic is to insert in each iteration of the building phase of the GRASP method the complete solution of one-machine scheduling problems solved by a branch-an-bound method, instead of insert one sequence of two individual operations as it is usual in other GRASP methods for this problem.

We have compared it with other similar methods also used as an initialization phase within more complex algorithms; namely a GRASP of Binato, Hery, Loewenstern and Resende [381], which is the base for a GRASP with path-relinking procedure of Aiex, Binato and Resende [378], and the Shifting Bottleneck procedure of Adams, Balas and Zawack [377], incorporated in the successful guided local search of Balas and Vazacopoulos [380]. The comparison to the GRASP of Binato, Hery, Loewenstern and Resende shows that the GRASP-B&B is much faster than theirs. The quality of their best solution is slightly better than ours in 60% of the instances tested. When comparing GRASP-B&B with the Shifting Bottleneck, the first one is still faster, and it achieves better solutions, except for 3 of the comparable instances. Therefore we can conclude, that the GRASP-B&B is a good method to use as the initialization phase of more elaborated and complex methods to solve the job-shop scheduling problem. As future research, we are working on this elaborated method also using OSH ideas, i.e. combing heuristic and exact methods procedures.

Acknowledgements

The work of Susana Fernandes is suported by the the programm POCI2010 of the Portuguese Fundação para a Ciência e Tecnologia. The work of Helena R. Lourenço is supported by Ministerio de Educacion y Ciencia, Spain, MEC-SEJ2006-12291.

14

Parallel Memetic Algorithms for Independent Job Scheduling in Computational Grids

Fatos Xhafa and Bernat Duran

Polytechnic University of Catalonia, Department of Languages and Informatics Systems, C/Jordi Girona 1-3, 08034 Barcelona, Spain
{fatos,bduran}@lsi.upc.edu

Summary. In this chapter we present parallel implementations of Memetic Algorithms (MAs) for the problem of scheduling independent jobs in computational grids. The problem of scheduling in computational grids is known for its high demanding computational time. In this work we exploit the intrinsic parallel nature of MAs as well as the fact that computational grids offer large amount of resources, a part of which could be used to compute the efficient allocation of jobs to grid resources.

The parallel models exploited in this work for MAs include both fine-grained and coarse-grained parallelization and their hybridization. The resulting schedulers have been tested through different grid scenarios generated by a grid simulator to match different possible configurations of computational grids in terms of size (number of jobs and resources) and computational characteristics of resources. All in all, the result of this work showed that Parallel MAs are very good alternatives in order to match different performance requirement on fast scheduling of jobs to grid resources.

Keywords: Memetic Algorithms, Independent Job Scheduling, Computational Grids, Parallelisation, Two Level Granulation.

14.1 Introduction

In this chapter we present several parallel implementation of Memetic Algorithms (MAs), both unstructured and structured, for the problem of independent job scheduling to grid resources. The scheduling problem is at the heart of any Computational Grid (CG) [415, 416]. Due to this, the job scheduling problem is increasingly receiving the attention of researchers from the grid computing community with the objective of designing schedulers for high performance grid-enabled applications. The independent job scheduling on computational grids is computationally hard. Therefore the use of heuristics is the *de facto* approach in order to cope in practice with its difficulty. Thus, the evolutionary computing research community has already started to examine this problem [407, 411, 420, 423, 413]. Yet, the parallelization of meta-heuristics, in particular of MAs [422], for the resolution of the problem has not been explored.

This work builds upon previous work on MAs for independent job scheduling [424]. One of the most advantageous characteristics of the family of Evolutionary Algorithms (EAs) is the intrinsic parallel nature of their structure.

C. Cotta and J. van Hemert (Eds.): Recent Advances in Evol. Comp., SCI 153, pp. 219–239, 2008.
springerlink.com   © Springer-Verlag Berlin Heidelberg 2008

Holland [418] in his early works introduced the first ideas for defining a parallel structure for EAs. The main observation here is that the algorithms based on populations of individuals could have a very complex structure, yet easily decomposable in smaller structures. This decomposition could be very useful to distribute the work of the algorithm to different processors. The objective of parallelizing EAs is essentially to distributed the burden of work during the search to different processors in such a way that the overall search time within the same (sequential) exploration is reduced. Therefore, the parallelization of EAs, in general, and MAs, in particular, could be beneficial for the resolution of the independent job scheduling in computational grids. Moreover, parallelizing EAs could imply not only the reduction of resolution time or better quality of solutions; it is also a source for new ideas in re-structuring the EA algorithm, differently from its original sequential structure, which could eventually lead to better performance of the algorithm.

In this chapter three different models of parallelizing unstructured MAs and Cellular MAs (here after refereed to as MAs and cMAs, resp.) are studied: (a) the model of independent searches, referred to as Independent Runs (IR) model for MAs; (b) the Master-Slave (MS) for MAs; and, (c) parallel hybridization among the coarse-grained and fine-grained models (also referred to as Two Level Granulation) for cMAs. The IR model consists of simultaneous execution of independent searches. In the MS model, the search algorithm is executed by a *master* processor, which delegates independent sub-tasks of high computational cost to the rest of processors (*slaves*). In the context of MAs, different slave processors could apply different local search procedures [421, 419] on the individuals of the population. Finally, hybrid parallel implementation is done for cMAs by combining the coarse-grained model, at a first hierarchical level, and the fine-grained model, at a second hierarchical level.

The proposed parallel MAs are implemented in C++ and MPI using a skeleton for MAs [409], extended for the purposes of this work. The implementations are extensively tested, on the one hand, to identify a set of appropriate values for the search parameters and, on the other, to compare the results for the makespan parameter obtained by different parallel implementations and the corresponding sequential versions. To this end we have used different grid scenarios generated using a grid simulator [425] to match different possible configurations of computational grids in terms of size (number of jobs and resources) and computational characteristics of resources.

The remainder of the chapter is organized as follows. We give in Section 14.2 the description of independent job scheduling problem. An overview on the taxonomy of parallel models for meta-heuristics and their application to MAs and cMAs is given in Section 14.3. The implementation of the parallel MAs is given in Section 14.4. The experimental study and some relevant computational results are presented in Section 14.5. We conclude in Section 14.6 with most important aspects of this work and indicate directions for future work.

14.2   Job Scheduling on Computational Grids

For the purposes of this work, we consider the following scheduling scenario: the tasks being submitted to the grid are independent and are not preemptive, that is, they cannot change the resource they has been assigned to once their execution is started, unless the resource is dropped from the grid.

The problem of job scheduling is formulated by using Expected Time to Compute (ETC) matrix model (see e.g. Braun et al. [410]). In this model, we have an estimation or prediction of the computational load of each task, the computing capacity of each resource, and an estimation of the prior load of the resources. In the Expected Time to Compute matrix $ETC[t][m]$ indicates the expected time to compute task t in resource m. The entries $ETC[t][m]$ could be computed, for instance, by dividing the workload of task t by the computing capacity of resource m. An instance of the problem is defined as follows: (a) A *number* of independent (user/application) *tasks* that must be scheduled. Any task has to be processed entirely in unique resource; (b) A *number* of heterogeneous *machines* candidates to participate in the planning; (c) The *workload* (in millions of instructions) of each task; (d) The *computing capacity* of each machine (in *mips*); (e) The time $ready_m$ when the machine will have finished the previously assigned tasks; and, (f) The expected time to compute ETC matrix of size $nb\_tasks \times nb\_machines$.

In this work the objective is to minimize the makespan parameter (the problem is multi-objective in its general formulation), that is, Makespan is a the time when finishes the latest task: $makespan : \min_{S_i \in Sched}\{\max_{j \in Jobs} F_j\}$, where F_j denotes the time when task j finalizes, $Sched$ is the set of all possible schedules and $Jobs$ the set of all jobs to be scheduled. For the purposes of computation, it is better to express makespan in terms of the *completion time* of machines. Let $completion[m]$ indicates the time in which machine m will finalize the processing of the previous assigned tasks as well as of those already planned tasks for the machine. Its value is calculated as follows:

$$completion[m] = ready\_times[m] + \sum_{\{j \in Tasks \mid schedule[j]=m\}} ETC[j][m].$$

Then, $makespan = \max\{completion[i] \mid i \in Machines\}$.

14.3   Overview of the Taxonomy of the Parallel Models for Memetic Algorithms

In this work we are based on the classification of the parallel models according to the taxonomy introduced by Crainic and Toulose [414] (see Fig. 14.1).

Low level parallelism: The model of low level parallelism is based on the simultaneous execution of the operations of the execution flow of the sequential method that can be executed in different processors. In this type of parallelism belongs the Master-Slave (MS) model, in which the search method is organized

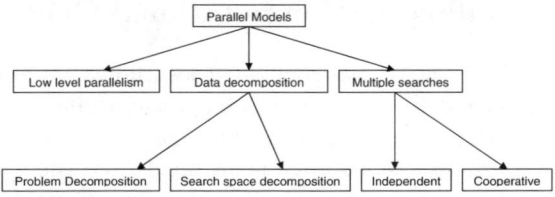

Fig. 14.1. Classification of Parallel Models

in a main process (*master*) that carries out the search and several processes (*slaves*), which execute the different sub-tasks of the search that the master delegates to them. This model is interesting for EAs given that in such algorithms several independent operations can be easily identified and performed in parallel.

Data decomposition: This type of parallelism considers the data decomposition for the resolution of the problem. Two classes of parallelism can be distinguished here according to whether the problem is decomposed into smaller problems or whether the space of solutions is divided into smaller regions (each processor solves exactly the same problem but treating a smaller data subset): (a) *Problem decomposition*: the problem is divided into subproblems that are assigned to different processes, the solutions of the subproblems are later combined into the solution of the original problem; and, (b) *Search space decomposition*: the decomposition is carried out in the space of solutions by assigning disjoint regions to slave processors.

14.3.1   Multiple Searches

The parallel schemes in this category are those that perform several simultaneous searches of the same space of solutions. The searches can be more or less independent among them, but in any case they work as independent entities in their own and thus different searches need not to have the same configuration. The strategy of multiple searches differs thus according to the degree of dependence established among them.

Independent searches: The independent searches consist of searches among which there is no communication at all. Notice that in this model we run a sequential version of the MAs /cMAs so this model doesn't introduce modifications to the sequential version of the algorithm. Yet, this model is particularly interesting when different searches use different configurations that is, different starting populations and values for the search parameters. In this setting larger regions of search space could be explored by different searches. We refer to this model as *Independent Runs - Different Strategies* (IR-DS). The advantage of this model is not in the reduction in the execution time, but in the possibility of each independent search to explore a region of the solution space within the same time and therefore the possibility to find a better solution is increased. The different configurations are automatically generated alleviating thus the burden to the user to manually introduce the values of the parameters. Moreover, the automatic configuration of different independent searches allows a much more versatile resolution of the problem, which is very suitable in the planning in real time of jobs in a computational grid due to its dynamics.

Cooperative searches: The cooperative search model consist of searches with a certain degree of cooperation and interdependence. The different strategies that fall into this category are distinguished through several parameters: a) the topology that establishes the relation among the processes; b) the communication mechanism; c) the information exchanged by the processes; d) the time period used to exchange information; and e) the synchronism among the processes (synchronized or asynchronous communication).

Regarding the Parallel Evolutionary Algorithms (PEAs), two cooperative search strategies are usually distinguished: the coarse-grained and the fine-grained strategy.

The *coarse-grained strategy* divides the global population of the EA in sub-populations assigned to different processors and each processor applies the EA on the corresponding sub-population. This sub-population, however, doesn't evolve independently but rather according to migration policy of individuals in such a way that the exchanged individuals influence the evolution of the sub-populations. Thus, new parameters that guide the search appear: a policy of selection of the individuals to be exchanged (*emigration policy*), the number of individuals to be exchanged (*migration rate*), the time interval when the exchange takes place (*migration interval*) and the policy that integrates the exchanged individuals into the corresponding sub-populations (*immigration policy*). Besides, depending on how the migratory flows are established among the sub-populations, in the coarse-grained strategy two models are distinguished: (a) *Island model*: an exchange of individuals among any couple of sub-populations (islands) can be done; and (b) *Steeping-stone model*: a relation among the sub-populations is established so that only exchanges of individuals among neighboring sub-populations can be done (it is necessary to establish a certain topology among the sub-populations).

In the *fine-grained strategy* the population of individuals is equally divided into small sub-populations, with trend of a cardinality 1, that is, just an individual is assigned to each processor. The subsets of individuals are related through a topology in neighboring groups influencing, thus, on the convergence of the population. The cMA designed in this work follows this model in a sequential setting, where the individuals of the population are structured in a 2D toroidal grid and the individuals are only combined with the neighboring individuals (according to the type of chosen neighborhood).

14.3.2 Hybrid Models

There are several strategies based on the hierarchical combination of the parallel models presented above, where different types of parallelism are established at different hierarchical levels. In [412] are surveyed some of the hybridization possibilities studied in the literature.

Coarse-grained + fine-grained: The coarse-grained model is combined at a first level, by dividing the population into sub-populations and establishing a migration mechanism among them, with the fine-grained model at a second level, in which each sub-population is structured in a topology (see Fig. 14.2, left). This strategy was originally proposed in [417].

Coarse-grained + Master-Slave: Different processors (coarse-grained model at a first level) divide the population into sub-populations and each master processor delegates tasks to slave processors following the model Master-Slave at a second level (see Fig. 14.2, right). This models simply reduces the computational cost of the coarse-grained model.

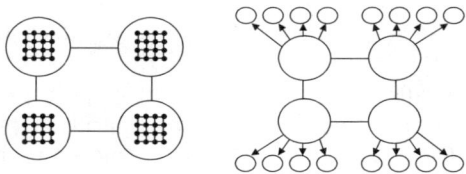

Fig. 14.2. Hierarchic PEA: coarse-grained model at 1st level and fine-grained model at 2nd level (left); Hierarchic PEA: coarse-grained at 1st level and MS at 2nd level (right)

Coarse-grained + coarse-grained: This model applies the coarse-grained at two hierarchical levels. The population is divided into sub-populations, and each sub-population on its own is divided into sub-populations; two different mechanisms of migration for the two hierarchical levels are applied (see Fig. 14.3, left).

Coarse-grained + fine-grained for cMA algorithm: In this model the grid structure of the population is distributed to different processors in such a way that two hierarchical levels are established: the first formed by the processors related among them in a grid topology (coarse-grained steeping-stone model) and the other one by the grid of individuals in each processor (see Fig. 14.3, right). The joint performance of all processors has to produce the evolution of only one population structured in a grid (fine-grained). Notice that the division of the population in sub-populations (coarse-grained model) at the first level establishes blocks of semi-isolated individuals and thus the behavior of a structured and parallelized population differs from the behavior of the same population in a sequential model.

14.4 Design and Implementation of the Parallel EAs (PEAs)

We show now the design and implementation of the parallel models for MAs using an algorithmic skeleton.

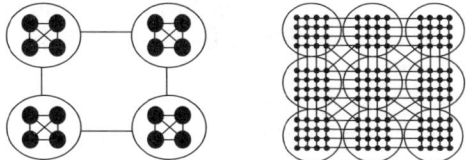

Fig. 14.3. Hierarchic PEA: coarse-grained model at 1st and 2nd level (left); Parallel cMA coarse-grained + fine-grained (right)

14.4.1 Extension of the MALLBA Skeletons

The implementation of the parallel models uses as a basis the MALLBA skeletons [408] adopted and extended here for the case of MAs and cMAs. The skeleton encapsulates the different models of parallelism in different classes through inheritance from the Solver base class (Solver_Seq, Solver_IR and Solver_MS, shown in Fig. 14.4). The design of the skeleton achieves the separation between the *provided* part, which implements the generic parallel models, from the *required* part, which implements the problem dependent features of the skeleton.

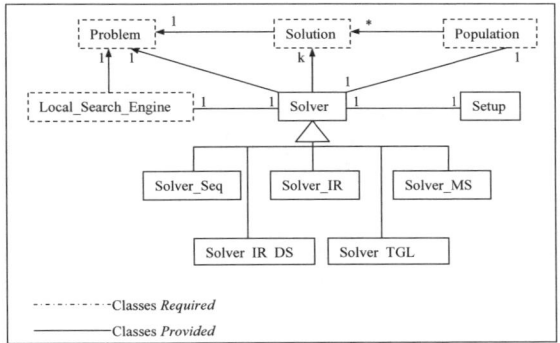

Fig. 14.4. The parallel skeleton for MAs and cMAs

As can be seen from Fig. 14.4, apart from IR and MS implementations, two new parallel models have been implemented, namely, a) the model IR_DS for the Parallel MAs (PMA) and Parallel Cellular MA (PCMA) and b) Two Level Granulation model for PCMA. Also, the Solver_MS class has been extended to support PMA algorithm.

Communication cost in the parallel models: A key aspect in the implementation of the different parallel models is to achieve a reduced communication time among the processors. Depending on the running environment the communication cost could suppose a bottleneck that reduces drastically the expected speed up. In order to minimize the communication time we have minimized the amount of information exchanged among the processors. More precisely, the data exchanged most frequently in PEAs is the solution to be manipulated by the processors. This is not critical in the IR and IR-DS models but could be critical in case of MS and TLG models since solutions are frequently exchanged. Instead of sending the whole solution, we implemented sending/receiving routines, which send only *changes* in the new solution, given that the processors know the previous solution. This approach reduces drastically the communication time since the changes are those produced by the local perturbation, which are small or very small changes.

Generation and use of different search strategies: A search strategy is defined as a starting population and a set of values for the search parameters. In order to establish the different configurations of the searches, which are automatically generated and assigned to processors. The user is required to only provide the range of values for each parameter ($liminf, limsup$) and a probability distribution (Constant, Normal, Uniform, Exponential or Triangular) for choosing the value of the parameter. Regarding the MS model, we have also introduced the novelty that each slave processor can use its own strategy for the search, that is, the individuals of the same population are locally perturbed using different local search methods. It is important to notice that this feature of the parallel implementation implies that the behavior of the MS model is different from the sequential algorithm.

14.4.2   Communication and Migration in TLG Model for the cMAs

As we already mentioned, in this model we have two hierarchical levels, the coarse-grained and fine-grained levels. Regarding the first we have to decide the exchange protocol for the individuals and regarding the second we have to specify the migration policy.

Communication in the coarse-grained level: Two types of communications can be established. a) *Synchronous communication*: the processors exchange the individuals in pre-established stages of the algorithm, precisely at the end of each iteration, in such a way that the neighboring processor will not have the modifications carried out until the exchange take place; and b) *Asynchronous communication*: any sub-population can dispose of the updated individuals of the neighboring sub-populations, that is, the modification of an individual is immediately available for the rest of sub-populations. The communication model used in this work is a) since it combines a synchronous model of semi-isolated sub-populations with the asynchronous model in each sub-population.

Fine-grained level: This second hierarchical level corresponds to the sub-population level, which preserves the grid topology but now a set of migration operators must be defined in order to "export" and "import" individuals in neighboring sub-populations. Due to this, now the 2D torodidal grid topology is not applicable; instead, a virtual topology is used (see Fig. 14.5).

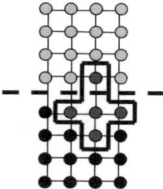

Fig. 14.5. Virtual topology of neighboring sub-populations

14.4.3   Design and Implementation of the Coarse-Grained Steeping Stone Level

Given the specifications of a global population, that is, the parameters of population height, population width, number of recombinations and number of mutations, the global population is divided into several sub-populations, assigned to the different available processors in such a way that cMA can be applied to the sub-populations and by achieving the global evolution of a unique population with the specified parameters. Two roles are distinguished for the processors: a *main processor* in charge of mapping the sub-populations to the processors and to carefully distribute the number of recombinations and mutations to be performed by the processors, and the *secondary processors* that work on sub-populations in a coordinated way among them and with the main processor. The algorithm run by the main processor is shown in Fig. 14.6.

```
Mapping of the population to p processors
For each independent_run do
    While stopping condition is not met do
        For i=0 to p do
            send signal of continuation to processor i
            calculate #recombinations and #mutations for processor i
            send #recombinations and #mutations to processor i
        EndFor
    EndWhile
    Receive the result from the execution of each processor
    Calculate the best solution and running time statistics
EndFor
```

Fig. 14.6. Main processors' algorithm in the coarse-grained steeping stone level

Regarding the secondary processors, these have to execute the cMA algorithm on their sub-population. In this case, the algorithm is essentially the sequential version of the algorithm except that new elements are added to perform the distribution of the population (see Fig. 14.7).

```
Receive the sub-population size and the set of
neighboring processors V.
/*Construct sub-population P*/
For each independent_run do
    Initialize the grid P(t=0) with n individuals
    /*Permutations containing the order used for
    the recombination and mutation of the individuals*/
    Initialize rec_order and mut_order
    For each individual i of P, apply local search on i
    Evaluate the individuals of P
    Exchange the shared individuals with V
    /*Reception of the stopping_signal */
    While not (stopping_signal) do
        Receive #recombinations and #mutations
        For j = 1 to #recombinations do
            SelectToRecombine S ⊆ N_{P[rec_order.current()]}
            i' := Recombination of S
            Local search on i'; Evaluation of i'
            Substitute P[rec_order.current()] by i'
            rec_order.next()
        EndFor
        For j = 1 to #mutations do
            i = P[mut_order.current()]
            i' := Mutation of i; Local search on i'; Evaluation of i'
            Substitute P[mut_order.current()] by i'
            mut_order.next()
        EndFor
        Update rec_order and mut_order
        Exchange the updates of the solutions shared with V
        Receive stopping_signal
    EndWhile
    Send results to the main processor
EndFor
```

Fig. 14.7. Secondary processors's algorithm in the coarse-grained steeping stone level

Next, we detail the set of operators used in the coarse-grained level, such as mapping of the population or the exchange of individuals (the rest of cMA operators are omitted here).

Mapping of the population: The mapping of the population consists in dividing it into p grids, where p is the number of secondary processors. Therefore, we have to fix the value p_h (number of cells per column), p_w (number of cells per row) and h_p (the cell width). The following conditions must hold $p = p_h \times p_w$, $h_p \times w_p \times p = h \times w$, where h and w are the height and the width of the global population, resp. It is attempted to preserve the proportions of the population (relation between the height and the width) in each of the small grids, in such a way that the subdivision of the population has to be the "squarest" possible, where the difference between p_h and p_w is minimum. The ideal case is that where p has an exact square root ($p_h = p_w = \sqrt{p}$), yielding to perfect proportions: $h_p = h/\sqrt{p}$ and $w_p = w/\sqrt{p}$, $\frac{h}{w} = \frac{h_p}{w_p}$.

The extreme case corresponds to p a prime number; in this case the population would remain in p portions in a same dimension (in our approach divided into p rows).

Next, the topology of the secondary processors is established as a 2D toroidal grid, in which each processor is assigned the corresponding neighbors. More precisely, a vector of 8 positions contains the identifiers of the 8 neighboring processors. The processors will exchange the individuals according to the neighborhood pattern established at fine-grained level.

Computing the number of recombination and mutations: Given the total number of recombination and mutations, it is necessary to distribute them uniformly for each sub-population in order to avoid possible *imbalances* (more evolved sub-populations). The distribution is thus done at random and is generated at each iteration in order to avoid the same scheme along the algorithm.

Exchange of the individuals: Two types of exchanges are distinguished: in the first, the exchange takes place just after having initialized the sub-populations and, in the second, the exchange takes place at the end of each iteration. The way the solutions are sent and received as well as the number and the set of exchanged solutions depend on the target and origin processor respectively, therefore, it depends on the neighborhood pattern established among the processors. These aspects are encapsulated in the fine-grained level, since it is at this level where the distribution of the individuals in space is known.

The exchange protocol applied here consists in sending and reception of solutions in an ordered way to avoid a deadlock situation. For each cardinal point, the processors initially export the corresponding solutions to the neighbor that it finds in that direction, next expect the reception of the solutions from the reverse direction in such a way that the exchange is guaranteed without blocking.

14.4.4   Design and Implementation of the Fine-Grained Level

Recall that the fine-grained level deals with the structure of the population. Each sub-population has to maintain a sequence of replicated individuals corresponding to the individuals the sub-population requires from the neighbors. The distribution of these replicas in the grid and the rules of how to inter-operate with the rest of individuals are established by the cMA algorithm. Besides, the sub-population has to provide other sub-population mechanisms to extract and to import individuals according to the position (north, south,...).

Extension of the population structure: In order to manage the replicas of the neighboring individuals in the population, and that these interact with the proper individuals of the population in a transparent way, the grid has been broadened in such a way that in the central part there is the real sub-population and in the peripheral parts are placed the replicas according to their origin. These replicas are useful only in the recombination of the neighboring individuals.

Note that the new region of replicas depends directly on the pattern of the neighborhood, avoiding to import replicas that will never be needed. We show in Fig. 14.8 the graphical representation of the extension of the grid with replicas for five most important topologies (panmixia, L5, L9, C9 and C13).

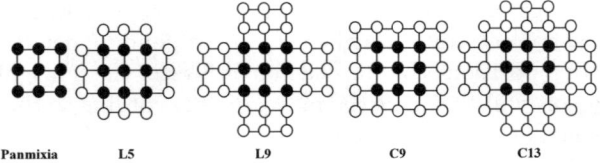

Panmixia   L5   L9   C9   C13

Fig. 14.8. Representation of a population of 3×3 individuals and the replicated region according to the neighborhood pattern

Importing replicas: Another issue is the design of the operators for importation of replicas and immigration operators. The former is used for initializing the population and thus complete solutions are imported, while the later serves for updating the replicas. In the first case the whole solution is copied for each replica and, in the second case, are applied just the changes to the solution.

The position from which the replicas will be imported depends on the cardinal point where the individuals come from (see Fig. 14.9, left). The individuals are imported following a fixed arbitrary order (left-right and up-down). It should be noted however that this order should be the same in exporting the central individuals, so as to not alter the virtual neighborhoods among the individuals of the neighboring sub-populations.

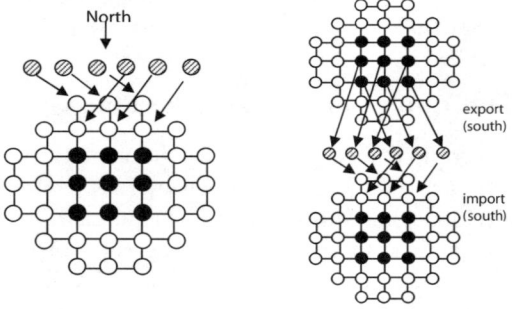

Fig. 14.9. Importing of individuals coming from the north neighbor in a C13 neighborhood pattern (left); Correspondence of the exporting and importing individuals between north-south neighbors (right)

Exporting replicas: The operators of exporting and emigration of individuals correspond to the opposite versions of the import operator. In the exportation operator whole copies corresponding to the central individuals are exported, while in emigration only the resulting modifications are exported as new changes of the same individuals only in case they have been modified. The export protocol coincides with that of import operator (left-right and up-down), as shown in Fig. 14.9 (right).

14.5   Experimental Study

In this section we present the experimental study to evaluate the performance and improvement of the implemented parallel models for MAs and cMAs. Notice that we are especially interested to evaluate the trade-off between the quality of solutions and the computation time given the dynamic nature of grid systems. Two criteria, directly related to the performance of the parallel implementations, have been considered: (a) the speed-up (or the reduction in computation time obtained by the parallelism) and (b) the performance of those parallel versions, which behave differently from their sequential version.

For the first criterion, the speed-up is computed in the usual way, that is, $S(p) = T_s/T_p$, where T_s and T_p are the sequential and parallel execution time, resp. Regarding the second criterion, as we already mentioned, some models of parallelism considered in this work introduce changes, which make the parallel algorithms to behave differently from their sequential versions; in a certain sense, we could speak of two different algorithms, in this case. More precisely, from the parallel models considered in this work, the MS for MAs –which differs from the sequential model in the fact that different mechanisms of local search can be applied on the individuals of the same population–, and the TLG for cMA, in which the division of the population into partially isolated small grids alters the behavior that would show the same population without the coarse-grained distribution.

We note that in all computational results shown in next subsections, the number of processors refers to the total number os processors participating in the execution. Thus, in the case of IR-DS model, having p processors means that $p - 1$ independent searches are executed and for the MS model there is a master processor and $p - 1$ slave processors. Finally, in the TLG model, p indicates the number of subdivisions of the population.

It should also be mentioned that no speedup is provided by the independent searches model, which consists of just running the sequential version in different processors.

14.5.1   Speed up Analysis

A group of four instances of semi-consistent ETC matrices has been generated using a grid simulator [425] and used for this experimental study; their sizes are shown in Table 14.1. For each instances, computational results are reported for MS implementation for the Parallel MA (PMA) and TLG implementation for the Parallel cMA (PcMA). The results are averaged over 20 executions using machines of our department PC cluster of usual configuration[1].

Performance of MS implementations

The values of the parameters used in PMA algorithm are shown in Table 14.2. The number of generations has been fixed to 100 and the size of the population

---

[1] All computers have the same configuration running Linux as operating system (Redhat distribution).

Table 14.1. Instance sizes used in the speedup analysis

	Small	Medium	Large	Very Large
Nb. of Tasks	512	1024	2048	4096
Nb. of Resources	32	64	128	256

Table 14.2. Parameter values used in PMA algorithm

nb_generations	100
nb_solutions_to_recombine	3
nb_recombinations	0.2× population_size
nb_mutations	0.8× population_size
start_choice	MCT and LJFR-SJFR
select_choice	Random Selection
recombine_choice	One point recombination
recombine_selection	Binary Tournament
rec_selection_extra	0.9
mutate_choice	Rebalance-Both
mutate_extra_parameter	0.6
mutate_selection	Best Selection
local_search_choice	LMCTS
nb_local_search_iterations	10
nb_ls_not_improving_iterations	4
add_only_if_better	false

Table 14.3. Size of the population established according to the instance size

	Small	Medium	Large	Very Large
Population size	50	60	70	75

has been settled to follow a gradual slow down according to the increase of the instance size (see Table 14.3).

In the table, MCT and LJFR-SJFR denote two deterministic methods used to generate initial solutions; they stand for Minimum Completion Time and Longest-Job to Fastest Resource -Shortest Job to Fastest Resource, resp. LMCTS (Local MCT Swap) is the local search methods used.

We present in Table 14.4 the execution and communication time among the processors for large and very large size instances. Notice that we have omitted the makespan value since, in order to measure the communication time the same local search mechanism is used in the slave processors.

Table 14.4. Execution and communication time for large size instances (left) and very large size instances (right) for MS implementation

Nprocs	t(s)	tcomm(s)
1	268.98	0
3	220.46	7.06
4	161.58	6.89
5	122.15	6.54
7	91.29	6.43
8	73.09	6.32

Nprocs	t(s)	tcomm(s)
1	602.64	0
3	499.06	14.1
4	353.51	13.74
5	268.21	13.19
7	191.97	13.25
8	176.94	13.13

Table 14.5. Parameter values used in the evaluation of the PCMA algorithm

nb_generations	200
nb_solutions_to_recombine	3
nb_recombinations	1×population_height×population_width
nb_mutations	0.5×population_height×population_width
start_choice	MCT and LJFR-SJFR
neighborhood_pattern	C9
recombination_order	FLS
mutation_order	NRS
recombine_choice	One point recombination
recombine_selection	N Tournament
rec_selection_extra	3
mutate_choice	Rebalance-Both
mutate_extra_parameter	0.6
local_search_choice	LMCTS
nb_local_search_iterations	5
nb_ls_not_improving_iterations	$+\infty$
lsearch_extra_parameter	0
add_only_if_better	true

Table 14.6. Population size according to instance size

	Small	Medium	Large	Very Large
Population size (height×width)	6×6	6×6	7×7	7×7

As can be observed, for all of instances the reduction in the execution time is very considerable with the increase of the number of processors. The reduction in time follows an evolution similar to $T_s/\log(p-1)$ function, for $p > 2$ processors.

Performance of Two Level Granulation implementation

The configuration of parameters used in TLG is shown in Table 14.5. The only parameter that varies is the population size whose value is fixed according to the instance size (see Table 14.6). It should be noticed that the population size value is fixed independently of the number of processors, therefore the population size has to be large enough to ensure admissible sub-divisions in TLG. The number of generations has been fixed to 200.

The makespan values[2], the execution and communication time in TLG implementation according to the instance size and number of processors used are given in Tables 14.7. Notice that the number of processors coincides with the number of sub-divisions of the population.

A qualitative improvement is observed in the value of makespan as the number of processors increases. This implies that for a same configuration and workload, the division of the cellular population in smaller grids benefits the search. Besides, a clear improvement is observed in the execution time with the increase in the number of processors.

However, TLG implementation doesn't behave as regularly as the MS, which could be explained by the fact that according to the number of processors a different structure of the sub-population is established at the coarse-grained level.

---

[2] In arbitrary time units.

Table 14.7. Makespan values, execution and communication time for small size instances (left) and medium size instances (right) in TLG implementation

Nprocs	Makespan	t(s)	tcomm(s)
1	1834405.18	81.45	0
3	1803143.55	53.9	1.01
4	1806527.72	32.48	1.56
6	1796988.74	33	1.51
8	1767354.44	24.74	1.7
9	1791683.11	32.61	1.68

Nprocs	Makespan	t(s)	tcomm(s)
1	1179876.25	186.26	0
3	1155363.36	102.08	1.07
4	1152594.67	71.02	1.59
6	1148556.35	61.26	1.54
8	1134427.8	39.88	1.76
9	1146915.11	47.81	1.74

Table 14.8. Makespan values, execution and communication time for large size instances (left) and very large size instances (right) in TLG implementation

Nprocs	Makespan	t(s)	tcomm(s)
1	768778.44	588.93	0
3	753870.58	239.17	1.46
4	748673.41	190.07	1.92
6	745782.64	147.1	2.22
8	743934.16	104.65	2.16
9	745861.82	92.5	2.06

Nprocs	Makespan	t(s)	tcomm(s)
1	490088.47	1242.77	0
3	481851.38	515.65	2.43
4	481153.9	393.14	2.66
6	478877.1	279.29	3.93
8	477989.05	191.69	3.79
9	476626.85	175.41	3.17

As a matter of fact, the makespan improvement as well as the communication time depend more on the concrete value of the number of processors (e.g. when it is a primer number) rather than the amount of processors used. The most evident case is that when the number of processors p is a prime number in which case the grid is divided into p rows. This phenomenon minimizes the communication time significantly due to the vertical cuts in the grid (lateral communication), but on the other hand it implies a deformation of the proportions given by the user to construct the sub-populations, without respecting the relation between the height and the width of the global population.

Again, the execution time follows the function $T_s/\log(p-1)$ although not as adjusted as in the case of MS implementation.

14.5.2   Qualitative Improvement of the Makespan Results

The experimental study presented in previous section aimed at identifying the relation of execution and communication time with the increasing number of processors, respectively. This is important to further analyze the qualitative improvement in the makespan of the schedule, which is the main objective of the considered parallel implementations. On the other hand, we also noticed that, independently of the execution time, there could be a qualitative improvement due to the introduction of new elements in the parallel versions of the algorithms, which yields to "*new*" algorithms. In the case of MS this behavior is due to the fact that each processor can apply different local search procedures to the individuals of the population. In the TLG model the new behavior comes from the fact that by dividing a population into sub-populations, these become partially isolated since they don't share the individuals in asynchronous way, as in the sequential cMA algorithm. Finally, the IR-DS model behaves essentially as the sequential algorithm, although qualitative improvements are possible by using the best configuration in each independent search.

We used different size instances (small, medium, large and very large) to observe the improvement obtained by each parallel model according to the particular algorithm that it has been applied to, namely, Parallel MA and Parallel cMA. Again, the results are averaged over 20 executions carried out in same PC cluster (see above). The maximum execution time has been fixed to 90 seconds, which is considered as a reasonable time interval for scheduling independent jobs in real grid environments.

Comparison of IR-DS, MS and sequential implementations of PMA

We present next the configuration used in PMA as well as the makespan values obtained with IR-DS, MS and sequential implementations, respectively, of the PMA. Based on the computational results we can compare the improvements by the IR-DS and MS implementations of PMA with respect to the sequential implementation according to instance size and number of processors.

Notice that, in order to generate the different strategies for the IR model, lower and upper bounds for the parameters should be fixed. These values are given in Fig. 14.10; the range of the population size according to instance size is shown in Table 14.9.

In Fig. 14.10, MCT (Minimum Completion Time) and LJFR-SJFR (Longest-Job to Fastest Resource -Shortest Job to Fastest Resource) are the deterministic methods used to generate initial solutions. LMCTS (Local MCT Swap) and LMCTM (Local MCT Move) are the local search methods used, resp.

Parameter	Inferior limit	Superior limit	Distribution
nb_generations	(max 90 s)		
nb_solutions_to_recombine	2	4	uniform
nb_recombinations	0,1× population_size	0,4× population_size	uniform
nb_mutations	0,7× population_size	0,9× population_size	uniform
start_choice	MCT and LJFR-SJFR		
select_choice	Random		
select_extra_parameter	∅		
recombine_choice	One point recombination		
recombine_selection	Binary Tournament		
rec_selection_extra	0,7	1	uniform
mutate_choice	Rebalance-Both		
mutate_extra_parameter	0,5	1	normal
mutate_selection	Best Selection		
mut_selection_extra	∅		
local_search_choice	LMCTS and LMCTM		
nb_local_search_iterations	5	20	uniform
nb_ls_not_improving_iterations	0,5×nb_local_search_iterations	1×nb_local_search_iterations	uniform
lsearch_extra_parameter	∅		
Boolean parameter	**Likelihood**		
add_only_if_better	0,1		

Fig. 14.10. Configuration of the IR-DS implementation for PMA

Table 14.9. Population size in IR-DS implementation for PMA

	Small	Medium	Large	Very Large
Population size (sequential mode)	50	60	70	75
Population size (IR-DS) [inf, sup], distr.	[35,65] uniform	[45,75] u. [45,75] u.	[55,85] u.	[60,90] u.

Table 14.10. Makespan values obtained by the PMA versions IR-DS and MS for small size instances (left) and medium size instances (right)

Nprocs	PMA(IR)	PMA(MS)
(Seq)	1632339.51	
3	1636429.04	**1634867.64**
4	1622557.25	**1620267.95**
5-6	1609863.35	**1609296.35**
7-8	1610404.06	**1601005.42**
9	1611584	**1594404.15**

Nprocs	PMA(IR)	PMA(MS)
(Seq)	1008864.68	
3	1021601.5	**989279.28**
4	992045.54	**975075.05**
5-6	986709.91	**973972.77**
7-8	976843.64	**960896.63**
9	979943.91	**952360.84**

Table 14.11. Makespan values obtained by the PMA versions IR-DS and MS for large size instances (left) and very large size instances (right)

Nprocs	PMA(IR)	PMA(MS)
(Seq)	654102.25	
3	663821.81	**658875.39**
4	**645805.22**	647980.86
5-6	648881.96	**644351.72**
7-8	641974.1	**639227.82**
9	640485.7	**633337.48**

Nprocs	PMA(IR)	PMA(MS)
(Seq)	441155.97	
3	**444951.86**	449175.51
4	**440938.81**	441038.38
5	**439542.5**	440562.97
7	438428.26	**436174.14**
9	437178.6	**432571.49**

The computational results for makespan parameter are given in Tables 14.10 and 14.11 (the best value of both implementations is shown in bold). As can be observed from the tables, the makespan value is improved, compared to the sequential implementation, as the number of processors increases. We exemplify the makespan reductions by both implementations in Fig. 14.11, where we can see that MS implementation outperforms the IR-DS implementation.

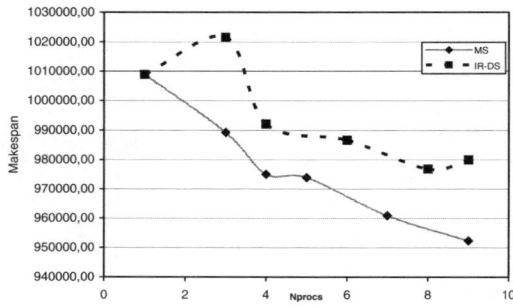

Fig. 14.11. Makespan reduction of IR-DS and MS implementations of the PMA for a medium size instance

Parameter	Inferior limit	Superior limit	Distrib.
nb_generations	(max 90 s)		
nb_solutions_to_recombine	2	4	uniform
nb_recombinations	0,8× population_height× population_width	1× population_height× population_width	uniform
nb_mutations	0,4× population_height× population_width	0,6× population_height× population_width	uniform
start_choice	MCT and LJFR-SJFR		
neighborhood_pattern	L5, L9, C9		uniform
recombination_order	FLS		
mutation_order	NRS		
recombine_choice	One point recombination		
recombine_selection	N Tournament		
rec_selection_extra	3		
mutate_choice	Rebalance-Both		
mutate_extra_parameter	0,5	1	normal
local_search_choice	LMCTS and LMCTM		uniform
nb_local_search_iterations	3	15	uniform
nb_ls_not_improving_iterations	0,3×nb_local_search_iterations	1×nb_local_search_iterations	uniform
lsearch_extra_parameter	∅		
Boolean parameter	Likelihood		
add_only_if_better	0,8		

Fig. 14.12. Configuration of the IR-DS implementation for PCMA

Table 14.12. Population size in IR-DS implementation for PCMA

	Small	Medium	Large	Very Large
Sizes sequential mode (height×width)	5×5	6×6	6×6	7×7
Height (IR-DS) [inf, sup], distr.	[4,6] uniform	[5,7] u.	[5,7] u.	[6,8] u.
Width (IR-DS) [inf, sup], distr.	[4,6] uniform	[5,7] u.	[5,7] u.	[6,8] u.

Comparison of IR-DS, TLG and sequential implementations of PCMA

The configuration used in PCMA as well as the makespan values obtained with IR-DS, TLG and sequential implementations, respectively, of the CPMA are presented next. Based on the computational results we can compare the improvements by the IR-DS and MS implementations of CPMA with respect to the sequential implementation according to instance size and number of processors participated in the search.

Again, in order to generate the different strategies for the IR model, lower and upper bounds for the parameters are fixed. These values are given in Fig. 14.12 and the range of the population size according to instance size is shown in Table 14.12.

As can be seen from the configuration of the IR-DS model, the proportions of the structured population are not fixed, therefore the independent searches not only could vary in the number of individuals but also in the structure of the population. The patterns of the neighborhood used are those that guarantee a more acceptable behavior, discarding the panmixia and C13 since both favor a too intense search policy. The most influential operators in the search have been fixed in order to avoid the negative effect of the bad configurations of searches.

Table 14.13. Sizes of the population used in TLG

Nprocs	Small	Medium	Large	Very Large
3	6×4	6×6	6×6	7×7
4	5×5	6×6	6×6	7×7
6	6×4	6×6	6×6	7×7
8	8×4	8×4	8×4	7×7
9	6×6	6×6	6×6	7×7

Table 14.14. Makespan values obtained by the PMA versions IR-DS and TLG for small size instances (left) and medium size instances (right)

Nprocs	PCMA(IR-DS)	PCMA(TLG)
(Seq)	1595730.87	
3	1618178.02	**1591373.7**
4	1614640.4	**1581589.48**
5-6	1579210.6	**1572990.76**
7-8	1580631.56	**1568097.94**
9	1574928.9	**1578471.46**

Nprocs	PCMA(IR-DS)	PCMA(TLG)
(Seq)	1087018.94	
3	1071528.14	**1044949.77**
4	1032718.9	**1012936.09**
5-6	1027370.45	**1000646.61**
7-8	1017213.54	**983838.58**
9	1018227.21	**989905.15**

The local search procedures used, are again LMCTS and LMCTM, as in the case of PMA algorithm.

An additional aspect to take into account in the context of the sub-populations is that of the their size. The decision has been to maintain the number of individuals and the pattern of the population as in the sequential model. For the TLG model better results are obtained with smaller populations; indeed, smaller size populations are computationally less expensive and the cellular structure seem to protect them from a premature convergence. We show in Table 14.13, the values of the height and the width according to the instance size and the number of processors used.

The computational results for the makespan parameter obtained from IR-DS and TLG implementations of PCMA are shown in Table 14.14 and Table 14.15 (the best value of both implementations is shown in bold). The IR-DS shows the same behavior: initially with the increasing number of processors, the makespan is reduced and later it's stagnated. In in a certain sense this behavior shows the usefulness but also the limitations of the independent searches with different strategies. Regarding TLG implementation, the results show a considerable improvement outperforming the IR-DS model. However, as in the case of MS implementation, the communication cost of the model shows to be considerable with the increase of the number of processors, which goes in detriment of the

Table 14.15. Makespan values obtained by the PMA versions IR-DS and TLG for large size instances (left) and very large size instances (right)

Nprocs	PCMA(IR)	PCMA(TLG)
(Seq)	767549.31	
3	750571.49	**741375.89**
4	746843.73	**727341.88**
5-6	734531	**717319.02**
7-8	736794.5	**701782.12**
9	733531.82	**695951.31**

Nprocs	PCMA(IR)	PCMA(TLG)
(Seq)	500354.57	
3	499817.34	**399661.47**
4	496147.31	**356350.92**
5-6	493774.6	**356528.66**
7-8	492910.19	**478943.6**
9	492125.28	**475351.07**

proper search time. On the other hand, it's not worth increasing the population size since it doesn't improve the quality of the results with less processors without increasing the total execution time.

14.6   Conclusions

In this chapter we have presented the implementation and the computational evaluation of three parallel models of Memetic Algorithms (MAs) for the Independent Job Scheduling problem in Computational Grids. The Independent Search with Different Strategies (IR-DS), the Master-Slave (MS) and a hybridization between the coarse-grained and fine-grained models (Two Level Granulation –TLG) are considered. The interest of these models is twofold. First, these models can contribute to finding better solutions, either by searching larger areas of the solution space or by speeding up the running time of the algorithms. Second, for the Cellular MAs (cMA), some parallel models, such as MS and TLG behave differently from their sequential versions. These structural differences could *per se* yield better performance of cMAs.

The extensive experimental results using different size instances (from small to very large) generated with a grid simulator showed that all these models provide important qualitative improvements for makespan parameter as compared to the sequential versions. From a practical perspective, the parallel implementations of MAs and cMAs are very good alternatives for scheduling in computational grids since they are able to obtain a very accentuated reduction in the makespan value of the schedule.

In our future work we plan to use the resulting Parallel MAs and parallel cMA-based schedulers as part of grid applications and evaluate their performance in a real grid environment.

Part V

Routing and Travelling Salesman Problems

Reducing the Size of Travelling Salesman Problem Instances by Fixing Edges

Thomas Fischer and Peter Merz

Distributed Algorithms Group, Department of Computer Science,
University of Kaiserslautern, Germany
{fischer,pmerz}@informatik.uni-kl.de

Summary. The Travelling Salesman Problem (TSP) is a well-known NP-hard combinatorial optimization problem, for which a large variety of evolutionary algorithms are known. However, with instance size the effort to find good solutions increases considerably. Here, we discuss a set of eight edge fixing heuristics to transform large TSP problems into smaller problems, which can be solved easily with existing algorithms. The edge fixing heuristics exploit properties of the TSP such as nearest neighbour relations or relative edge length. We argue, that after expanding a reduced tour back to the original instance, the result is nearly as good as applying the used solver to the original problem instance, but requires significantly less time to be achieved.

Keywords: Travelling Salesman Problem, Search Space Reduction, Minimum Spanning Tree, Nearest Neighbour.

19.1 Introduction

The Travelling Salesman Problem (TSP) is a widely studied combinatorial optimization problem, which is known to be NP-hard [248].

Let $G = (V, E, d)$ be an edge-weighted, directed graph, where V is the set of $n = |V|$ vertices, $E \subseteq V \times V$ the set of (directed) edges and $d : E \to \mathbb{R}^+$ a *distance function* assigning a distance $d(e)$ to each edge $e \in E$. A *path* is a list $(u_1, \ldots, u_k)$ of vertices $u_i \in V$ $(i = 1, \ldots k)$ holding $(u_i, u_{i+1}) \in E$ for $i = 1, \ldots, k - 1$. A *Hamiltonian cycle* in G is a path $p = (u_1, \ldots, u_k, u_1)$ in G, where $k = n$ and $\bigcup_{i=1}^{k} u_i = V$ (each vertex is visited exactly once except for u_1). The TSP's objective is to find a Hamiltonian cycle t for G that minimizes the cost function $C(t) = \sum_{i=1}^{k-1} d((u_i, u_{i+1})) + d((u_k, u_1))$ (sum of weights of the edges in t). Depending on the distance function d, a TSP instance may be either *symmetric* (for all $u_1, u_2 \in V$ holds $d((u_1, u_2)) = d(u_2, u_1))$) or *asymmetric* (otherwise). Most applications and benchmark problems are *Euclidean*, i. e., the vertices V correspond to points in a Euclidean space (mostly 2-dimensional) and the distance function represents a Euclidean distance metric. The following discussion focuses on symmetric, Euclidean problem instances.

C. Cotta and J. van Hemert (Eds.): Recent Advances in Evol. Comp., SCI 153, pp. 243–258, 2008.
springerlink.com   © Springer-Verlag Berlin Heidelberg 2008

Different types of algorithms for the TSP are known, such as exact algorithms [249, 250, 251] or local search algorithms. Local search algorithms require the construction of an initial solution before starting to improve this solution. Common construction heuristics for the TSP are the nearest neighbour heuristics, greedy heuristics, and insertion heuristics. Well-known local search algorithms for the TSP are k-exchange operators, which remove k edges from a tour and insert k new edges ($k \in \{2, 3\}$ for practical cases). A tour is k-opt, if there is no k-exchange which improves the tour's quality. To improve the speed of this type of local search, the search of exchanges can be restricted to a limited set of candidate edges or the concept of don't look bits can be used, which allows to skip exchange moves when no improvement is expected. An extension to k-opt algorithms is the LK algorithm [252], which keeps k variable. Here, a given tour is transferred into a 1-tree on which a sequence of 1-exchange moves is performed. The algorithm allows to backtrack and thus undo intermediate exchange moves. Finally when fulfilling the gain criterion, the 1-tree can be transferred back into a valid TSP tour by a final 1-exchange move. The Lin-Kernighan algorithm is known for its complexity and open design choices.

Among the best performing TSP algorithms are those utilizing Lin-Kernighan local search within an evolutionary framework such as Iterated Lin-Kernighan [253] or Memetic Algorithms [254, 255]. Even exact algorithms like Branch & Cut rely on these heuristics. The heuristic used in Concorde [256] to find near-optimum solutions to large TSP instances is essentially a memetic algorithm using the Lin-Kernighan (LK) heuristic as local search and tour-merging [257] for recombination [258].

Other algorithms have been proposed for the TSP such as Simulated Annealing, Tabu Search, Ant Colony Optimization [259], or Evolutionary Algorithms [260, 261].

As the TSP is NP-hard, computation time is expected to grow exponentially with the instance size. E. g. for a TSP instance with 24 978 cities, an exact algorithm required 85 CPU years to prove a known tour's optimality [262]. Using a comparable environment, sophisticated heuristic algorithms such as Helsgaun's LK (LK-H) [263] are expected to find solutions with 1 % excess within a few hours at most.

The problem of time consumption can be approached by distributing the computation among a set of computers using distributed evolutionary algorithms (DEA) [264, 265]. Another problem when solving extremely large TSP instances such as the World TSP [266] is the memory consumption, as data structures such as neighbour or candidate lists have to be maintained. We address this problem in this paper by proposing different *edge fixing heuristics*, which may reduce the problem to a size suitable for standard TSP solvers. In the general fixing scheme, some heuristic selects edges of an existing tour for fixing; paths of fixed edges are merged into a single fixed edge reducing the instance size. Fixed edges are 'tabu' for the TSP solver, which is applied to the reduced instance in a second step. Finally, the optimized tour is expanded back to a valid solution for the original problem by releasing fixed edges and paths.

In the remainder of this section related work is presented. In Sec. 19.2 problem reduction techniques based on fixing edges are discussed. A set of TSP instances is analyzed in Sec. 19.3 regarding the discussed fixing heuristics. In Sec. 19.4 the results when applying the fixing heuristics to an evolutionary local search are discussed. Finally, our findings are summarized in Sec. 19.5.

19.1.1 Related Work

Only limited research regarding the reduction of TSP instances in combination with evolutionary local search has been done. The primary related work to our concept is the *multilevel approach* by Walshaw [267], which has been applied to several graph problems including the TSP [268]. Basically, multilevel algorithms work as follows: In the first phase a given graph is recursively coarsened by matching and merging node pairs generating smaller graphs at each level. The coarsening stops with a minimum size graph, for which an optimal solution can easily be found. In the second phase, the recursion backtracks, uncoarsening each intermediate graph and finally resulting in a valid solution of the original problem. In each uncoarsening step, the current solution is refined by some optimization algorithm. It has been reported that this strategy results in better solutions compared to applying the optimization algorithm to the original graph only. When uncoarsening again, the optimization algorithm can improve the current level's solution based on an already good solution found in the previous level. As the coarsening step defines the solution space of a recursion level, its strategy is crucial for the quality of the multilevel algorithm.

In [268] the multilevel approach has been applied to the TSP using a CLK algorithm [269] for optimization. Here, a multilevel variant (called $\mathrm{MLC}^{N/10}\mathrm{LK}$) of CLK gains better results than the unmodified CLK, being nearly 4 times faster. The coarsening heuristic applied to the TSP's graph matches node pairs by adding a fixed edge connecting both nodes In each step, nodes are selected and matched with their nearest neighbour, if feasible. Nodes involved in an (unsuccessful) matching may not be used in another matching at the same recursion level to prevent the generation of sub-tours. Recursion stops when only two nodes and one connecting edge are left. The difference to our approach will be evident after the categorization in the following section.

19.2 Problem Reduction by Fixing Edges

To reduce a TSP instance's size different approaches can be taken. Approaches can be either node-based or edge-based. Both can be divided further whether they operate on a TSP instance or use an existing solution, respectively.

A *node-based approach* may work as follows: Subsets of nodes can be merged into meta-nodes (cluster) thus generating a smaller TSP instance. Within a meta-node a cost-effective path connecting all nodes has to be found. The path's end nodes will be connected to the edges connecting the meta-node to its neighbour nodes building a tour through all meta-nodes. Problems for this approach

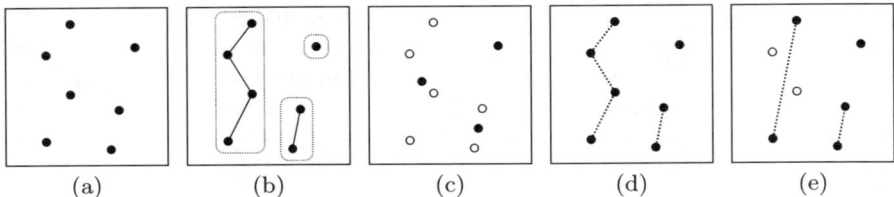

Fig. 19.1. Examples of node-based and edge-based problem reductions. Starting from the original problem instance (a), the node-based approach assigns node sets to clusters (marked by dashed boxes) and defines a spanning path within each cluster (b). Subsequently, in (c) only representatives of the clusters have to be considered (black nodes, here arbitrarily located at each cluster's center), whereas the original nodes (white) can be ignored. For the edge-based approach, edges to be fixed have to be selected (dotted lines in (d)). Subsequently, paths can be merged to single edges and inner path nodes (white) may be ignored (e).

are (i) how to group nodes into meta-nodes (ii) how to define distances between meta-nodes (iii) which two nodes of a cluster will have outbound edges. In an *edge-based approach*, a sequence of edges can be merged into a meta-edge, called a fixed path. Subsequently, the inner edges and nodes are no longer visible and this meta-edge has to occur in every valid tour for this instance. Compared to the node-based approach, problems (ii) and (iii) do not apply, as the original node distances are still valid and a fixed path has exactly two nodes with outbound edges. So, the central problem is how to select edges merged into a meta-edge. Examples for both node-based and edge-based problem reductions are shown in Fig. 19.1.

Edges selected for merging into meta-edges may be chosen based on instance information only or on a candidate tour's structure. The former approach may select from an instance with n nodes any of the $n(n-1)/2$ edges for a merging step, the latter approach reuses only edges from a given tour (n edges). The *tour-based approach*'s advantage is a smaller search space and the reuse of an existing tour's inherent knowledge. Additionally, this approach can easily be integrated into memetic algorithms. A disadvantage is that the restriction to tour edges will limit the fixing effect especially in early stages of a local search when the tour quality is not sufficient.

Walshaw's multilevel TSP approach focuses on an edge-based approach considering the TSP instance only. In this paper, we will discuss edge-based approaches, too, but focus on the following tour-based edge fixing heuristics:

Minimum Spanning Tree (MST): Tour edges get fixed when they occur in a *minimum spanning tree* (MST) for the tour's instance. This can be motivated by the affinity between the TSP and the MST problem [270], as the latter can be used to establish a lower bound for the TSP. However, global instance knowledge in form of an MST (complexity of $\mathcal{O}(m + n \log n)$ for m edges using Fibonacci heaps) has to be available in advance.

Alternate Minimum Spanning Tree (altMST): Here, a second minimum spanning tree is built where edges from the previous MST are tabu. This edge selection uses edges from both spanning trees as it is guaranteed that each node in the combined spanning tree graph has at least two adjacent edges like a TSP tour.

Nearest Neighbour (NN): As already exploited by the *nearest neighbour* tour construction heuristic, edges between a node and it's nearest neighbour are likely to occur in optimal tours thus being promising fixing candidates, too. Determining nearest neighbour lists may be computationally expensive (complexity of $\mathcal{O}(n^2 \log n)$), but can be sped up e.g. by using kd-trees [271, 272].

Second Nearest Neighbour (NN2): This edge selection heuristic considers both edges to the nearest and the *second nearest neighbours*. Motivation for this selection is that a lower bound for TSP tours uses the distances of each node to its two nearest neighbours.

Lighter than Median (<M): Edges whose length is *less* than the median over all edges' lengths in a tour are selected, as it is beneficial to keep short edges by fixing them and leaving longer edges for further optimization. The most expensive operation of this approach is the necessary sorting of all tour edges (complexity of $\mathcal{O}(n \log n)$). There may be tours that have very few different edge lengths resulting in a small number of edges that are strictly shorter than the median.

Lighter than or equal to Median ($\leq$M): Similar to the previous strategy (<M), but edges whose length is *equal* to the median are also fixed.

Close Pairs (CP): Here, a tour edge's length is compared to the lengths of the two neighbouring edges. The edge will be fixed if and only if it is shorter than both neighbouring edges and the edge's nodes therefore form a *close pair*. This approach considers only local knowledge (edge and its two neighbour edges) allowing it to be applied even on large instances. It is expected to work well in graphs with both sparse and dense regions.

Relaxed Close Pairs (CPR): Like the Close Pairs selection approach, an edge's length is compared to the edge's neighbour edges. Here, however, the edge under consideration will be fixed if it is shorter than *any* of both neighbouring edges.

The selection heuristics described above can be grouped into four pairs, where one heuristic of each pair represents a 'pure' approach (MST, NN, <M, CP), whereas the other heuristic represents a 'relaxed' approach (Alternate MST, Second NN, $\leq$M, Relaxed CP), which will select more edges in addition to the selected edges from the pure approach. This allows the user to select between two strength levels for each concept of edge selection.

For the Lighter than Median variant (<M) and the Close Pairs variant (CP), at most half of all edges may be selected. This statement, however, does not hold for their respective counterparts $\leq$M and Relaxed CP.

In the example in Fig. 19.2, solid lines represent tour edges and dotted lines represent edges of the minimal spanning tree (MST). Out of 7 tour edges, 5 edges

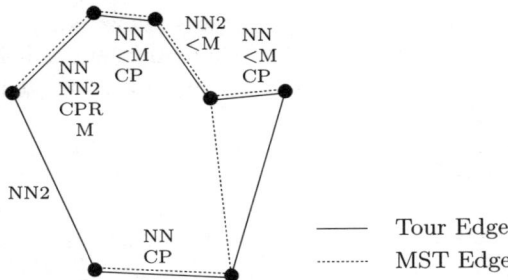

Fig. 19.2. Example tour and minimum spanning tree with different edge properties highlighted. Tour and MST edges are drawn in different line styles, properties *Nearest Neighbour* (NN), *Second Nearest Neighbour* (NN2), *Lighter than Median* ($<$M), *Lighter than or equal to Median* ($\leq$M), *Close Pairs* (CP), and *Relaxed Close Pairs* (CPR) are shown as markings next to the edges.

are MST edges, too, 4 edges connect nearest neighbours (NN), 3 edges connect close node pairs (CP) and 3 edges are lighter than the edge weight median ($<$M).

For asymmetric TSP instances (ATSP) edge fixation heuristics can be developed, too, but this approach has not been pursued here. Applying the fixation heuristics presented here to ATSP instances poses new questions such as determining the direction when the fixation heuristic is based on undirected knowledge (e. g. from an MST).

19.3   Analysis of TSP Instances

The tests in this section were performed to evaluate if the edge selection strategies discussed in this paper are applicable for edge-fixation heuristics. Probabilities of each selection strategy to select a tour edge and the probabilities if the selected edge is part of an optimal tour were evaluated. These criteria describe the quantity and quality, respectively, of a selection strategy.

For our analysis, six TSP instances have been selected: From the TSPLIB collection [273] instances ja9847, brd14051, d15112, and d18512 and from a collection of national TSPs [274] instances it16862, and sw24978 were taken (numbers in the instance names denote the problem sizes). These instances were selected, because they are among the largest instances with known optimal solutions. The choice of instances was limited by the availability of optimal or at least high quality tours which were used to evaluate the tours found in our experiments.

For each TSP instance 32 nearest-neighbour tours were constructed. Each of these tours was optimized to a local optimum by the local search heuristics 2-opt, 3-opt, Lin-Kernighan (LK-opt), and LK-Helsgaun (LK-H), respectively. For Helsgaun's LK parameter MAX_TRIALS was set to 100 (instead of number

Table 19.1. Probabilities (in percent) for edges in a tour to match certain criteria. The data is grouped by TSP instance and tour type. The eight right most columns contain the original and the conditional probabilities for the edge fixation heuristics MST, altMST, NN, and NN2.

	Tour Type	P(OPT)	P(MST)	P(OPT|MST)	P(altMST)	P(OPT|altMST)	P(NN)	P(OPT|NN)	P(NN2)	P(OPT|NN2)
ja9847	NN	64.90	75.70	75.93	88.26	71.72	61.91	77.51	83.33	72.34
	2-opt	70.06	76.68	78.98	90.28	75.21	62.55	80.23	84.14	76.21
	3-opt	74.81	78.39	81.49	92.20	78.38	64.34	82.30	85.42	79.67
	LK-opt	79.73	78.41	84.79	93.29	81.96	63.93	85.40	85.46	83.64
	LK-H	93.28	76.98	94.74	92.88	93.86	61.85	94.70	84.41	94.43
	optimal	100.00	75.92	100.00	92.47	100.00	60.30	100.00	83.72	100.00
brd14051	NN	59.34	70.44	73.38	85.64	67.60	58.38	74.75	83.46	66.58
	2-opt	65.34	73.71	75.95	89.82	70.81	60.34	77.40	84.70	71.58
	3-opt	70.79	76.41	78.72	92.51	74.42	62.75	79.81	86.39	75.78
	LK-opt	77.22	77.49	82.69	94.42	79.20	62.88	83.56	86.92	81.16
	LK-H	92.56	76.75	93.98	94.53	93.02	61.07	94.08	86.53	93.76
	optimal	100.00	75.50	100.00	93.94	100.00	59.28	100.00	85.65	100.00
d15112	NN	61.38	71.35	74.31	86.95	69.01	59.09	76.08	83.48	68.83
	2-opt	66.05	73.63	76.12	90.19	71.28	60.47	77.78	84.56	72.39
	3-opt	71.03	76.06	78.60	92.52	74.55	63.04	79.86	86.33	76.06
	LK-opt	76.88	77.13	82.17	94.27	78.85	63.17	83.31	86.83	80.96
	LK-H	92.46	76.25	93.78	94.37	92.87	61.33	94.07	86.40	93.81
	optimal	100.00	74.85	100.00	93.85	100.00	59.69	100 00	85.63	100.00
it16862	NN	61.44	72.35	74.39	86.89	68.87	60.12	76.05	82.87	69.15
	2-opt	66.86	74.78	76.73	89.88	72.06	61.96	78.14	84.23	73.19
	3-opt	71.89	76.95	79.29	92.06	75.49	64.23	80 38	85.80	76.96
	LK-opt	77.27	77.63	82.57	93.60	79.36	64.22	83 58	86.13	81.37
	LK-H	91.23	76.63	92.83	93.50	91.77	62.63	92.98	85.51	92.66
	optimal	100.00	74.98	100.00	92.79	100.00	60.58	100.00	84.43	100.00
d18512	NN	60.91	71.81	74.01	87.10	68.41	59.31	75.72	83.92	68.22
	2-opt	65.83	74.24	76.02	90.43	70.95	60.67	77.62	84.76	72.14
	3-opt	70.79	76.76	78.48	92.79	74.28	62.96	79.78	86.46	75.86
	LK-opt	77.17	77.68	82.53	94.59	79.10	62.91	83.58	86.83	81.22
	LK-H	92.68	76.83	94.06	94.67	93.16	61.10	94.19	86.41	93.99
	optimal	100.00	75.48	100.00	94.16	100.00	59.32	100.00	85.60	100.00
sw24978	NN	65.46	76.23	75.26	88.55	72.12	65.59	75.79	85.27	71.94
	2-opt	68.22	75.62	76.95	90.23	73.18	64.25	77.71	85.26	73.86
	3-opt	72.29	77.01	79.11	92.06	75.83	65.54	79.65	86.31	76.88
	LK-opt	76.72	77.04	82.03	93.33	78.92	64.53	82.67	86.14	80.72
	LK-H	90.32	76.16	92.11	93.28	90.94	62.93	92.14	85.52	91.86
	optimal	100.00	74.58	100.00	92.57	100.00	60.89	100.00	84.49	100.00

Table 19.2. Probabilities (in percent) for edges in a tour to match certain criteria. The data is grouped by TSP instance and tour type. The eight right most columns contain the original and the conditional probabilities for the edge fixation heuristics <M, ≤M, CP, and CPR.

	Tour Type	$P(OPT)$	$P(<M)$	$P(OPT|<M)$	$P(\leq M)$	$P(OPT|\leq M)$	$P(CP)$	$P(OPT|CP)$	$P(CPR)$	$P(OPT|CPR)$
ja9847	NN	64.90	46.02	74.20	56.95	74.13	34.49	75.72	71.99	71.28
	2-opt	70.06	45.42	78.05	56.80	77.63	32.46	78.90	71.04	75.55
	3-opt	74.81	45.94	80.89	57.68	80.46	31.28	81.76	69.93	79.41
	LK-opt	79.73	45.68	84.43	57.35	84.02	29.90	84.94	68.44	82.95
	LK-H	93.28	44.45	93.81	56.02	93.84	27.00	94.81	64.68	94.17
	optimal	100.00	43.75	100.00	55.22	100.00	25.93	100.00	63.11	100.00
brd14051	NN	59.34	54.85	71.31	57.09	70.98	40.01	71.94	75.21	67.11
	2-opt	65.34	53.38	74.98	55.60	74.68	38.40	75.70	74.60	72.16
	3-opt	70.79	52.35	78.10	54.68	77.90	37.22	78.79	73.69	76.41
	LK-opt	77.22	52.15	82.51	54.60	82.27	35.18	83.10	71.91	81.02
	LK-H	92.56	48.57	93.73	51.21	93.72	31.73	93.93	68.21	93.38
	optimal	100.00	47.83	100.00	50.45	100.00	30.34	100.00	66.46	100.00
d15112	NN	61.38	54.65	72.14	55.42	72.02	39.61	73.51	75.62	68.53
	2-opt	66.05	54.13	74.67	54.87	74.55	38.55	76.14	75.33	72.41
	3-opt	71.03	53.60	77.72	54.44	77.64	37.68	79.05	74.70	76.33
	LK-opt	76.88	52.43	81.73	53.27	81.69	35.92	82.72	73.24	80.58
	LK-H	92.46	50.63	93.27	51.59	93.28	32.56	93.84	69.62	93.26
	optimal	100.00	49.79	100.00	50.75	100.00	31.25	100.00	67.79	100.00
it16862	NN	61.44	52.58	73.39	64.02	70.60	32.78	73.52	70.53	68.86
	2-opt	66.86	53.85	76.32	64.34	74.45	30.64	76.44	69.26	73.23
	3-opt	71.89	54.57	79.23	54.86	79.14	29.09	79.29	67.75	77.23
	LK-opt	77.27	53.45	82.71	53.90	82.62	27.38	82.88	65.56	81.12
	LK-H	91.23	51.22	92.48	51.80	92.46	24.15	92.92	61.49	92.24
	optimal	100.00	49.92	100.00	50.49	100.00	22.88	100.00	59.23	100.00
d18512	NN	60.91	55.37	72.60	58.62	72.38	39.19	73.23	74.74	68.50
	2-opt	65.83	53.91	75.31	55.75	75.12	38.07	76.09	74.34	72.52
	3-opt	70.79	53.20	78.22	55.07	78.04	37.13	78.90	73.50	76.45
	LK-opt	77.17	53.01	82.36	54.94	82.22	35.26	82.94	71.68	81.05
	LK-H	92.68	49.99	93.77	51.76	93.75	31.67	94.08	67.81	93.52
	optimal	100.00	49.06	100.00	50.83	100.00	30.34	100.00	66.10	100.00
sw24978	NN	65.46	42.58	74.52	58.33	74.34	24.92	74.93	66.03	71.12
	2-opt	68.22	41.17	76.71	58.38	75.34	25.05	76.98	66.19	73.79
	3-opt	72.29	41.80	78.86	59.56	77.58	24.31	79.16	65.15	77.07
	LK-opt	76.72	41.19	81.89	59.48	80.55	23.34	82.04	63.14	80.30
	LK-H	90.32	40.30	91.26	58.55	90.77	20.56	92.43	59.19	91.47
	optimal	100.00	39.15	100.00	57.46	100.00	19.36	100.00	57.07	100.00

of cities). Totally, 960 tours were constructed to test the fixing heuristic when applied to tours of different quality levels.

Each heuristic's selection scheme was applied to the set of 960 tours. Average values over each set of 32 tours with the same setup were taken and summarized in Tables 19.1 and 19.2. Here, the first column sets the instance under consideration. The 'Type' column determines in which local optimum the tours are located. Column 'P(OPT)' shows for a local optimal tour the percentage of edges that also occur in the known global optimal tour. Columns 'P(MST)', 'P(altMST)', 'P(NN)', 'P(NN2)', 'P(<M)', 'P($\leq$M)', 'P(CP)', and 'P(CPR)' contain the probability that an edge from a local optimal tour matches the given property. Columns 'P(OPT|MST)', 'P(OPT|altMST)', 'P(OPT|NN)', 'P(OPT|NN2)', 'P(OPT|<M)', 'P(OPT|$\leq$M)', 'P(OPT|CP)', and 'P(OPT|CPR)' contain conditional probabilities, that an edge in a local optimal tour is part of the global optimal tour, given that it matches the properties MST, altMST, NN, NN2, <M, $\leq$M, CP, or CPR, respectively.

19.3.1 Results of the Analysis

Column 'P(OPT)' shows the percentage of edges from a local optimal tour occurring in the global optimal tour. The better a tour construction and improvement heuristic works, the more edges the resulting tour has in common with the optimal tour. Whereas tours constructed by a nearest neighbour construction heuristic share about 60–65 % of all edges with optimal tours (depending on the instance), tours found by subsequently applied local search algorithms are better, ranging from 65–70 % for 2-opt to more than 90 % for LK-H-opt tours.

As each edge selection strategy has different criteria how to select edges, they differ in the number of optimal edges. The most edges get chosen by the altMST strategy (column 'P(altMST)' in Table 19.1) selecting about 85–95 % of all tour edges. Other strategies select less edges: From NN2 with 80–90 % and MST with 70–80 % via CPR with 60–75 % and NN with 60–65 % down to $\leq$M, <M, and CP with 50–65 %, 40–55 %, and 20–40 % of all edges, respectively. The number of selected edges is highest in most cases for strategies MST, altMST, NN, and NN2 when applied to 3-opt or LK-opt tours regardless of the TSP instance, whereas it is highest for strategies $\leq$M, <M, CP, and CPR when applied to nearest neighbour and 2-opt tours.

However, the quantity (number of edges selected for fixing) is not the only criterion to rate a selection strategy. Selected edges were checked whether they occur in the corresponding known optimal tour, too. When applying a fixing heuristic to a sub-optimal tour, a good heuristic should more likely select edges that occur in the optimal tour, too, rather than edges that do not occur in the optimal tour. Therefore we were interested in the probability that a sub-optimal tour's edge selected by an edge selection strategy would actually be contained in an optimal tour ('true positive') rather than being a 'false positive'.

The quality of an edge selection strategy can be estimated by comparing its conditional probability (e. g. column 'P(OPT|MST)' for the MST heuristic) with the probability of an edge to part of an optimal tour (column 'P(OPT)'). A selection strategy that randomly selects edges would have a conditional probability equal to 'P(OPT)', whereas a purposive strategy is expected to have higher probability values. This property holds for each of our strategies. Furthermore, it can be observed that strategies with higher conditional probabilities select less edges for fixation compared to selection strategies with lower conditional probabilities. E.g. for 2-opt tours of the instance d15112, the probability of an edge to be part of an optimal tour is 66.1%. For the edge selection strategy NN, 60.5% of all edges would be selected, of which 77.8% occur also in the optimal tour. By contrast, the edge selection strategy MST selects 90.2% of the edges, but only 71.3% of those are part of an optimal tour.

Generally, the value of 'P(OPT)' increases with tour quality. E.g. for instance sw24978, nearest neighbour tours share about 65% of all edges with the optimal tour. This proportion increases with better tours up to LK-H optimized tours which share 90% of all edges with the optimal tour.

19.4   Experimental Setup and Results

For the experimental evaluation, the fixing heuristics have been integrated into a simple TSP solver written in Java. The solver works as follows: Each tour was reduced using one of the fixing heuristics and subsequently improved by an iterated local search (ILS) algorithm. In each iteration of the algorithm the current tour was perturbed and locally optimized by an LK implementation, enriched to handle fixed edges. For the perturbation a variable-strength *double-bridge move* (DBM) was used increasing the number of DBMs each two non-improving iterations. At the end of each iteration the new tour was compared to the previous tour and discarded if no improvement was found, otherwise kept for subsequent iterations. The iterations would stop after 2 non-improving iterations. Finally, the improved tour was expanded back to a solution for the original TSP instance. This reducing and expansion procedure was repeated 10 times on each solution. For comparison, all tours were optimized by the iterated local search algorithm without any reduction, too. This solver was not designed to compete with state-of-the-art solvers, but merely to evaluate our fixation heuristics. Each parameter setup was tested by applying it to the tours described in Sec. 19.3; average values were used for the following discussion. Computation times are utilized CPU time on a 3.0 GHz Pentium 4 system with 512 MB memory running Linux.

Tables 19.3 and 19.4 hold the results for the different setups applied to the ILS and are structured as follows: Rows are grouped by instance (ja9847 to sw24978) and by starting tour for the ILS ('NN' to 'LK-H'). The instances are ordered by number of cities, the starting tours are ordered by descending tour length. Columns are grouped into blocks, where the first block summarizes an ILS setup without any fixation and the subsequent blocks summarize ILS setups with each fixation heuristics ('MST' to 'Second Nearest Neighbour' and 'Light Median <' to 'Relaxed Close Pairs'). Each column block consists of four columns: Improvement found when applying the ILS, required CPU time until

Table 19.3. Results for different setups of instance, start tour, and fixing heuristics. The data is grouped by TSP instance and tour type. The result columns are grouped into five blocks for the fixing heuristics 'No Fixing', 'MST', 'Alternate MST', 'Nearest Neighbour', and 'Second Nearest Neighbour' consisting of four columns each: Achieved improvement from starting tour, utilized CPU time, fraction of free edges and size of the reduced instance (based on number of nodes).

Instance	Start Tour Type	No Fixing Impr. [%]	No Fixing CPU Time [s]	No Fixing Free Edges [%]	No Fixing Red. Size [%]	MST Impr. [%]	MST CPU Time [s]	MST Free Edges [%]	MST Red. Size [%]	Alternate MST Impr. [%]	Alternate MST CPU Time [s]	Alternate MST Free Edges [%]	Alternate MST Red. Size [%]	Nearest Neighbour Impr. [%]	Nearest Neighbour CPU Time [s]	Nearest Neighbour Free Edges [%]	Nearest Neighbour Red. Size [%]	Second Nearest Neighbour Impr. [%]	Second Nearest Neighbour CPU Time [s]	Second Nearest Neighbour Free Edges [%]	Second Nearest Neighbour Red. Size [%]
ja9847	NN	22.7	164.2	—	—	20.8	19.3	55.9	40.9	17.2	8.1	55.4	22.6	22.0	40.6	54.4	69.8	19.9	13.2	53.9	32.2
	2-opt	13.6	171.3	—	—	11.9	18.7	55.3	41.9	9.2	7.0	53.5	21.1	13.1	41.2	54.4	70.3	11.0	13.1	53.1	31.6
	3-opt	6.1	162.6	—	—	4.7	16.4	54.7	43.0	3.1	4.9	52.4	19.4	5.6	38.0	54.3	70.9	4.0	10.4	52.3	31.3
	LK-opt	1.0	139.1	—	—	1.1	13.2	54.7	45.3	0.9	3.0	52.1	16.8	0.9	32.4	55.1	73.0	1.0	7.6	52.2	30.9
	LK-H	0.0	77.0	—	—	0.0	7.7	55.1	44.5	0.0	1.6	51.7	15.0	0.0	19.5	54.8	72.4	0.0	4.2	51.7	28.6
brd14051	NN	23.2	128.5	—	—	20.9	21.4	56.6	46.5	16.0	7.9	55.4	24.2	22.4	45.3	55.3	72.5	19.2	13.8	54.5	34.4
	2-opt	12.5	135.6	—	—	9.9	21.5	55.6	46.6	6.2	6.0	53.0	22.2	11.6	46.9	54.9	72.5	8.5	12.3	53.2	33.7
	3-opt	6.0	139.8	—	—	3.9	19.1	55.1	47.3	2.0	4.0	51.9	19.6	5.1	46.2	54.8	72.6	2.9	9.9	52.4	33.3
	LK-opt	0.4	122.1	—	—	0.1	12.7	55.1	48.9	0.0	1.5	51.5	14.5	0.2	33.6	55.8	75.1	0.0	4.9	51.9	30.7
	LK-H	0.0	56.7	—	—	0.0	7.7	54.3	45.4	0.0	1.0	50.9	11.8	0.0	20.4	55.1	73.8	0.0	2.9	51.5	25.6
d15112	NN	21.7	163.7	—	—	19.0	22.9	57.0	46.4	14.3	7.8	55.6	23.9	20.7	49.1	55.5	71.4	17.2	12.8	54.5	33.4
	2-opt	11.7	171.1	—	—	9.0	23.4	56.0	47.2	5.4	5.8	53.1	22.0	10.6	51.7	55.2	71.8	7.5	12.1	53.2	33.2
	3-opt	5.8	179.0	—	—	3.5	21.2	55.3	48.1	1.8	3.7	52.2	19.7	4.8	51.6	51.6	72.0	2.6	9.5	52.5	32.6
	LK-opt	0.5	160.9	—	—	0.0	13.6	55.2	49.5	0.0	1.4	51.7	15.0	0.1	37.6	56.0	74.6	0.0	5.1	52.1	30.2
	LK-H	0.0	66.9	—	—	0.0	8.4	54.3	46.3	0.0	1.0	51.3	12.1	0.0	22.6	55.2	73.3	0.0	2.8	51.5	25.3
fnl16862	NN	22.3	241.1	—	—	19.7	35.0	56.2	45.1	15.3	13.1	54.9	24.2	21.4	70.2	55.1	70.0	18.1	21.7	53.7	33.7
	2-opt	12.4	249.3	—	—	9.8	33.7	55.2	45.7	6.4	10.6	53.0	22.5	11.5	73.0	54.9	70.3	8.5	20.1	52.8	33.3
	3-opt	5.9	250.4	—	—	3.8	30.7	54.9	46.4	2.0	7.1	52.1	20.5	5.1	71.8	54.7	70.4	2.9	16.0	52.2	32.9
	LK-opt	0.7	223.7	—	—	0.2	21.8	55.0	48.5	0.1	3.4	51.6	16.7	0.4	57.8	55.4	72.9	0.1	5.9	51.9	31.6
	LK-H	0.0	96.6	—	—	0.0	11.7	54.4	46.0	0.0	1.8	51.3	14.2	0.0	30.9	55.0	71.8	0.0	5.0	51.5	27.7
d18512	NN	21.0	191.8	—	—	19.0	31.1	56.6	46.0	14.2	9.9	55.6	23.5	20.7	67.6	55.2	72.4	17.4	18.5	54.5	33.7
	2-opt	11.4	202.0	—	—	8.8	30.8	55.7	46.8	5.1	7.4	53.0	21.9	10.5	72.0	55.0	72.7	7.4	17.5	53.2	33.6
	3-opt	5.8	212.0	—	—	3.6	27.8	55.2	47.7	1.8	4.8	52.0	19.3	4.9	74.5	54.8	73.0	2.7	13.6	52.4	33.2
	LK-opt	5.4	187.9	—	—	0.0	18.8	55.0	49.0	0.0	1.8	51.6	14.4	0.1	51.5	55.8	75.4	0.0	7.5	52.0	30.7
	LK-H	0.0	85.4	—	—	0.0	11.3	54.1	45.3	0.0	1.2	51.1	11.4	0.0	31.3	55.1	73.8	0.0	4.0	51.5	25.5
sw24978	NN	19.9	540.6	—	—	10.8	58.5	56.7	42.8	13.3	20.1	55.8	22.8	18.7	132.7	54.9	65.8	15.5	33.8	54.5	31.5
	2-opt	11.6	588.5	—	—	9.0	65.5	55.9	43.0	6.1	16.6	53.4	21.9	10.5	147.8	54.9	67.5	7.9	35.5	53.2	32.4
	3-opt	5.8	590.7	—	—	3.8	60.2	55.3	46.6	2.1	11.7	52.2	20.5	4.9	143.5	54.9	68.4	3.0	28.2	52.4	32.4
	LK-opt	0.5	474.0	—	—	0.1	39.3	55.3	49.2	0.0	5.0	51.8	17.4	0.3	103.7	55.9	71.7	0.1	16.2	52.1	31.7
	LK-H	0.0	196.3	—	—	0.0	22.2	54.6	47.0	0.0	2.9	51.4	14.7	0.0	54.5	55.3	70.8	0.0	8.3	51.5	28.0

Table 19.4. Results for different setups of instance, start tour, and fixing heuristics. The data is grouped by TSP instance and tour type. The result columns are grouped into five blocks for the fixing heuristics 'No Fixing', 'Light Median <', 'Light Median ≤', 'Close Pairs', and 'Relaxed Close Pairs', consisting of four columns each: Achieved improvement from starting tour, utilized CPU time, fraction of free edges and size of the reduced instance (based on number of nodes).

Instance	Start Tour Type	No Fixing				Light Median <				Light Median ≤				Close Pairs				Relaxed Close Pairs			
		Impr. [%]	CPU Time [s]	Free Edges [%]	Red. Size [%]	Impr. [%]	CPU Time [s]	Free Edges [%]	Red. Size [%]	Impr. [%]	CPU Time [s]	Free Edges [%]	Red. Size [%]	Impr. [%]	CPU Time [s]	Free Edges [%]	Red. Size [%]	Impr. [%]	CPU Time [s]	Free Edges [%]	Red. Size [%]
ja9847	NN	22.7	164.2	—	—	22.2	59.7	76.5	73.3	21.8	42.9	71.6	63.2	22.3	110.5	73.2	100.0	20.8	42.6	51.6	68.0
	2-opt	13.6	171.3	—	—	13.2	55.1	76.4	73.4	13.1	44.1	71.5	62.9	13.5	111.6	73.3	100.0	11.9	42.6	51.7	68.4
	3-opt	6.1	162.6	—	—	5.9	55.9	76.4	73.4	5.7	40.6	71.4	62.7	6.2	107.8	73.4	100.0	5.0	36.8	51.9	69.1
	LK-opt	1.0	139.1	—	—	1.2	48.7	76.5	73.7	1.5	33.1	71.6	63.0	1.1	94.3	73.6	100.0	0.7	25.5	52.3	69.7
	LK-H	0.0	77.0	—	—	0.0	27.4	76.5	73.5	0.0	19.4	71.5	62.7	0.0	53.0	74.1	100.0	0.0	17.9	52.6	70.0
brd14051	NN	23.2	128.5	—	—	21.7	50.1	73.7	68.7	21.6	47.6	72.7	66.6	23.1	105.3	68.9	100.0	20.9	48.8	50.4	64.4
	2-opt	12.5	135.6	—	—	11.2	49.9	73.8	68.4	11.0	47.4	72.8	66.0	12.4	105.5	68.9	100.0	10.2	46.6	50.4	64.9
	3-opt	6.0	139.8	—	—	5.1	49.2	74.2	68.6	5.0	43.8	72.9	65.8	5.9	109.4	68.9	100.0	4.2	38.9	50.5	65.6
	LK-opt	0.4	122.1	—	—	0.3	36.8	74.3	67.9	0.2	33.7	73.3	65.6	0.4	82.6	69.2	100.0	0.1	21.3	50.6	66.4
	LK-H	0.0	56.7	—	—	0.0	23.0	75.7	69.1	0.0	21.5	74.5	66.7	0.0	43.8	69.6	100.0	0.0	14.8	50.6	66.2
d15112	NN	21.7	163.7	—	—	20.1	52.3	75.4	66.7	20.0	50.5	75.2	65.8	21.5	118.3	68.3	100.0	19.0	50.4	50.1	62.0
	2-opt	11.7	171.1	—	—	10.3	53.4	75.4	66.8	10.2	51.6	75.1	65.8	11.5	128.6	68.3	100.0	9.0	47.9	50.1	62.6
	3-opt	5.8	179.0	—	—	4.7	51.5	75.5	66.8	4.7	49.0	75.2	65.9	5.5	121.4	68.3	100.0	3.7	38.2	50.2	63.4
	LK-opt	0.5	160.9	—	—	0.2	37.6	75.8	66.4	0.2	35.7	75.5	65.5	0.3	97.3	68.5	100.0	0.0	21.0	50.2	64.1
	LK-H	0.0	66.9	—	—	0.0	22.2	76.3	65.9	0.0	21.5	76.0	64.9	0.0	49.3	68.9	100.0	0.0	15.4	50.2	63.8
fl14662	NN	22.3	241.1	—	—	21.3	89.4	71.5	70.7	19.5	55.0	67.7	55.6	22.3	190.9	75.9	100.0	20.4	93.5	52.6	72.1
	2-opt	12.4	249.3	—	—	11.7	89.1	71.5	70.5	11.6	86.4	71.0	68.8	12.4	203.8	76.1	100.0	10.6	90.4	52.8	72.7
	3-opt	5.9	250.4	—	—	5.4	85.8	71.6	70.7	5.3	85.6	71.4	69.7	5.9	198.2	76.2	100.0	4.6	80.8	53.1	73.6
	LK-opt	0.7	223.7	—	—	0.6	69.0	71.5	70.2	0.5	67.6	71.5	69.6	0.7	176.9	76.4	100.0	0.2	50.1	53.7	74.8
	LK-H	0.0	96.6	—	—	0.0	35.2	71.6	70.1	0.0	34.6	71.5	69.4	0.0	77.4	77.2	100.0	0.0	31.2	54.1	75.2
d18512	NN	21.6	191.8	—	—	20.2	77.8	73.0	70.1	20.0	71.7	71.7	67.4	21.5	162.1	68.9	100.0	19.3	75.4	50.5	64.9
	2-opt	11.4	202.0	—	—	10.2	78.5	73.0	70.1	10.1	73.7	72.3	68.3	11.3	168.8	68.9	100.0	9.1	73.9	50.5	65.4
	3-opt	5.8	212.0	—	—	4.9	73.1	73.3	69.9	4.8	72.4	72.4	68.1	5.6	166.7	68.9	100.0	4.0	59.9	50.5	66.2
	LK-opt	0.4	187.9	—	—	0.2	56.5	73.3	69.8	0.2	53.5	72.5	67.9	0.3	133.0	69.2	100.0	0.0	32.7	50.7	67.0
	LK-H	0.0	85.4	—	—	0.0	32.4	73.5	69.3	0.0	30.7	72.8	67.5	0.0	67.3	69.7	100.0	0.0	23.2	50.8	66.7
sw24978	NN	19.9	540.6	—	—	19.2	219.7	77.4	77.3	18.5	128.9	71.4	61.7	19.9	462.8	79.8	100.0	18.2	213.0	54.1	74.5
	2-opt	11.6	588.5	—	—	11.1	244.4	77.4	78.5	10.3	135.4	71.1	61.4	11.7	472.2	79.7	100.0	9.9	213.3	53.7	74.6
	3-opt	5.8	590.7	—	—	5.4	234.0	77.0	78.5	4.8	126.3	71.0	61.2	5.9	481.3	79.8	100.0	4.6	176.3	54.5	75.5
	LK-opt	0.5	474.0	—	—	0.4	186.6	77.1	79.6	0.3	92.0	71.1	61.3	0.6	389.5	80.0	100.0	0.1	108.3	54.5	76.6
	LK-H	0.0	196.3	—	—	0.0	87.4	76.9	79.2	0.0	48.1	71.1	60.0	0.0	163.5	80.6	100.0	0.0	64.7	55.7	77.1

termination, size of the reduced instance (normalized to number of cities), and fraction of edges that are fixed (tabu for any ILS operation).

For every instance and each fixing heuristics (including no fixing) holds that the better the starting tour is, the smaller are the improvements found by the ILS. Applying our TSP solver to nearest neighbour tours (Tables 19.3 and 19.4, rows with start tour type 'NN') results improvements of about 15–20 % for most setups (columns 'Impr. [%]'). For better starting tours, less improvement is achieved, down to improvements of 0 % for starting tours coming from LK-H.

Each fixing ILS setup can be compared with the corresponding ILS setup without fixing regarding improvement on the given start tour and the required CPU time. The following observations can be drawn from Tables 19.3 and 19.4:

- For non-fixing setups, the CPU time is always higher compared to fixing setups, as the effective problem size is larger for the non-fixing setup. However, time consumption does not directly map to better tour quality.
- The Close Pairs (CP) fixing heuristic yields in improvements as good as for non-fixing ILS, but requires significantly less time to reach these quality levels. E. g. for instance **brd14051** starting with 3-opt tours, both the non-fixing ILS and the CP fixing ILS improve the start tour by about 6 %, but the CP variant requires only 109.4 s, whereas the non-fixing ILS requires 139.8 s.
- For the other fixing heuristics hold that they consume both less CPU time and result in less improvements compared to the Close Pairs heuristic. Although this makes comparing the different fixing strategies hard, improvements are still competitive while requiring significantly less CPU time compared to the non-fixing ILS.
- Among all fixation-based ILS setups, the Alternate MST heuristic results in both the smallest improvements and lowest running times compared to the other fixation heuristics. E. g. for instance **sw24978** starting with 2-opt tours, the Alternate MST heuristic results in an improvement of 6.1 % consuming 16.6 s, whereas the next closest heuristic (Second Nearest Neighbour) finds improvements of 7.9 % consuming 41.0 s.

Comparing CPU time consumption versus possible improvement, the fixation heuristics can be put into three groups as visualized Fig. 19.3, which compares the improvements for each fixing heuristics applied to 3-opt tours. Both time and improvement are normalized for each of the six TSP instances by the values from the runs without fixing.

1. The 'expensive', but good Close Pairs heuristic reaching improvements as good as the non-fixing ILS requiring only 3/4 of the time.
2. The 'medium' heuristics Relaxed Close Pairs, NN, $\leq$M and $<$M with normalized improvement above 75 % and normalized time of less than 50 %.
3. The 'cheap', but not so good heuristics (Alternate MST, Second NN, and MST) each having a normalized time of less than 25 % (quarter of the CPU time) and a normalized improvement of about 35 %, 50 %, and 65%, respectively. The outlier of each of the three heuristics are due to instance **ja9847**.

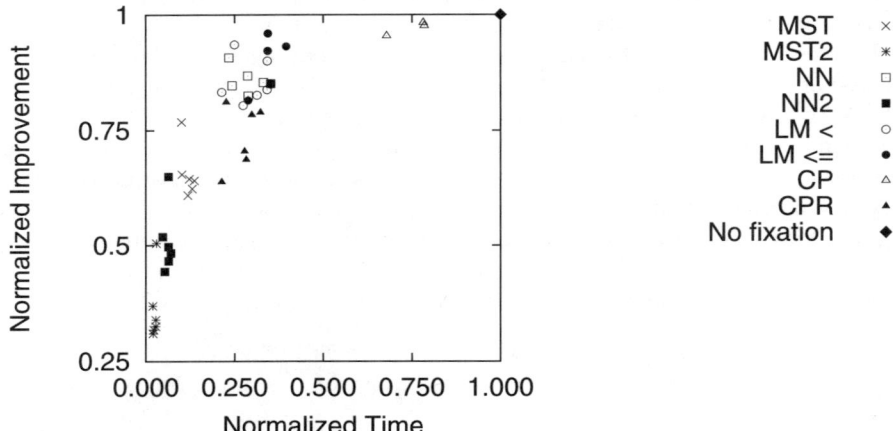

Fig. 19.3. For each of the six instances using 3-opt starting tours, improvements and CPU time of the ILS with fixation (MST, Alternate MST, NN, Second NN, <M, ≤M, CP, and Relaxed CP) are normalized with the results from ILS runs without fixation

Relaxed Close Pairs performs worst in the group of 'medium' heuristics bridging the gap to the 'cheap' heuristics.

For all fixing heuristics the size of the original instance has been compared to the corresponding reduced instances' size (in percent, columns 'Red. Size [%]' in Tables 19.3 and 19.4) and the number of free edges within the reduced instance (in percent, columns 'Free Edges [%]').

- In most cases, the number of free edges and the reduced instances' size is about the same for each combination of instance and fixing heuristic, regardless of the starting tour type.
- For Close Pairs (CP) fixations holds that the reduced instance's size equals always the original instance's size, as fixed edges can not have neighbouring edges that are fixed, too, as this would contradict the selection criterion. Thus, no nodes are redundant.
- For all combinations of instance and start tour, the Alternate MST fixing heuristic is the most progressive one resulting in the smallest instances. Here, fixed instances have on average less than the quarter the number of cities compared to the original instances. Within these reduced instances, more than half of the edges are free for the local search. E. g. for instance d15112 and 3-opt tours, only 19.7 % of the original nodes are left, whereas comparable heuristics leave e. g. 32.6 % (Second NN) or 48.1 % (MST).
- The Nearest Neighbour heuristic reduces most instance types for about the same level (to 70–75 % of the original size); instance sw24978 has a relative size of 65.8 % to 70.8 %. Similar observations can be made for the Second Nearest Neighbour heuristic which reduces that instance size to about

30–35 % except for setups starting from LK-H tours, which get reduced to less than 30 % of their original size.

- The Nearest Neighbour, Second Nearest Neighbour, MST and Alternate MST heuristics have very stable percentages of free edges. For the Nearest Neighbour and the MST heuristics, the values range only between 54.3 % to 56.0 % and 54.1 % to 57.0 %, respectively. For the Second Nearest Neighbour heuristic and Alternate MST, these values are only marginally smaller ranging between 51.5 % to 54, 5 % and 50.9 % to 55.8%, respectively.

- The Lighter than Median heuristic leaves about 70–75 % of all edges free for modification. The similar, but more greedy Lighter or equal to Median heuristic leaves, depending on instance, about the same percentage of edges available (instances d15112, it16862, and d18512) or a smaller fraction (instances ja9847, brd14051, sw24978).

- The Close Pairs fixing heuristic leaves between 68–81 % of all edges free (depending on instance), which results in the largest absolute number of free edges among all fixing heuristics considering that reduced instances have the same size as the original instance.

- The Relaxed Close Pairs fixing heuristic has the smallest proportion of free edges among all fixing heuristics (about 50–55 %). The reduced instances' size differs among instances from 62.0–64.1 % (instance d15112) to 74.5–77.1 % (instance sw24978).

As discussed, the proposed edge selection heuristics offer a range of choices for reducing the search space and thus affecting the expected solution quality. Selection heuristics which are very progressively (Alternate MST, Second NN, and MST) may considerably reduce the computation time to find a local optimum but this greediness will notably reduce the gain of the local search. Other heuristics that are more selective (e. g. Close Pairs) fix less edges and thus allow the local search to exploit a larger search space which still requires less time compared to a local search in the original instance. Accepting a minor decrease but achieving a considerable saving in time is interesting when using the problem reduction algorithm as part of a metaheuristic such as a memetic algorithm or an ant colony optimization algorithm.

19.5 Conclusion

In order to extend current memetic algorithms for the TSP to find close to optimum solutions for large TSP instances, we studied several edge-based problem reduction techniques that can easily be incorporated into an evolutionary local search framework. We have shown that fixing edges in TSP tours can considerably reduce the computation time of a TSP solver compared to applying the same solver to the unmodified problem instance. Still, the solutions found when using fixing heuristics are nearly as good as the solutions found without fixing. Therefore, edge fixing is a feasible approach to tackle tours that are otherwise too large for solvers regarding memory or time consumption.

When selecting one of the proposed fixing heuristics, a trade-off between expected solution quality, computation time, or required preprocessing steps has to be made. E. g. the Alternate MST heuristic is expected to consume the least time, but requires building two MSTs in advance and has the worst results among all fixing heuristics. Better results can be expected e. g. from the MST heuristic, which requires building only one MST. The Close Pairs strategy can be used if no global knowledge is available, but here too few edges get fixed to decrease an instance's size considerably. As a compromise regarding time and quality, either the Nearest Neighbour (NN), the Lighter or equal to Median ($\leq$M), or the Lighter than Median ($<$M) heuristics can be applied.

16

Algorithms for Large Directed Capacitated Arc Routing Problem Instances

Urban Solid Waste Collection Operational Support

Vittorio Maniezzo and Matteo Roffilli

Department of Computer Science, University of Bologna, Italy
{vittorio.maniezzo,roffilli}@unibo.it

Summary. Solid waste collection in urban areas is a central topic for local environmental agencies. The operational problem, the definition of collection routes given the vehicle fleet, can greatly benefit of computerized support already for medium sized town. While the operational constraints can vary, the core problem can be identified as a capacitated arc routing problem on large directed graphs (DCARP). This paper reports about the effectiveness of different metaheuristics on large DCARP instances derived from real-world applications.

Keywords: Combinatorial Optimization, Capacitated Arc Routing, Problem Reduction, Metaheuristics, Data Perturbation, Garbage Collection.

16.1 Introduction

The widely acknowledged increase in solid waste production, together with the increased concern about environmental issues, have led local governments and agencies to devote resources to solid waste collection policy planning. The problem, difficult *per se*, is further complicated by differential collection and recycling policies: this leads to scenarios where manual planning of waste collection can easily yield severely inefficient solutions. It is known in fact that an efficient planning can have a significant impact on the overall costs of solid waste collection logistics [275].

Waste collection, as most logistic activities, can be studied at different levels: strategical, tactical, operational. In this work we concentrate on operational planning, where a vehicle fleet and the service demand are given and the objective is to design the vehicle trips in order to minimize operational costs subject to service constraints.

Specifically, this work derives from an experience of decision support for the waste collection sectors of municipalities of towns with about 100000 inhabitants, with the objective of designing vehicle collection routes subject to a number of operational constraints. The reported results are for an abstraction level which does not consider several very specific issues, such as union agreements, third-party or personal contracts, etc. The problem to solve is modeled as a Capacitated Arc Routing Problem (CARP) on a directed graph and solved accordingly.

C. Cotta and J. van Hemert (Eds.): Recent Advances in Evol. Comp., SCI 153, pp. 259–274, 2008.
springerlink.com

The instances to be solved are bigger than the state of the art ones. Original heuristic approaches had to be designed in order to meet the solution quality and the allowed computation time specifications. The reported results have a relevance beyond the specific application, as several activities of real-world relevance can be modeled as CARP: foremost among them are mail collection or delivery, snow removal, street sweeping. The CARP is in fact a powerful problem model, which was originally proposed by Golden and Wong [276], and which, given its actual interest, have then been studied by many researches. Dror [277] collected a significant number of applications of CARP variants and of corresponding solution methodologies. For a survey the reader is also referred to [278].

In the literature, several heuristic approaches have been proposed for the CARP, while the only exact techniques so far were published by Hirabayashi et al. [279], who proposed a Branch and Bound algorithm and by Baldacci and Maniezzo [280], who transformed CARP instances into Capacitated Vehicle Routing Problem (CVRP) equivalents and proceeded solving these last. Moreover, Belenguer and Benavent [281] and recently Wøhlk [282] proposed effective lower bounds.

However, being CARP NP-hard, exact techniques can seldom cope with the dimension of real-world instances. These call for heuristic and metaheuristic approaches: among the most recent proposals we remind the tabu search of Belenguer et al. [283], the tabu search of Hertz et al. [284], the Variable Neighborhood Descent *VND-CARP* of the same authors [285] and the genetic algorithm of Lacomme et al. [286]. For a recent review, see [287].

As for more real-world oriented researches, we remind the work of Amberg et al. [288], who studied the M-CARP, that is a multidepot CARP with heterogeneous vehicle fleet, by means of a tabu search, of Xin [289], who implemented a decision support system for waste collection in rural areas based on a simple Augment-Merge heuristic, of Snizek et al. [290] dealing with urban solid waste collection and of Bodin and Levy [291].

The general strategy adopted in this work is that of translating CARP into node routing problem instances and then studying solution methodologies for the resulting data. Different approaches were compared, both directly applied to CARP data and to node routing translations. Among those, the most effective one is based on data perturbation [292] coupled with Variable Neighborhood Search (VNS) and acting on a node routing problem reformulation.

The paper is structured as follows.

Section 16.2 introduces the problem of urban solid state waste collection and its reduction to directed CARP. Section 16.3 presents the methodology we used, first transforming DCARP into Asymmetric CVRP (ACVRP) instances, then applying three variants of VNS to the resulting instances. Section 16.4 reports about the computational results we obtained, both on standard CARP test sets and on DCARP instances derived from real-world applications. Finally, Section 16.5 contains the conclusions of this work.

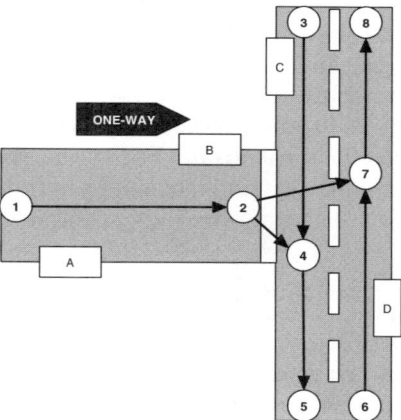

Fig. 16.1. Conversion of a road network into a graph. Bin C can be emptied only by a vehicle which travels the directed arc $(3, 4)$. Bin D can be emptied only by a vehicle which travels the directed arc $(6, 7)$. A vehicle travelling along the one-way arc $(1, 2)$ can empty both bin A and B.

16.2 Urban Solid State Waste Collection

The problem is defined over the road network of the town to support. The network is used as a basis for geocoding all problem elements, which are:

- one or more depots, where vehicles are stationed;
- landfills and disposal plants, where waste has to be brought;
- waste bins, which are usually specialized for differential collections. In our examples, we had the following waste types: i) undifferentiated; ii) plastics; iii) glass; iv) paper; v)biologic or organic.

A number of vehicles are available, where each vehicle can load only bins of given types and has a capacity specified as a maximum number of bins. The actual problem faced asks for designing the routes of the available vehicles, in order to collect all specified bins, satisfying capacity and bin compatibility constraints, forbidding partial bin collection, with the objective of minimizing the total travel distance. Moreover, bins can be collected only within given time widows.

Bins are set in an urban context, which can be modeled by means of a directed graph. One way streets and forbidden turns have to be considered. Each bin is moreover set on one specific side of the street. A vehicle driving along a street can usually collect all bins of a compatible type on the right side of that street. In case of one-way streets, the vehicle can instead collect both bins on the right and on the left side of the street, if adequately equipped.

Figure 16.1 hints at some issues of urban bin collection when modeled on a graph.

This work reports about an optimization procedure for an abstract version of the general problem, in that we consider only one depot and one disposal facility per waste type. Moreover, time windows are not considered and the number of vehicles is an input parameter and not a decision variable, thus the objective function is only given by the total distance traveled. However, the extension to actual settings is a lesser effort when compared to the ability of solving the core problem for actual dimension. This problem can be formulated as an extension of a well known problem from the literature: the Capacitated Arc Routing Problem.

16.2.1 Capacitated Arc Routing Problem

CARP was originally proposed by Golden and Wong [276] and is usually studied in a version which would correspond to an instance of the waste collection problem with a single depot, a single disposal unit coincident with the depot, only one bin type, a non-heterogeneous fleet and no temporal constraints. Moreover, the graph is undirected and bins are positioned on an arc, not on one side of an arc. Different mathematical formulations have been proposed for this version of the CARP ([276], [281], [278]), all of them for the undirected case. The problem we have to solve is however characterized by the fact that the road network of interest must be modeled as a directed graph, thus the problem to solve is a Directed CARP (DCARP). We can take advantage of this in the mathematical formulation of the problem.

16.2.2 Transformation into an ACVRP

The huge dimension of the graphs we had to deal with, rules out the possibility to apply either exact methods or sophisticated heuristics. Our approach for solving the DCARP instances is based on their transformation into Asymmetric Capacitated Vehicle Routing Problem instances. CARP and CVRP are in fact closely related, the main difference being that in the CARP clients are set on arcs while in the CVRP clients are set on the nodes.

The CVRP has been more thoroughly studied in the literature and, even though instances of our size are still out of the current state of the art, efficient heuristics are available for instances bigger than the CARP ones.

The possibility to transform a CARP instance into a CVRP one is well-known, having been first proposed in the literature by Pearn et al. [293]. This transformation, when applied to a CARP defined on a graph with m mandatory edges, results in a CVRP with $3m + 1$ nodes. More recently, Baldacci and Maniezzo [280] presented an alternative transformation which results in a CVRP with $2m + 1$ nodes.

This transformation is of general validity, however, we can take advantage of the need to work on digraphs and use a modified approach which transforms a DCARP instance in an ACVRP one, by simply associating with each DCARP mandatory arc an ACVRP node which inherits the arc demand, by maintaining the depot node, and by redefining distances according to the formulation of section 16.2.1.

Specifically, we compute the all pairs shortest paths matrix among every out-going vertex j, including depot node 0, and every incoming vertex h, including depot node 0, for each pair of nodes $(ij), (hl) \in R$, being R the set of mandatory arcs. We then add the cost c_{ij}, $(ij) \in R$, to the cost $dist(ij)$ of each path exiting from j. This results in a graph with $m + 1$ nodes and $O(m^2)$ arcs, which defines an instance of ACVRP having the solution set which can be mapped 1 to 1 into equivalent DCARP solutions.

In the case of our application, the transformation function parses a DCARP instance and rewrites it as an equivalent ACVRP instance having:

- the same vehicle set with the same capacity constraints;
- the objective function defined over the cost matrix as computed above;
- multiple routes per vehicle, where the first one starts from the depot and ends at the disposal, all other ones start and end at the disposal except for the last one, which starts at the disposal and ends at the depot;
- vehicle objections associated to each client: clients associated to nodes corresponding to two-ways roads can be serviced by any vehicle, while clients associated to nodes corresponding to one-way roads can be serviced only by vehicles which can load bins both from their left and from their right side.

Figure 16.2 shows the transformation on a simple graph.

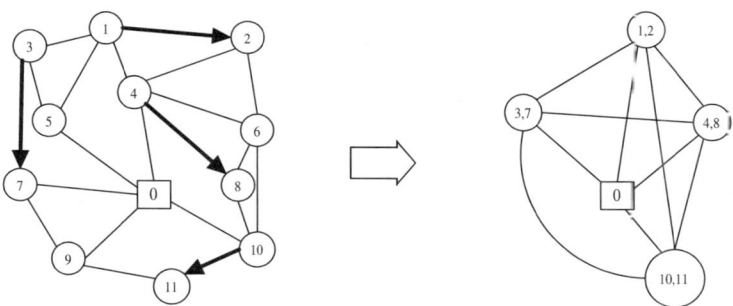

Fig. 16.2. CARP→CVRP. a) DCARP graph. b) ACVRP graph (all edges are pairs of arcs).

Since the ACVRP graph has a number of nodes corresponding to the number of clients of the DCARP instance, in the waste collection case we get a smaller graph than the original road network. The computational complexity is not determined by the road network graph, but by the number of clients.

16.3 New Metaheuristics

Unfortunately, the size of the ACVRP instances of interest for us are very big (hundreds or thousands of nodes) thus all but very simple heuristics are ruled out. We implemented three such heuristics; all of them make use of a local search procedure based on the following neighborhoods:

1. move of a client within one same route;
2. move of a client from a route to another;
3. swap of two clients within one same route;
4. swap of two clients of different routes;
5. 3-opt within one same route;
6. single point route crossover [286].

The basic approach we used for solving the ACVRP instances was a modified Iterated Variable Neighborhood Search. Variable Neighborhood Search (VNS) is an approach proposed by Mladenović and Hansen [294], whose basic framework is the following:

Algorithm 16.1. Variable Neighborhood Search

(Initialization)
Define the set of neighborhood structures $N_k, k = 1, ..., k_{max}$ to be used during search.
Generate an initial solution x.
repeat
  $k = 1$
  repeat
    (Shaking)
    Randomly generate a solution x' in the k-th neighborhood of x, $x' \in N_k(x)$
    k=1
    (Local search)
    Find the local optimum x' using neighborhood k.
    if x' is better than x **then**
      $x = x'$
      $k = 1$
    else
      $k = k + 1$
    end if
  until $k > k_{max}$
until (termination_condition)

Notice that the *Shaking* at line 7 implements a search diversification which permits to escape form cyclic dynamics. The way shaking is implemented deeply affects the algorithm performance. We implemented three different shaking strategies: i) random multistart, ii) a genetic algorithm and iii) data perturbation.

Before describing the three variants, we present the initial solution construction procedure, which is common to the three of them. To construct initial solutions we used a two phase algorithm [295] because it is fast and consistently produces solutions of reasonable quality.

16.3.1   Two Phase Construction Algorithm

The two phase algorithm [295], constructs a solution in two different ways, one per phase, starting from the same pivots. We remind that a *pivot* is the first client assigned to a vehicle, its proper choice is essential for the construction of a good route as it determines the area where the vehicle will operate. Usually, pivots are chosen as the customers hardest to service, therefore nodes which have high requests, which are far from the depot, which are far from previously considered routes, etc. The two phases share only the pivot set and are otherwise independent from one another. The solution proposed is the best among the two obtained. The first phase is a standard parallel construction based on extramileage criteria. The pseudocode is as follows.

Algorithm 16.2. Phase 1

---

for each vehicle $k = 1, \ldots, K$, choose a pivot and initialize one route
$q = n - K$ /*number of unserviced clients*/
while $(q > 0)$ **do**
  Compute the extra mileage for inserting each client in each position of each route
  Insert the best client in the best position of the best route
  $q = q - 1$
end while

The second phase, which tries to use implicit dual information obtained from the solution of the first, is more involved and is as follows.

Algorithm 16.3. Phase 2

Use the same K pivots as in phase 1
Initialize a route for each pivot: let $p(h)$ be the pivot of route h, $h = 1 \ldots, K$
Compute the extramileage $e(i, h)$ for inserting client i in route h, for each unserviced client i and for each route h:
$e(i, h) = c_{0i} + c_{ip(h)} - c_{p(h)0}$
let $hmin$ be the route for which $e(i, h)$ is minimized
let $hsec$ be the route for which $e(i, h)$ is the second minimum
For each client i, compute the regret $g(i)$ which estimates the cost for inserting i in his second-best tour:
$g(i) = e(i, hsec) - e(i, hmin)$
Order the clients by decreasing regrets $g(i), i = 1, \ldots, n$
repeat
  take the client with highest regret still unserviced and insert it in the best possible route $(hmin, hsec, \ldots)$
until all clients are serviced

16.3.2 Multistart Algorithm

This is a straightforward repeated local search algorithm, where at each iteration search is started from a different seed solution, constructed independently from those constructed before. The pseudocode is as follows.

Algorithm 16.4. Multistart algorithm

$\bar{S} = \emptyset$, $f(\bar{S}) = \infty$
repeat
  Construct solution S_1 with the Phase 1 of the Two Phase algorithm
  Apply VNS to S_1 obtaining S_1'
  if $(f(S_1') < f(\bar{S}))$ **then**
    $\bar{S} = S_1'$
  end if
  Construct solution S_2 with the Phase 2 of the Two Phase algorithm
  Apply VNS to S_2, obtaining S_2'
  if $(f(S_2') < f(\bar{S}))$ **then**
    $\bar{S} = S_2'$
  end if
until termination condition
return $\bar{S}$

The key to the Multistart (MS) algorithm is its ability to start from different seed solutions. The seed solutions are constructed by the multistart algorithm, which in turns depends on the pivots choice. We used a probabilistic node difficulty quantification function in order to get different seed solutions. The function is given by the following formula:

$$\text{(difficulty degree)} \quad q(i) = \alpha d_i + \beta \left(c_{0i} + \sum_{k \in U} c_{ik} \right) \tag{16.1}$$

where d_i is the weight of node i, c_{ij} is the distance between nodes i and j and U is the set of already chosen pivots. α and β are two user-defined parameters.

Given the difficulty degrees, the actual pivot choice is demanded to a montecarlo extraction.

The termination condition is the disjunction of:

- Elapsed CPU time is grater than a given limit;
- Number of iterations is greater than a given limit;
- The best solution cost is equal to a (known) optimal value.

The user parameters for this algorithm are the following:

1. limit to maximum CPU time;
2. limit to maximum number of iterations;
3. maximum number of vehicles;
4. vehicle capacity;
5. parameters in the pivot choice function.

16.3.3   Genetic Algorithm

The multistart algorithm has no dependencies among the different seed solutions. As an alternative, we designed a Genetic Algorithm (GA) [296] to let the evolution learn the best pivot set. The implemented GA has a very standard structure, based on selection (roulette wheel), crossover and mutation. The individuals encode which, among the different clients, are chosen to be pivot in the resulting solution. The individual is encoded as a binary string, with as many bits as clients and a bit set to 1 if the corresponding client is a pivot. The population is randomly initialized with individuals with a number of 1s, that is a number of routes, estimated as follows:

$$\text{(number of pivot)} \quad numpivot = \left\lceil \frac{K \sum_{i=1}^{n} d_i}{\sum_{j=1}^{K} C_j} \right\rceil \tag{16.2}$$

where C_j is the capacity of vehicle j. Mutation and crossover operators are standard (random mutation, two points crossover), with the only caution that the number of 1s of each individual is restored to $numpivot$ in case it is changed. Having defined the pivots, the remaining of the algorithm is as described for the multistart, with two phase to construct solutions and VNS to optimize them.

Algorithm 16.5. Genetic Algorithm

Estimate the number of pivots: $numpivot$
Generate a population P of $dimPop$ individuals
Each individual has $numpivot$ randomly chosen bits set to 1
repeat
  for all individual $i \in P$ **do**
    Construct a solution s with the Two Phase algorithm using the pivots in i
    Apply VNS to s obtaining s'
  end for
  for all individual $i \in P$ **do**
    compute the fitness as a function of the cost of s'
  end for
  Generate a new population P' from P by selection, crossover and mutation
  $P = P'$
until termination condition

Termination condition and parameters are the same as in MS. with three more:

1. population dimension;
2. crossover probability;
3. mutation probability.

To transform costs (to be minimized) into fitness (to be maximized) we used Dynamic Fitness Scaling [297].

16.3.4   Data Perturbation

Data Perturbation (DP) is a technique originally proposed by Codenotti et al. [292] which can be framed within an Iterated Local Search approach [298] with the objective of modifying in a controlled way the successive starting solutions. Whereas the more usual approach in VNS is to modify a solution in order to get the new starting one, DP suggests to modify input data. Several works adopted this approach, see for example [299], [300], [301].

If P' is the result of the perturbation of instance P, it is possible to define an injective function $f : P \rightarrow P'$ which associates each element of P with an element of P'. Given $f(S)$, where S is a local optimum of P obtained by some local search, a perturbated solution S' can be obtained as $S' = f(S)$. It is now possible to apply to S', which is not necessarily a local optimum for P', the local search on the perturbed data, thus obtaining S''. Finally, going back to the original instance P it is possible to construct the solution T as $T = f^{-1}(S'')$. Again, T usually is not a local optimum for P, thus applying the local search we get a new local optimum T' for the original instance. This procedure can be applied iteratively. Perturbation was originally presented for the TSP [292], and a suggested data perturbation functions was $\varepsilon - move$, where each datum i is perturbed of a value ε_i: in the case of a geocoded Euclidean TSP this could be implemented by displacing each town in a random direction and recomputing the distance matrix.

This idea can obviously be applied to other problems, the general structure of a data perturbation algorithm is the following:

Algorithm 16.6. Data perturbation

Find an initial locally optimal solution S for instance P
repeat
    perturb P obtaining a new instance P'
    find the solution S' corresponding to S in P'
    find a new solution S'' for P' applying a local search starting from S'
    transform S'' into the corresponding solution T for P
    apply a local search to T obtaining T'
    if $z(T') < z(S)$ **then**
        $S = T'$
    end if
until termination condition
return S

This general structure was made easier in our case by the fact that steps 4 and 6 were not necessary, since any solution is feasible both in the original and in the perturbed settings. Our procedure is as follows.

First we constructed an initial solution for the input instance P by means of phase one of the two phases algorithm. Then we applied to it the VNS obtaining a solution S, which is a local optimum for P. The perturbation of P into P' is then obtained by perturbing each entry c_{ij} of the client distance matrix, specifically by computing:

$$c'_{ij} = c_{ij} \pm \tfrac{k}{100} c_{ij}, \quad k \text{ random in } [0, \delta]$$

where δ is a parameter (maximum percentage variation of original value).

Then we apply to S only a local search based on the swap of nodes between different routes and we obtain a solution S' for P'. Actually, S' is also a solution for P, albeit usually not a local optimum. We can thus apply to S' the complete VNS obtaining solution S'', which is a local optimum for P.

Algorithm 16.7. Data perturbation for CVRP-CARP

Construct an initial solution for the input instance P (parallel extramileage) and apply to it VNS algorithm obtaining S, local optimum for P
repeat
    Perturb the distance matrix of P obtaining P'
    Apply to S the 1-1 swap local search on P' obtaining S', local optimum for P'
    /* S' is also a solution for P */
    Apply to S' VNS algorithm on P obtaining S'', local optimum for P
    if $z(S'') < z(S)$ **then**
        $S = S''$
    end if
until termination condition
return S

Notice that there is no need to transform solution S', computed on P', into a feasible solution for P, as the perturbation used does not impact on the problem constraints. All feasible solutions for P are such also for P' and vice-versa. Both the algorithm parameters and the termination conditions are the same as for MS.

16.4 Computational Results

We have implemented the three variants of VNS (MS, GA and DP). All algorithms were tested on the same test instances and on the same computer, a Pentium 4, 2.8GHz with 512Mbyte of RAM, running under Windows XP.
To test the algorithms we used four sets of instances:

DeArmon instances [302]. Presented in 1981, these are relatively easy, small-sized undirected CARP instances widely used in the literature.
Benavent instances [303]. Proposed in 1997, again small-sized undirected CARP with 24 to 50 nodes and from 34 to 97 arcs, all mandatory.
Set M, real world. These are four instances adapted from real-world problems. The first two ones have 246 nodes and 611 arcs, 14 or 179 of which mandatory; the second two ones have 1222 nodes and 1842 arcs, 5 and 200 of which are mandatory.

Set L, real world. These are five different instances, modeling the centers of the two towns of about 100000 and 150000 inhabitants, respectively. The two C instances, after some preprocessing of the CARP graph, have 1222 nodes, 1842 arcs, 300 of which are mandatory (two different client samples of 300 bins), the R ones have 12388 nodes, 19045 arcs, 1400 of which are mandatory (three different samples).

The DeArmon and the Benavent instances are undirected CARP instances, the M and L are directed CARP instances. We first present results on the first two sets, then on the two real-world ones. All tables in this Section show the following columns:

Instance: instance identifier.
$|V|$: number of nodes of the instance graph.
$|E|$: number of arcs of the instance graph.
$|A|$: number of selected clients.
$|K|$: number of vehicles.
MS: best solution found by algorithm MS.
GA: best solution found by algorithm GA.
DP: best solution found by algorithm DP.
VND: best solution found by algorithm VND-CARP.
T_{alg}: CPU time, in seconds, used by algorithm alg to find its best solution.

16.4.1 Algorithm Setting

Algorithm MS was run using the following parameters setting: Max number of iterations: 10000, Max CPU time: 180 seconds The most influential parameter is Nv, number of available vehicles. We ran the algorithm 3 times for each Nv value, for a number of different values, and report the best result obtained.

Algorithm GA was run using the following parameters settings: Max number of iterations: 10000, Max CPU time: 180 seconds, $\alpha = 2$, $\beta = 1$, $\mu = 50$, $N = 40$, $p_c = 0.6$, $\tau = 3$, $p_m = \frac{3}{n}$

Table 16.1. Analysis of the impact of δ (DeArmon and Benavent instances). Values are the solution costs.

Instance	DP,$\delta = 5$	DP,$\delta = 10$	DP,$\delta = 20$	DP,$\delta = 40$
1	316	316	316	316
10	352	350	350	352
11	323	319	319	325
23	156	156	156	156
1.B	187	184	183	187
2.C	457	457	457	459
3.B	87	87	87	87
5.A	431	425	423	435
8.C	563	561	559	563

Algorithm DP was run using the following parameters settings: Max number of iterations: 10000, Max CPU time: 180 seconds, $\alpha = 2$, $\beta = 1$, $\mu = 50$, $\delta = 20$.

Table 16.1 shows some results obtained on some DeArmon and Benavent instances for setting parameter δ, thus the sensitivity to this value.

16.4.2 Algorithm Comparison

Tables 16.2 and 16.3 report the best results obtained by the different algorithms on the DeArmon and Benavent instances.

Table 16.2. Comparison on the DeArmon instances

Instance	VND	MS	T_{MS}	GA	T_{GA}	DP	T_{DP}
1	316	316	0.01	316	0.01	316	0.01
2	339	339	0.01	339	0.03	339	0.01
3	275	275	0.01	275	0.01	275	0.01
4	287	287	0.01	287	0.01	287	0.01
5	377	377	0.03	377	0.01	377	0.02
6	298	298	0.01	298	0.01	298	0.01
7	325	325	0.01	325	0.01	325	0.01
10	350	350	0.55	350	2.60	350	1.92
11	315	315	11.78	315	14.98	315	2.35
12	275	275	0.01	275	0.06	275	0.01
14	458	458	0.06	458	0.01	458	0.01
15	544	544	0.01	544	0.01	544	0.01
16	100	100	0.01	100	0.01	100	0.01
17	58	58	0.01	58	0.01	58	0.01
18	127	127	0.06	127	0.11	127	0.01
19	91	91	0.01	91	0.01	91	0.01
20	164	164	0.01	164	0.01	164	0.01
21	55	55	0.21	55	0.48	55	0.07
22	121	121	0.24	121	0.53	121	0.02
23	156	156	0.66	156	2.79	156	0.04
24	200	200	0.14	200	14.98	200	1.30
25	235	235	1.06	235	1.01	235	0.88

Given the transformation UCARP $\rightarrow$ DCARP, which was based on the VND-CARP solution, the reference value is that found by VND-CARP itself. When compared to it, the three VNS variants show comparable performance on simple DeArmon instances, both with respect to the quality of the solution and to the time to get it.

The Benavent instances are more challenging and permit to discriminate more among the algorithms. On the bigger instances, MS encountered some difficulties; GA performed better than MS and DP was the best of the three, being able to get in 19 cases out of 26 the same result of VND-CARP in comparable CPU time. Given these results, we used only DP for further testing on harder instances.

Table 16.3. Comparison on the Benavent instances

Instance	VND	MS	T_{MS}	GA	T_{GA}	DP	T_{DP}
1.A	173	173	0.12	173	0.02	173	0.09
1.B	178	183	14.98	183	14.98	179	4.84
1.C	248	248	0.01	248	0.10	248	0.02
2.A	227	227	0.02	227	0.01	227	0.45
2.B	259	259	8.40	259	10.49	259	1.02
2.C	457	457	2.47	457	1.52	457	0.06
3.A	81	81	0.59	81	0.17	81	0.03
3.B	87	87	0.03	87	0.68	87	0.02
3.C	140	140	0.08	140	0.96	140	0.06
4.A	400	408	14.98	406	14.98	400	3.17
4.B	414	426	14.99	422	14.99	418	14.98
4.C	428	444	14.98	440	14.99	436	14.98
4.D	544	560	14.99	558	14.98	556	14.98
5.A	423	427	14.98	423	6.14	423	5.18
5.B	449	457	15.00	451	14.99	451	14.98
5.C	474	483	14.98	482	14.99	480	14.98
5.D	599	611	14.98	608	14.98	599	4.66
6.A	223	223	1.76	223	4.67	223	0.12
6.B	233	233	1.34	233	5.84	233	0.13
6.C	325	331	14.98	331	14.98	325	0.57
7.A	279	279	2.22	279	0.28	279	3.30
7.B	283	283	0.05	283	0.44	283	0.50
7.C	335	345	14.98	343	14.98	335	0.05
8.A	386	386	2.52	386	9.00	386	2.57
8.B	403	403	1.99	403	3.32	403	7.77
8.C	538	563	14.98	559	14.98	544	14.98
9.A	323	323	0.78	323	0.76	323	0.25
9.B	331	331	3.09	331	2.61	331	2.13
9.C	339	339	4.63	339	3.13	339	2.21
9.D	413	413	21.22	413	19.64	413	7.84
10.A	428	428	0.88	428	0.96	428	0.25
10.B	436	436	2.45	436	2.72	436	2.83
10.C	451	451	6.79	453	7.48	451	8.39
10.D	556	556	11.13	552	12.07	556	7.28

16.4.3   Real World Cases

The section presents results on bigger instances derived from real-world problems, much bigger than those analyzed so far, and for which the set R of clients is a subset of the set of arcs.The first set of tests was aimed at ascertaining the applicability of the VND-CARP to bigger instances. We used testset M for this. Being it composed of DCARP instances, the opposite transformation of that described in Section 16.2.2 was in order. This is fortunately very straightforward, requiring only the redefinition of arcs as edges since the instances were already

Table 16.4. VND-CARP applied to testset M

Instance	$|V|$	$|E|$	$|A|$	$\mathbf{T}_{VND}$
C mini	246	611	14	18.53
C mini	246	611	179	118.72
C	1222	1842	5	116.39
C	1222	1842	200	>3600

defined with the depot coinciding with the disposal. We ran VND-CARP on the four instances of testset M obtaining the results shown in Table 16.4.

As appears from Table 16.4, VND-CARP cannot scale up to real-world instances. Only the smaller instances could be solved in reasonable time, even though the number of clients was kept small in all instances. This is mainly because the solution is represented as a list of all nodes traversed by each vehicle, whose number grows with the number of arcs of the graph, and not as a list of the clients as in the case of DP. For this reason, on the bigger instances of set L we applied only the DP algorithm.

16.4.4 Results of DP Algorithm

Table 16.5 shows the results obtained by DP algorithm applied to the large-scale instances of testset L. For each instance, we report the number of vehicles K, the number of selected clients A, the cost of the solution of the 2 phase initialization $2p$ and the time in seconds to produce it $\mathbf{T}_{2p}$, the best found solution DP, the time in seconds to produce it $\mathbf{T}_{DP}$ and the total CPU time allowed $\mathbf{T}_{tot}$. In all tests the termination condition was set on the maximum number of iterations $NumIter$, which was set to 20000 for the C instances and to 10000 for the R ones.

Table 16.5. DP applied to real world instances

Instance	$|V|$	$|E|$	$|A|$	$|K|$	2p	$\mathbf{T}_{2p}$	DP	$\mathbf{T}_{DP}$	$\mathbf{T}_{tot}$
C1	1222	1842	300	3	43475	0.23	38245	8924	24073
C2	1222	1842	300	3	65996	0.50	56273	16939	17464
R1	12388	19045	1400	13	826358	7.20	574389	16664	22521
R2	12388	19045	1400	13	768036	5.19	528974	17856	18620
R3	12388	19045	1400	13	779252	5.27	533121	16939	17464

Notice how the instance dimensions, coupled with the computational resources available, induced a high CPU time for processing the requested number of iterations. The two C instances could be improved of 12.03% and of 14.73%, respectively, while the R instances got an improvement of 30.49%, 31.13% and of 31.59%, respectively.

The high CPU time needed to produce the best solution found in all cases testifies that search is actively going on and that probably an even higher number of iterations would produce still better results.

16.5 Conclusions

The paper reported about an approach for the management of urban waste collection. A transformation of the problem into a node routing equivalent is shown and different metaheuristics are tested on it, both from the literature and original. Computational results testify the effectiveness of a perturbation-based VNS approach. This technique proved also to be able to scale to real-world instances, having dimensions much greater than those so far attacked in the literature. Considering that waste collection in bigger towns is usually planned by first partitioning the town into zones, and then solving the routing for each zone, we believe that the proposed approach can be considered an option for actual routing also for larger municipalities.

An Evolutionary Algorithm with Distance Measure for the Split Delivery Capacitated Arc Routing Problem

Nacima Labadi, Christian Prins, and Mohamed Reghioui

Institute Charles Delaunay, University of Technology of Troyes, BP 2060, 10010
Troyes Cedex, France
{nacima.labadi,christian.prins,mohamed.reghioui_hamzaoui}@utt.fr

Summary. This chapter deals with the Split Delivery Capacitated Arc Routing Problem or SDCARP. Contrary to the classical Capacitated Arc Routing Problem or CARP, an edge may be serviced by more than one vehicle. An integer linear program, an insertion heuristic and a memetic algorithm with population management are proposed for this problem seldom studied. The algorithm calls an effective local search procedure, which contains new moves able to split demands. Compared to one of the best meta-heuristics for the CARP, we show that splitting demands can bring important savings and reduce significantly the number of required vehicles.

Keywords: Capacitated Arc Routing Problem, Split Delivery, Memetic Algorithm, Population Management, Distance Measure.

17.1 Introduction

The Capacitated Arc Routing Problem (CARP) has various applications like mail delivery, waste collection or street maintenance. It is usually defined on an undirected weighted graph, with identical vehicles at a depot node. Some edges with known demands must be serviced by vehicles. The goal is to determine a set of trips to process all these required edges, knowing that each edge must be serviced by one vehicle. This constraint is relaxed in the Split Delivery CARP (SDCARP): each edge may be served by different trips. This problem can be met in urban refuse collection: indeed, a vehicle may become full in the middle of an edge (street) and using a CARP model is not realistic.

Most of works dealing with split demands concern the split delivery vehicle routing problem (SDVRP), introduced by Dror and Trudeau. [304, 305]. Later, these authors showed that allowing split deliveries results in significant savings in total cost and number of vehicles used, especially when the average demand per customer exceeds 10% of vehicle capacity (Dror et al. [306]).

Archetti et al. [307] have shown recently that the SDVRP is NP-hard, even for unit demands and a vehicle capacity greater than 2, provided edge costs are symmetric and satisfy the triangle inequality: this proof can be transposed to the SDCARP. Two versions of the SDVRP are studied in literature. In the first

C. Cotta and J. van Hemert (Eds.): Recent Advances in Evol. Comp., SCI 153, pp. 275–294, 2008.
springerlink.com   © Springer-Verlag Berlin Heidelberg 2008

version, the maximum number of vehicles which may be used is imposed. In the second one, this number is left as a decision variable.

Concerning the first version, Belenguer et al. [308] proposed a cutting plane algorithm, studied lower bounds and succeeded in solving small instances to optimality. Very recently, Mingzhou et al. [309] have presented a two-stage algorithm with valid inequalities, while Mota et al. [310] have designed the first metaheuristic, a scatter search. Concerning the second version, the first heuristic was presented by Dror and Trudeau. [304]. The results obtained by these authors were improved by Archetti et al. [311], using a tabu search algorithm. Frizzell and Giffin [312,313] studied an SDVRP variant with time windows and a grid network. They developed constructive heuristics that use certain structural features of the problem.

The SDCARP literature is more scarce: to the best of our knowledge, only two studies are publicly available. Mullaseril et al. [314] proposed an adaptation of the SDVRP heuristic of Dror and Trudeau for the SDCARP with time windows. This method was used to solve a livestock feed distribution problem. In his Ph.D., Guéguen [315] transforms the SDCARP with time windows into a node routing formulation solvable by a column generation approach. Contrary to the VRP, there are still relatively few metaheuristics in the CARP literature, and even less for its variants. For instance, two metaheuristics are available for the CARP with time windows or CARPTW (Labadi et al. [316] and Wøhlk [317]), but none for the SDCARP.

This chapter bridges the gap by presenting a memetic algorithm (MA) for the SDCARP. It is organized as follows. Section 17.2 introduces an integer linear model for the SDCARP. Three initial heuristics and the main components of the memetic algorithm are presented respectively in Section 17.3 and 17.4. In Section 17.5, numerical experiments are conducted to evaluate the effectiveness of the approach. Some conclusions are given in Section 17.6.

17.2 Problem Statement and Linear Formulation

The SDCARP can be defined on an undirected graph $G = (V, E)$, where V is a set of n nodes, including one depot at node 1, and E a set of m edges. Each edge $[i, j] \in E$ has a cost (length) c_{ij}. $R \subseteq E$ is a subset of τ required edges. Each required edge $[i, j]$ has a positive demand d_{ij}. F is a fleet of K vehicles with a limited capacity W, located at the depot.

The SDCARP is the problem of finding a minimum-length traversal of required edges, such that each required edge is serviced by *at least* one vehicle. In our version, all data are integer, U-turns on an edge $[i, j]$ are forbidden after a partial service, and no demand exceeds vehicle capacity. U-turns are not handled because they require the knowledge of the distribution of demand along the edge, to compute the fraction of demand satisfied. As for large demands, consider for instance one edge with $d_{ij} = 2.7 \times W$: it can be satisfied using three vehicles, including two full ones. By pre-computing such full trips and removing satisfied demands from the data, one can assume without loss of generality that $d_{ij} \leq W$ for each required edge $[i, j]$.

The SDCARP can be formulated as the following integer linear program. Let x_{ijk} be the number of times vehicle k traverses edge $[i, j]$ from i to j, l_{ijk} a binary variable equal to 1 if and only if vehicle k serves $[i, j]$ from i to j, and r_{ijk} the amount of demand of $[i, j]$ serviced by vehicle k. The u_S^k and y_S^k are additional binary variables used in subtour elimination constraints.

Minimize :
$$\sum_{k \in F} \sum_{(i,j) \in A} c_{ij} \cdot x_{ijk} \tag{1}$$

s.t.

$$\sum_{j \in \Gamma(i)} (x_{ijk} - x_{jik}) = 0 \qquad \forall i \in V, \forall k \in F \tag{2}$$

$$x_{ijk} \geq l_{ijk} \qquad \forall (i,j) \in A_R, \forall k \in F \tag{3}$$

$$\sum_{k \in F} (l_{ijk} + l_{jik}) \geq 1 \qquad \forall [i,j] \in R \tag{4}$$

$$\sum_{[i,j] \in R} r_{ijk} \leq W \qquad \forall k \in F \tag{5}$$

$$\sum_{k \in F} r_{ijk} = d_{ij} \qquad \forall [i,j] \in R \tag{6}$$

$$d_{ij} \cdot (l_{ijk} + l_{jik}) \geq r_{ijk} \qquad \forall [i,j] \in R, \forall k \in F \tag{7}$$

$$\left.\begin{array}{l} \sum_{i \in S} \sum_{j \in S} x_{ijk} - n^2 y_S^k \leq |S| - 1 \\ \sum_{i \in S} \sum_{j \in \bar{S}} x_{ijk} + u_S^k \geq 1 \\ u_S^k + y_S^k \leq 1 \end{array}\right\} \quad \forall S \neq \emptyset, S \subseteq V \setminus \{1\}, \forall k \in F \tag{8}$$

$$x_{ijk} \in I\!N \qquad \forall (i,j) \in A, \forall k \in F \tag{9}$$

$$l_{ijk} \in \{0,1\} \qquad \forall (i,j) \in A_R, \forall k \in F \tag{10}$$

$$r_{ijk} \in I\!N \qquad \forall [i,j] \in R, \forall k \in F \tag{11}$$

$$u_s^k, y_s^k \in \{0,1\} \qquad \forall S \neq \emptyset, S \subseteq V \setminus \{1\}, \forall k \in F \tag{12}$$

Variables x and l depend on edge directions, contrary to variables r. To better handle these directions, the formulation uses the set of arcs A obtained by replacing each edge $[i, j]$ by two arcs (i, j) and (j, i), and the set of arcs $A_R \subseteq A$ corresponding to the required edges. In reality, a vehicle may service one given edge several times. But since the shortest paths between required edges satisfy the triangle inequality, such solutions are dominated by the ones in which each required edge is serviced at most once by one given trip. This explains the use of binary variables l_{ijk}.

The objective function (1), to be minimized, is the total length of the trips. Constraints (2) ensure trip continuity ($\Gamma(i)$ is the set of successors of node i). Constraints (3) state that each edge serviced by this vehicle is also traversed by the same vehicle. Constraints (4) guarantee that each required edge is serviced by at least one vehicle, in any direction. Vehicles capacity is respected thanks to constraints (5). The demand of each edge must be completely satisfied, which is enforced by constraints (6). Constraints (7) ensure that a partial service of edge $[i, j]$ by vehicle k implies that the edge is serviced by k. Constraints (8) are subtour elimination constraints introduced by Golden and Wong [318] in an integer linear program for the CARP. The remaining constraints define the variables.

We have tried to solve the LP model with the commercial solver CPLEX 9. However, the software was not able to solve problems with more than 10 required edges. This justifies the need of an effective heuristic method, able to solve larger instances to near-optimality in reasonable time.

17.3 Heuristics

Let $\mu = \sum_{[i,j] \in R} \sum_{k \in F} (l_{ijk} + l_{jik})$ be the total number of edge services in an SDCARP solution. The minimum value being the number of required edges τ, we can define the *splitting level* as the ratio μ/τ. This ratio is sensitive to $\bar{d}/W$, where $\bar{d}$ denotes the average demand per required edge. When demands are small compared to vehicle capacity, the CARP and SDCARP optimal solutions are quite close in terms of total cost and number of vehicles. But the gap in favour of the SDCARP solution increases quickly with $\bar{d}/W$. An extreme case is when $\bar{d}/W > 0.5$: in that case, any CARP solution needs one trip per required edge and splitting demands can lead to important savings.

This section presents three heuristics used for the initial population of our memetic algorithm. The two first are classical CARP heuristics, Augment-Merge and Path-Scanning: they are not affected by the ratio $\bar{d}/W$. The third one is a new insertion heuristic able to split demands and called Split-Insertion Heuristic or SIH. SIH is expected to return better solutions than Augment-Merge and Path-Scanning, especially for large ratios $\bar{d}/W$.

17.3.1   Augment-Merge and Path-Scanning

Augment-Merge (AM) is a greedy heuristic for the CARP proposed by Golden and Wong [318]. The heuristic is composed of two phases. In the first phase, called *Augment*, each edge is serviced by a unique trip. The number of trips is then reduced by absorbing the trips with small demands: if a trip T traverses the unique required edge e of another trip T', e may be serviced by T and T' may be suppressed.

The second phase called *Merge* is similar to the famous savings heuristic introduced by Clarke and Wright [319] for the Vehicle Routing Problem (VRP): pairs of trips are merged if vehicle capacity allows it and if this decreases the total cost. In each iteration, all feasible mergers are evaluated and the one which results in the largest saving is executed. The algorithm stops when no additional improvement is possible.

The second heuristic is the Path-Scanning heuristic (PS), which was first presented for the classical CARP by Golden et al. [320]. PS builds one trip at a time using the nearest neighbour principle. Each iteration computes the set of edges not yet serviced and closest to the last edge of the incumbent trip. For a trip T built up to an edge e, one iteration determines the edge f closest to e and adds it at the end of T. Ties are broken using five priority rules. One solution is computed for each rule and the best one is returned at the end.

17.3.2 Split-Insertion Heuristic for the SDCARP

SIH is inspired by the best-insertion heuristic (BIH) for the VRP. BIH starts with one empty trip. In each iteration, it computes the minimum insertion cost for each unrouted customer and performs the best insertion. When the residual trip capacity becomes too small to accept additional customers, a new empty trip is created.

SIH transposes this principle to the SDCARP (required edges are inserted), while adding the possibility of splitting a demand d_e to insert it into a subset of trips S offering enough capacity, i.e., $\sum_{k \in S}(W - \sum_{[i,j] \in R} r_{ijk}) \geq d_e$. We call such an operation a *split-insertion*. SIH performs ordinary insertions to fill the first trip, like BIH. After the creation of the second trip, each iteration evaluates the cost variations of all feasible insertions and split-insertions and executes the best one.

The idea of split-insertions was introduced by Dror and Trudeau [304] for a local search for the SDVRP. However, to limit running time, these authors enumerate only the possible insertions into a limited number of trips k, explaining why they call such moves k-*split interchanges*. In fact, we show in the sequel that it is possible to determine efficiently the optimal split-insertion, without fixing k a priori.

Let e be the required edge to be inserted into existing trips, S the set of trips in which e can be inserted (the vehicle is not full), a_j the residual capacity of trip j, z_j the cost of a best insertion of e into j, y_j the amount of demand inserted into j and x_j a binary variable equal to 1 if trip j is used for insertion. The optimal split-insertion for e corresponds to the following integer linear program.

$$\min \sum_{j \in S} z_j x_j \qquad (17.13)$$

$$\sum_{j \in S} y_j = d_e \qquad (17.14)$$

$$0 \leq y_j \leq a_j x_j \qquad \forall j \in S \qquad (17.15)$$

$$y_j \in IN, x_j \in \{0, 1\} \qquad \forall j \in S \qquad (17.16)$$

In fact, this integer linear formulation can be replaced by the 0-1 knapsack below. Indeed, constraints (17.14) and (17.15) imply constraints (17.18), the optimal solution x^* of the knapsack problem tells us which trips must be used, and the demand d_e may be assigned as we wish to these trips, for instance using the very simple Algorithm 17.1.

$$\min \sum_{j \in S} z_j x_j \qquad (17.17)$$

$$\sum_{j \in S} a_j x_j \geq d_e \qquad (17.18)$$

$$x_j \in \{0, 1\} \qquad \forall j \in S \qquad (17.19)$$

Using dynamic programming (DP), the knapsack problem can be solved pseudo-polynomially in $O(|S| \cdot (\sum_{j \in S} a_j - d_e))$. In practice, this technique is fast enough

---

Algorithm 17.1. Insertion of demand d_e using the knapsack solution

---

1: **repeat**
2:    $j \leftarrow \arg\min \ \{z_p | p \in S \wedge x_p^* = 1 \wedge a_p \neq 0\}$
3:    $w \leftarrow \min\{d_e, a_j\}$
4:    Insert one copy of edge e in trip j, with an amount w
5:    $d_e \leftarrow d_e - w; \ a_j := a_j - w$
6: **until** $d_e = 0$

---

because the number of trips $|S|$ is much smaller than the number of required edges τ. We actually implemented this DP method and the MA running time increased by 30% on average, which is tolerable.

However, we tried also the greedy heuristic of Algorithm 17.2, with insertions in increasing order of the z_j/a_j. Preliminary tests have shown that this heuristic is much faster than the DP method: it can be implemented in $O(|S|(\tau + \log |S|))$. Moreover, it is optimal in 66% of cases for the small knapsacks generated by our SDCARP instances. On average, the substitution of the DP method by the heuristic does not degrade the best solution cost returned by the MA. This is why the greedy heuristic is implemented in the final version of our MA.

Figure 17.1 shows a small example solved optimally in three iterations by SIH. The numbers are edge lengths and there are three required edges a, b and c. All

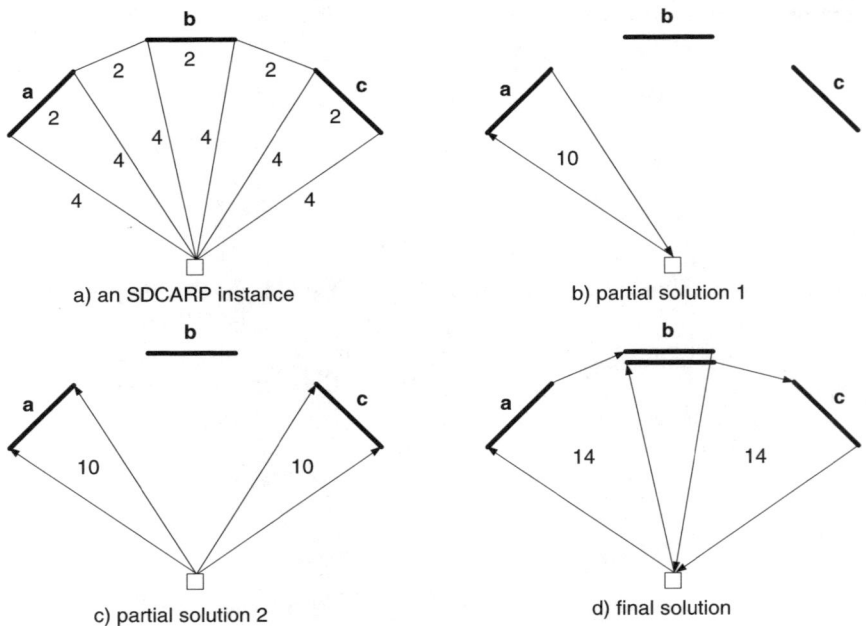

Fig. 17.1. An instance solved by SIH in three iterations

Algorithm 17.2. Insertion of demand d_e using the greedy heuristic

---

1: Sort the $|S|$ trips in increasing order of z_j/a_j in a list λ
2: $i \leftarrow 0$
3: **repeat**
4:    $i \leftarrow i + 1$
5:    $w \leftarrow \min\{d_e, a_{\lambda_i}\}$
6:    Perform a best insertion of edge e in trip λ_i, with an amount w
7:    $d_e \leftarrow d_e - w$
8: **until** $d_e = 0$

---

demands are equal to 2 and $W = 3$. At iteration 1, any edge can be inserted in the initial trip, with a cost variation $+10$. Assume a is selected. Since no other insertion is possible, a new empty trip is created. Edges b and c have the same insertion cost of $+10$, assume c is inserted in the second trip. An ordinary insertion is impossible for b without creating a new trip, but a split-insertion into the two existing trips is possible, with a cost variation $+8$.

17.4 Memetic Algorithm with Population Management

A memetic algorithm (MA) is a genetic algorithm, in which intensification is performed by applying an improvement procedure (local search) to new solutions. The concept was introduced by Moscato (see [321] for a tutorial) and has proven its effectiveness on many combinatorial optimization problems.

The memetic algorithm with population management (MA|PM) is a more powerful version recently proposed by Sörensen and Sevaux [322]. In addition to the traditional MA components, MA|PM uses a distance measure in solution space to control population diversity. More precisely, let $D(A, B)$ be the distance between two solutions A and B in solution space and, by extension, $D(A, Pop) = \min_{S \in Pop}(D(A, S))$ the minimum distance between a solution A and a population Pop. MA|PM accepts a new solution A only if its distance to Pop is not smaller than a given threshold value Δ, i.e., if $D(A, Pop) \geq \Delta$. The threshold may be constant or dynamically adjusted during the search.

17.4.1 Chromosomes and Evaluation

A chromosome X is defined by two lists. In the first list T, each of the τ required edges appears once or more, depending on the splitting level. An edge e may occur as one of its two possible directions, denoted by e and e' in the sequel (in practice $e' = \tau + e$). Implicit shortest paths are assumed between the depot and the first edge, between two consecutive edges, and between the last edge and the depot. A second list Q specifies the amount of demand satisfied for each edge occurrence in the first list. No trip delimiter is used: the chromosome can be viewed as a giant tour performed by a vehicle with infinite capacity, allowed to visit each required edge several times.

List T: 2 1 5 4 3 2' 1 2
List Q: 3 4 2 2 4 6 5 2

Fig. 17.2. An SDCARP chromosome

Figure 17.2 gives a simple example with $\tau = 5$ required edges: the only split edges are 1 and 2, with respective demands 9 and 11. Both are serviced twice, edge 1 always in the same direction and edge 2 in two directions. In the sequel, the lists belonging to several chromosomes are distinguished using the chromosome name as prefix, e.g., $X.T$.

To convert a chromosome X into a feasible SDCARP solution, we extend an optimal splitting procedure called *Split* and proposed by Lacomme et al. for the CARP [323]. *Split* for the CARP builds an auxiliary graph $H = (V, A, Z)$ which describes all ways of splitting the giant tour $X.T$ into capacity-feasible trips. V contains $\tau + 1$ nodes indexed from 0 onwards. A contains one arc $(i - 1, j)$ for each feasible trip (subsequence of edges) $(T_i, T_{i+1}, \ldots, T_j)$. The weight $z_{i-1,j}$ of arc $(i - 1, j)$ is the cost of the associated trip.

The optimal splitting, subject to the sequence imposed by T, is obtained by computing a shortest path from node 0 to node τ in H. The trips in the resulting solution $S = Split(X)$ correspond to the arcs of the shortest path. The cost $F(S)$ of this solution is the cost of the shortest path and also the fitness adopted for chromosome X. Since H contains $O(\tau^2)$ arcs in the worst case, the shortest path can be computed with the same complexity, using the version of Bellman's algorithm for directed acyclic graphs.

For the SDCARP, due to split demands, T may contain more than τ edge services and a trip $(T_i, T_{i+1}, \ldots, T_j)$ may visit several times some edges. Since the triangle inequality holds, the trip cost does not increase if all visits except one are suppressed for each split edge. The problem of setting the best possible cost on the arc which models the trip in H is equivalent to removing superfluous visits to minimize the cost of the resulting trip. This problem is sometimes easy. For instance, if T contains a trip $(1, 5, 1, 3, 1)$, the only split edge is 1 and there are only three ways of removing redundant visits: keep either the first, second or third occurrence. However, the running time increases quickly for SDCARP solutions with high splitting levels.

We decided to use a simpler but faster method: for each split edge in a trip, we keep the first visit and delete the others. This method is not always optimal but *Split* can still be implemented in $O(\tau^2)$. In fact, preliminary tests have shown that this sub-optimal policy is not a real drawback: in most cases, the local search is able to relocate the edge at a better position in the trip, if any.

17.4.2 Distance Measure

The distance measure proposed for the SDCARP extends a metric proposed by Martí et al. [324] for permutation problems like the TSP. Given two permutations

$$A \qquad\qquad\qquad B$$

A.T: 3' 2 1 3 1' 2'   B.T: 1' 2' 3 1 2

A.Q: 1 1 1 1 1 1   B.Q: 1 1 2 1 1

Matrix M^A							Matrix M^B							$\|M^A - M^B\|$					
	1	2	3	1'	2'	3'		1	2	3	1'	2'	3'		1	2	3	1'	2'	3'
1	0	0	1	0	0	1	1	0	1	0	0	0	0	1	0	1	1	0	0	1
2	2	0	0	0	0	0	2	1	0	0	0	0	0	2	1	0	0	0	0	0
3	0	0	0	1	0	0	3	1	0	0	0	0	0	3	1	0	0	1	0	0
1'	0	0	0	0	2	0	1'	0	0	0	0	1	1	1'	0	0	0	0	1	1
2'	0	0	1	0	0	0	2'	0	0	1	1	0	0	2'	0	0	0	1	0	0
3'	0	1	0	1	0	0	3'	0	1	0	0	0	0	3'	0	0	0	1	0	0

Fig. 17.3. Example of distance computation

of n elements A and B, Martí's distance is the number of consecutive pairs (A_i, A_{i+1}), $i < n$, which are "broken" (no longer consecutive) in B.

For the SDCARP, A and B may have different lengths and each edge u may occur several times in each chromosome, using traversal directions u and u'. Martí's distance is no longer a metric if we try to transpose it for the SDCARP, as shown by the following example. Let $A = (1, 2, 3, 1', 2)$ and $B = (2, 3, 1, 2')$. If e and e' are considered as two distinct entities, we find $D(A, B) = 3$ and $D(B, A) = 2$: the symmetry property does not hold. If directions are ignored, i.e., if e and e' are interpreted as two equivalent symbols for the same edge, we get $D(A, B) = D(B, A) = 0$, although A and B are visibly quite different.

To compute a suitable distance, we define for a chromosome X a matrix M^X, $2\tau \times 2\tau$, in which M^X_{uv} is the number of times the pair (u, v) or (v', u') occurs in X. This matrix is easily built by browsing the chromosome. The distance is then defined as $D(A, B) = \sum_{u=1}^{2\tau} \sum_{v=1}^{2\tau} |M^A_{uv} - M^B_{uv}|/2$. Figure 17.3 gives an example of distance between two chromosomes A and B, for an instance with $\tau = 3$ required edges. We find $D(A, B) = 5$.

This measure is a *semi-metric* with integer values. Indeed, for any three chromosomes A, B and C, we have $A = B \Rightarrow D(A, B) = 0$, $D(A, B) = D(B, A)$ and $D(A, B) + D(B, C) \geq D(A, C)$. In fact, D is also *reversal-independent*, i.e., $D(A, \overline{A}) = 0$ and $D(A, B) = D(\overline{A}, B)$, where $\overline{A}$ denotes the mirror of A (inverted string). This property is interesting because A and $\overline{A}$ give two equivalent SD-CARP solutions after evaluatation by *Split*: the trips are performed backwards by the vehicles in the second solution, but the total cost does not change. In fact, if we consider that A and $\overline{A}$ are two equivalent chromosomes, D is a true metric because we have $D(A, B) = 0 \Leftrightarrow (A = B) \vee (\overline{A} = B)$.

17.4.3 Initial Population

The population Pop consists of a fixed number nc of chromosomes, always kept sorted in increasing order of cost. To build the initial population, we first generate three good solutions using the Augment-Merge, Path-Scanning and the SDCARP insertion heuristic SIH described in Section 17.3. The population is then completed with random solutions which include a few split demands.

Each random solution is generated as follows. First, a giant tour T containing the τ required edges is generated, with a random order and a random traversal direction for each edge. Starting from the first edge, T is then partitioned into successive trips with a greedy heuristic. Each trip is completed when the vehicle gets full or if adding one more edge would violate vehicle capacity. In the second case, the incumbent edge is split to fill the vehicle completely and its residual demand is used to start a new trip. Hence, only the first and last edges of each trip may be split in the resulting solution.

Contrary to new solutions resulting from crossovers, we do not require that $D(A, B) \geq \Delta$ for any two initial solutions A and B, but all initial solutions must be distinct, i.e., $D(A, B) \neq 0$. In practice, starting from an empty population, the three heuristic solutions and then the random solutions are generated one by one and converted into chromosomes by concatenating the sequences of edges of their trips. The resulting chromosome A is added to the population if $D(A, Pop) \neq 0$, otherwise it is discarded. Therefore, in general, more than $nc-3$ random solutions must be generated to fill Pop.

17.4.4   Selection, Crossover and Replacement

At each MA iteration, two parents A and B are randomly selected using the binary tournament method: two solutions are randomly selected from the population and the best one is taken for A. The same process is repeated to get B.

The recombination operator is a cyclic, one-point crossover for chromosomes with different lengths. Only one child-solution C is generated. The position k after which A and B are cut is randomly drawn in the integer interval $[1, \min\{|A|, |B|\} - 1]$. Child C is initialized with the first k edges of A and the associated amounts. The child solution is completed by browsing circularly the second parent B, from position $k+1$ onwards. There are two cases when inspecting the current edge $B.T_i = e$ in B. If the demand d_e is already satisfied in C, we skip the edge. Otherwise, it is appended to C and its amount $C.Q_i = B.Q_i$ is eventually truncated to guarantee that the total amount serviced in C does not exceed its demand d_e.

Figure 17.4 depicts one example of crossover for 5 customers and $k = 3$. The vertical bars indicate the cutting points. Subsequence $(2, 1, 5)$ is copied into C and then B is scanned circularly from position 4 onward. Edge 3 is not present in C and can be appended with its demand: $C.T_4 \leftarrow 3$ and $C.Q_4 \leftarrow 4$. Edge 1 is copied but its amount is truncated to 5, because 4 units of demand are already serviced in C. Edge 5 is ignored because d_5 is already satisfied. Edge 2 is copied, with an amount reduced to 8. Edge 4 is entirely copied. Finally, the last edge browsed (2) is ignored.

The resulting child is finally evaluated with *Split*. A chromosome R is randomly selected to be replaced by child C. R is chosen in the worst half of Pop.

Parent A	Parent B
A.T: 2 1 5 | 4 3 2' 1 2	B.T: 2 4 2 | 3 1 5
A.Q: 3 4 2 | 2 4 6 5 2	B.Q: 9 2 2 | 4 9 2

Child C
C.T: 2 1 5 \| 3 1 2 4
C.Q: 3 4 2 \| 4 5 8 2

Fig. 17.4. Example of crossover for the SDCARP

If $D(C, Pop \setminus \{R\}) \geq \Delta$, R is replaced by C in Pop. However, to avoid missing a new best solution, C is also accepted if its cost is smaller than the cost of the current best solution, $Pop(1)$. In all other cases, C is discarded.

In the MA|PM framework initially introduced by Sörensen and Sevaux [322], when the diversity test fails for a new solution C, this solution is mutated until the test is successful. In general, this technique degrades the cost of C and is time-consuming, because the repeated application of a random mutation operator has often a limited impact on $D(C, Pop)$. The MA|PM provides better solutions on average if C is rejected, provided the population is rather small ($nc \leq 30$ for instance), to avoid excessive rejection rates.

17.4.5   Local Search

Local search is used in memetic algorithms to intensify the search, since classical genetic algorithms are in general not aggressive enough on combinatorial optimization problems, compared to tabu search for instance. In our MA, recall that the child C generated by a crossover is converted by *Split* into an SDCARP solution S. The local search is applied with a given probability p_{LS} to S and the trips of the resulting solution are concatenated to rebuild C. In fact, this local search is a loop in which two local search procedures are called.

The first internal local search evaluates some moves already developed for the CARP and two new moves able to split edges. Classical moves for the CARP include the transfer of a string of one or two required edges (Or-opt moves), the exchange of two strings with up to two edges each (2-interchanges), and 2-opt moves. All these moves are similar to the well-known moves designed for the Travelling Salesman Problem (TSP), but with three main differences:

- They can be applied to one or two trips.
- The two traversal directions are evaluated when reinserting one edge.
- If one edge e is partially treated by a trip and if another partial service of e is moved to this trip, the amount transferred is aggregated with the first visit. This guarantees the dominance property already mentioned for the linear model: all services in a trip must concern distinct required edges.

The first new move, depicted in the upper part of figure 17.5, exchanges two required edges e and f pertaining to two distinct trips and split the one with

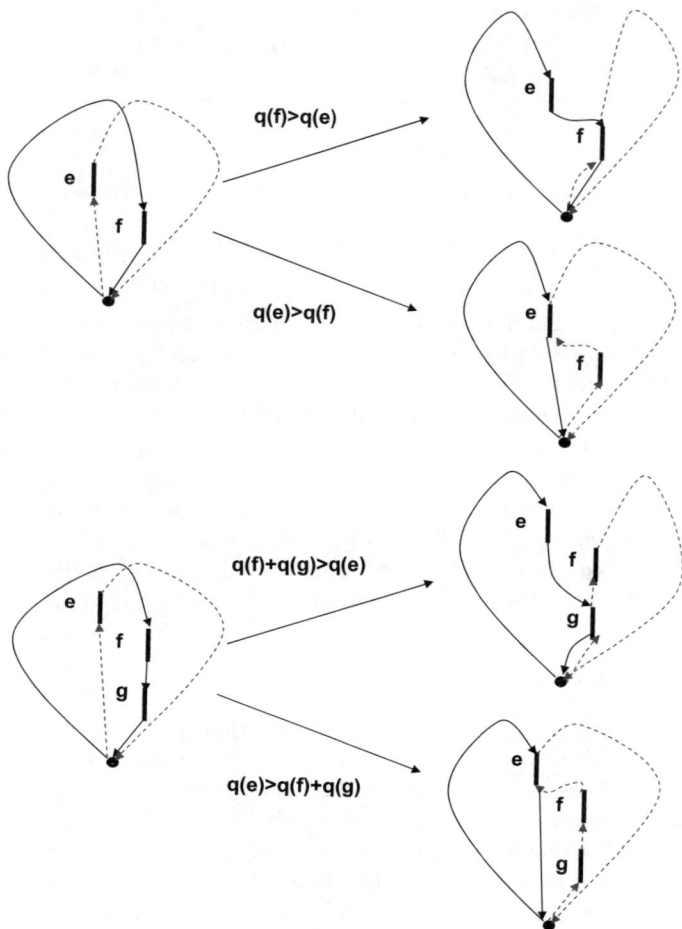

Fig. 17.5. New moves for the SDCARP

the largest amount. Let T_e, p_e, s_e, and q_e respectively denote the trip containing one given service of e, the predecessor and successor of e in this trip and the amount received. If $q_e > q_f$, f is removed with its amount from T_f and inserted before or after e in T_e (the best position is selected).

In parallel, a copy of e with an amount $q_e - q_f$ is created in T_f, to replace the visit to f which has been moved. If $q_e < q_f$, the move is similar, except that the roles of e and f are swapped. The case $q_e = q_f$ is ignored: it corresponds to a standard exchange tested in the CARP moves. Note that the new move does not affect trip loads: in particular, it works on two full vehicles.

In the second new move, e is swapped with a pair (f, g) of consecutive required edges (lower part of Figure 17.5). If $q_e > q_f + q_g$, the pair (f, g) is completely removed from T_f and inserted before or after e in T_e (whichever is the best), while a copy of e with a quantity $q_e - q_f - q_g$ is inserted at the old location. If $q_e < q_f + q_g$ and $q_e \geq q_f$, e and f are exchanged without splitting but a copy of g is moved to T_e, with an amount $q_f + q_g - q_e$. The case $q_e = q_f + q_g$ is a classical interchange of e with (f, g), evaluated by the CARP moves.

The moves in the second internal local search consists of removing one required edge from all the routes in which it is serviced and performing a best split-insertion, exactly like in the SIH initial heuristic of Section 17.3.

Algorithm 17.3. Local search

1: **repeat**
2:      $\overline{F} \leftarrow F(S)$
3:      $S \leftarrow$ CARPandNewMoves (S)
4:      $S \leftarrow$ SplitInsertions (S)
5: **until** $(F(S) = \overline{F})$

Algorithm 17.3 gives the structure of the whole local search. Each iteration of the first internal local search, *CARPandNewMoves*, searches for the first improving CARP move and executes it if found. Otherwise, the new moves are evaluated and the first improving one, if any, is executed. This process is repeated until solution S cannot be improved. The second internal local search, *SplitInsertions*, is then called. Each of its iterations checks all feasible split-insertions and performs the *best* improving move, contrary to *CARPandNewMoves*. The main loop stops when no further cost decrease is possible.

17.4.6   General Structure of the Algorithm

The structure of the resulting MA|PM is summarized by algorithm 17.4. It starts by building the initial population *Pop* of *nc* distinct chromosomes. These chromosomes are evaluated and the population is then sorted in increasing order of costs to facilitate its management. The *for* loop performs a fixed number of phases *np*, each consisting in a short MA (*repeat* loop) followed by a partial replacement procedure.

Each iteration of the short MA selects two parent-chromosomes P_1 and P_2 using the binary tournament technique. The crossover is applied to get one new chromosome C, which is converted by *Split* into an SDCARP solution S. With a given probability p_{LS}, solution S is improved by the local search LS and the trips of the resulting solution are concatenated by the *Concat* procedure, to return to a chromosome format.

The diversity check of the MA|PM is then applied to decide whether C is going to replace a chromosome R randomly selected in the worst half of *Pop*. The selection rule for R guarantees the *elitism* property, i.e., the best solution is preserved. C replaces R if its distance to $Pop\backslash\{T\}$ is at least Δ. However, to avoid loosing a new best solution, C is also accepted if it improves upon $Pop(1)$. Due to this exception and to the initial population structure, the distance between any two solutions in Pop may be smaller than Δ.

Algorithm 17.4. General structure of the MA|PM

1: Generate an initial population Pop of nc distinct chromosomes
2: Evaluate each chromosome using $Split$
3: Sort Pop in ascending order of costs
4: **for** $p \leftarrow 1$ **to** np **do**
5:     Initialize Δ
6:     $\alpha, \beta \leftarrow 0$
7:     **repeat**
8:         $\alpha \leftarrow \alpha + 1$
9:         Select two parents P_1 and P_2 in Pop by binary tournament
10:         $C \leftarrow Crossover(P_1, P_2)$
11:         $S \leftarrow Split(C)$
12:         **if** $random < p_{LS}$ **then**
13:             $S \leftarrow LS(S)$
14:             $C \leftarrow Concat(S)$
15:         **end if**
16:         Choose a chromosome R to be replaced in $Pop(\lfloor nc/2 \rfloor) \ldots Pop(nc)$
17:         **if** $(D(C, Pop \setminus \{R\}) \geq \Delta)$ or $(F(C) < F(Pop[1]))$ **then**
18:             $\beta \leftarrow 0$
19:             Replace R with C
20:             Shift C to keep Pop sorted
21:         **else**
22:             $\beta \leftarrow \beta + 1$
23:         **end if**
24:         Update Δ
25:     **until** $(\alpha = \alpha_{max})$ or $(\beta = \beta_{max})$
26:     $Pop \leftarrow PartialReplacement(Pop)$
27: **end for**

The short MA stops after a given number of iterations α_{max} or after a given number of iterations β_{max} without improving the best solution. The current phase ends by a partial renewal of the population (*PartialReplacement* procedure): the nk best solutions are kept (nk is a given parameter), while the other are replaced by new random chromosomes.

It is important to note that *no mutation operator* is used in this MA, contrary to the general MA model proposed by Moscato [321]. The role of mutation (increase diversity) is here played by the population management based on a distance measure. Moreover, the local search is not systematically applied to new solutions, for two reasons: a) more than 90% of the MA running time is spent in LS and a systematic local search would be too time-consuming, b) since the crossover used is able to change the splitting level, the generation of new solutions not improved by LS brings another source of diversification.

Concerning the diversity threshold Δ, four strategies were evaluated. The simplest one is the *constant policy*. For instance, $\Delta = 0$ corresponds to an ordinary MA since clones (identical solutions) are allowed, while $\Delta = 1$ guarantees a population without clones. Larger constant values improve diversity at the expense of higher rejection rates of children. The *increasing policy* is more elaborate: starting from an initial value $\Delta_{min} \geq 1$ for which clones are already prohibited, the threshold is increased linearly during iterations, until reaching a final value Δ_{max} for which clones are still forbidden. The *decreasing policy* is similar, Δ being linearly decreased. Finally, we have tested an *increasing policy with resets*: it corresponds to the basic increasing policy, but Δ is reset to Δ_{min} each time the current best solution is improved.

17.5 Computational Results

17.5.1 Instances, Implementation and Parameters

The computational evaluation is based on two sets of 42 instances each, generated by modifying the demands in standard CARP instances which can be downloaded at `http://www.uv.es/~belengue/carp.html`. The other data of the CARP instances are not changed: network, edge costs, position of depot and vehicle capacity. Each set contains 6 original CARP instances plus 6 groups of 6 new instances. Like Dror and Trudeau [304] for SDVRP instances, each group is obtained by taking the original instances and redefining each edge demand by randomly drawing one integer in a given interval defined by two fractions of vehicle capacity. The intervals for the 6 groups are [0.01,0.1], [0.1,0.3] [0.1,0.5], [0.1,0.9], [0.3,0.7] and [0.7,0.9].

The first set called *sdgdb* is derived from rather small classical CARP instances, used for instance in [323]: gdb8, gdb11, gdb14, gdb17, gdb20 and gdb23. The number of nodes n ranges from 7 to 27 and the number of edges m from 21 to 55, all required ($\tau = m$). The second set or *sdval* files contains larger problems, generated from the six classical CARP instances val1A, val1B, val1C, val7C, val8C, val9D, in which n and m vary respectively from 24 to 50 and from 39 to 92. Like in the *sdgdb* files, all edges are required. The selection of original instances aims at having different problem sizes, while discarding some instances with many unit demands, which cannot be split.

All algorithmic components are implemented in Delphi and executed on a 3 GHz PC with Windows XP Pro. After several preliminary tests, the MA|PM parameters were tuned as follows to get the best results on average: small population of $nc = 30$ chromosomes, number of phases $np = 10$, maximum number of iterations $\alpha_{max} = 3000$, maximum number of iterations without improvement $\beta_{max} = 2000$, local search rate $p_{LS} = 0.3$, number of solutions kept in partial replacements $nk = 1$ (i.e., only the best solution is kept).

Concerning the diversity control, the decreasing and increasing policies clearly dominate the constant policy and the increasing policy with resets explained in Section 17.4, but the decreasing policy slightly outperforms the increasing one. The threshold Δ is initialized to 25% of the average inter-solution distance in the population at the beginning of each phase (initial population or after a partial replacement), i.e. $\Delta_{max} = 1/4 \cdot (\sum_{X,Y \in Pop, X \neq Y} D(X,Y))/(nc \cdot (nc-1)/2)$, and decremented every $(\Delta_{max} - 1)/\alpha_{max}$ iterations to end with $\Delta_{min} = 1$.

It is important to note that the population at the beginning of each phase displays a strong diversity: most solutions are randomly generated and local search is not applied to them.

17.5.2   Results

Since no good lower bound is available for our instances, the MA|PM is compared to one of the best metaheuristics for the CARP, the memetic algorithm from Lacomme et al. [323]. Tables 1 and 2 share a common format to compare the two algorithms instance by instance, on both sets of benchmarks.

The first four columns indicate respectively the instance name, the interval of demands, the number of nodes n and the number of edges m, always equal to the number of required edges τ. Column 5 gives the results obtained by the CARP MA and listed in [323]. Solution costs found by our method are given in the *SDCARP MA|PM* column. The next two columns present respectively the savings obtained when splitting is allowed, in terms of total distance and number of vehicles used. The *Time* column gives the running time in seconds for the MA|PM. The last row displays the average value of each column.

The solution values reported are all obtained for one run and one single setting of parameters, the one detailed in previous subsection. The asterisks for the CARP instances (first group of each set) and the CARP MA indicate optimal solutions. The optimality for all *gdb* files except gdb08 and for val1A and val1B was already proven in [323], because the CARP MA reached a tight lower bound designed by Belenguer and Benavent [325]. The optimality for gdb08, val1C and val7C has been proven more recently by Longo et al. [326], using a transformation into a VRP problem solved by branch-and-cut.

The results for CARP instances indicate that the MA|PM in which splitting demands is allowed does not improve the solution values achieved by the CARP MA. This can be explained by the rather small demands in these instances. An exception is val1C, which is improved by splitting because the average demand per customer is larger. It is reassuring to see that MA|PM is always as good as the CARP MA, except on one instance (val8C).

Concerning the new *sgdb* instances, tables 1 and 2 show that MA|PM starts improving the total cost and the number of vehicles used when the average demand per customer reaches 30% of vehicle capacity (group C). An exception is sdgdb23-A: in spite of small demands, the distance is not improved but MA|PM saves two vehicles. The savings increase with the splitting level and reach 15.36% for the distance (sdgdb08-F) and 11 vehicles (sdgdb23-F).

Lacomme et al. report an average running time of 5.29 seconds for the CARP MA, on a 1 GHz Pentium III PC. This small duration is possible using an additional stopping criterion based on excellent lower bounds: on all gdb instances except one, these bounds are quickly achieved by the CARP MA. The average duration of MA|PM for the SDCARP is much larger (86.55 seconds) for three

Table 17.1. Results for SDGDB files

Instance name	Demand interval	Nodes n	Edges m	CARP MA	SDCARP MA|PM	Distance saved %	Vehicles saved	Time (s)
gdb08	0.04-0.33	27	46	*348	348	0.00	0	78.37
gdb11	0.02-0.18	22	45	*395	395	0.00	0	78.34
gdb14	0.05-0.38	7	21	*100	100	0.00	0	40.14
gdb17	0.05-0.19	8	28	*91	91	0.00	0	43.39
gdb20	0.04-0.33	11	22	*121	121	0.00	0	55.58
gdb23	0.04-0.33	11	55	*233	233	0.00	0	93.88
sdgdb08-A	0.01-0.1	27	46	250	250	0.00	0	73.49
sdgdb11-A	0.01-0.1	22	45	387	387	0.00	0	74.86
sdgdb14-A	0.01-0.1	7	21	96	96	0.00	0	37.81
sdgdb17-A	0.01-0.1	8	28	91	91	0.00	0	40.75
sdgdb20-A	0.01-0.1	11	22	121	121	0.00	0	38.83
sdgdb23-A	0.01-0.1	11	55	223	223	0.00	2	94.76
sdgdb08-B	0.1-0.3	27	46	334	334	0.00	0	79.00
sdgdb11-B	0.1-0.3	22	45	467	467	0.00	0	137.09
sdgdb14-B	0.1-0.3	7	21	100	100	0.00	0	42.23
sdgdb17-B	0.1-0.3	8	28	91	91	0.00	0	46.74
sdgdb20-B	0.1-0.3	11	22	123	123	0.00	0	40.11
sdgdb23-B	0.1-0.3	11	55	235	235	0.00	0	93.94
sdgdb08-C	0.1-0.5	27	46	412	404	1.94	1	208.59
sdgdb11-C	0.1-0.5	22	45	550	544	1.09	0	139.84
sdgdb14-C	0.1-0.5	7	21	104	104	0.00	0	40.66
sdgdb17-C	0.1-0.5	8	28	95	95	0.00	0	49.39
sdgdb20-C	0.1-0.5	11	22	128	128	0.00	0	42.36
sdgdb23-C	0.1-0.5	11	55	245	245	0.00	0	143.08
sdgdb08-D	0.1-0.9	27	46	742	673	9.30	1	89.45
sdgdb11-D	0.1-0.9	22	45	834	773	7.31	2	117.06
sdgdb14-D	0.1-0.9	7	21	120	118	1.66	0	51.39
sdgdb17-D	0.1-0.9	8	28	111	107	3.60	2	64.08
sdgdb20-D	0.1-0.9	11	22	163	153	6.13	2	44.06
sdgdb23-D	0.1-0.9	11	55	283	277	2.12	1	103.77
sdgdb08-E	0.3-0.7	27	46	727	665	8.53	2	222.55
sdgdb11-E	0.3-0.7	22	45	799	750	6.13	2	146.27
sdgdb14-E	0.3-0.7	7	21	144	138	4.16	0	41.05
sdgdb17-E	0.3-0.7	8	28	109	107	1.83	1	62.06
sdgdb20-E	0.3-0.7	11	22	176	159	9.66	3	42.95
sdgdb23-E	0.3-0.7	11	55	311	297	4.50	3	102.72
sdgdb08-F	0.7-0.9	27	46	1165	986	15.36	9	179.50
sdgdb11-F	0.7-0.9	22	45	1301	1133	12.91	8	211.14
sdgdb14-F	0.7-0.9	7	21	210	178	15.24	4	79.75
sdgdb17-F	0.7-0.9	8	28	133	125	6.02	4	50.01
sdgdb20-F	0.7-0.9	11	22	217	201	7.37	3	44.11
sdgdb23-F	0.7-0.9	11	55	403	361	10.42	11	170.05
Average				316.24	298.26	3.32	1.45	86.55

Table 17.2. Results for SDVAL files

Instance name	Demand interval	Nodes n	Edges m	CARP MA	SDCARP MA|PM	Distance saved %	Vehicles saved	Time (s)
val1A	0.01-0.09	24	39	*173	173	0.00	0	41.19
val1B	0.02-0.16	24	39	*173	173	0.00	0	58.58
val1C	0.04-0.42	24	39	*245	239	2.45	0	93.63
val7C	0.05-0.25	40	66	*334	334	0.00	0	97.31
val8C	0.06-0.28	30	63	527	535	-1.52	0	97.58
val9D	0.03-0.23	50	92	391	391	0.00	0	161.14
sdval1A-A	0.01-0.1	24	39	173	173	0.00	0	41.98
sdval1B-A	0.01-0.1	24	39	173	173	0.00	0	42.19
sdval1C-A	0.01-0.1	24	39	173	173	0.00	0	41.67
sdval7C-A	0.01-0.1	40	66	283	283	0.00	0	92.20
sdval8C-A	0.01-0.1	30	63	386	386	0.00	0	87.58
sdval9D-A	0.01-0.1	50	92	332	332	0.00	0	189.00
sdval1A-B	0.1-0.3	24	39	237	235	0.84	0	58.23
sdval1B-B	0.1-0.3	24	39	245	245	0.00	0	42.98
sdval1C-B	0.1-0.3	24	39	236	235	0.42	0	45.14
sdval7C-B	0.1-0.3	40	66	406	402	0.99	0	117.48
sdval8C-B	0.1-0.3	30	63	631	637	-0.95	0	81.50
sdval9D-B	0.1-0.3	50	92	561	557	0.71	1	310.64
sdval1A-C	0.1-0.5	24	39	306	298	2.61	0	59.94
sdval1B-C	0.1-0.5	24	39	326	314	3.68	0	69.74
sdval1C-C	0.1-0.5	24	39	314	306	2.55	1	45.47
sdval7C-C	0.1-0.5	40	66	542	533	1.66	0	123.78
sdval8C-C	0.1-0.5	30	63	912	909	0.33	1	215.09
sdval9D-C	0.1-0.5	50	92	752	739	1.73	1	514.81
sdval1A-D	0.1-0.9	24	39	499	444	11.02	2	77.64
sdval1B-D	0.1-0.9	24	39	496	450	9.27	2	75.36
sdval1C-D	0.1-0.9	24	39	456	432	5.26	1	45.62
sdval7C-D	0.1-0.9	40	66	899	817	9.12	6	129.51
sdval8C-D	0.1-0.9	30	63	1395	1325	5.02	3	258.75
sdval9D-D	0.1-0.9	50	92	1159	1087	6.21	3	655.98
sdval1A-E	0.3-0.7	24	39	504	450	10.71	2	62.67
sdval1B-E	0.3-0.7	24	39	481	446	7.28	1	73.64
sdval1C-E	0.3-0.7	24	39	485	447	7.84	2	46.73
sdval7C-E	0.3-0.7	40	66	978	839	14.21	8	282.51
sdval8C-E	0.3-0.7	30	63	1372	1319	3.86	2	233.84
sdval9D-E	0.3-0.7	50	92	1236	1103	10.76	3	685.33
sdval1A-F	0.7-0.9	24	39	798	687	13.91	7	85.55
sdval1B-F	0.7-0.9	24	39	798	689	13.66	7	58.41
sdval1C-F	0.7-0.9	24	39	798	706	11.53	7	51.08
sdval7C-F	0.7-0.9	40	66	1387	1218	12.18	11	447.39
sdval8C-F	0.7-0.9	30	63	2413	2064	14.46	12	139.78
sdval9D-F	0.7-0.9	50	92	1979	1680	15.11	17	693.24
Average				642.10	594.71	4.69%	2.38	162.66

main reasons: a) no lower bound is available to stop the algorithm before its planned number of iterations, b) the local search is more involved with its splitting moves and c) some iterations are not productive because the new solution is rejected by the distance test. If the MA running time is scaled for our 3 GHz PC (1.76 s approximately), the SDCARP algorithm is 49 slower than the CARP MA.

Similar conclusions can be deduced from table 2 for the *sdval* instances. This time, compared to the CARP MA, some instances are improved as from an average demand per customer equal to 0.2 (group B). The distance saved reaches 15.11% and the decrease in the number of vehicles 17 for sdval9D-F. For this problem, note that the CARP solution requires 92 vehicles, since the demands in $[0.7W, 09.W]$ imply one trip per required edge.

Compared to *gdb* files, the average running time increases, due to instance size: 2.7 minutes instead of 1.4. However, the time ratio MA|PM/MA is much smaller: in [323], the CARP MA requires 38.35 seconds on average on a 1 GHz PC, corresponding approximately to 12.78 seconds on our machine: the ratio is close to 13, versus 49 for *sdgdb* instances. The savings are more substantial on *sdval* problems: this can be explained by larger vehicle capacities.

17.6   Conclusion

This article investigates an NP-hard vehicle routing problem problem never studied before, to the best of our knowledge: the Split Delivery CARP or SDCARP. The proposed solution method is a memetic algorithm with population management (MA|PM). This metaheuristic is based on a non-trivial distance measure to control population diversity, and on new moves able to split edges and to group partial services. Compared to the case witout splitting (standard CARP), significant savings can be obtained, especially when the average customer demand represents an important fraction of vehicle capacity.

The advantage of allowing splitting was not obvious. The papers devoted to the SDVRP consider instances without service times at customers. In the SDCARP, the length of a split edge is counted each time the edge is traversed for a partial service and split deliveries seem less appealing. In spite of this important difference with the SDVRP, splitting can be an interesting option.

The current implementation could be perfected, the weak point being the rather large running times which can be explained by the lack of tight lower bounds to stop the algorithm earlier. The design of such bound is in progress, using a cutting plane algorithm based on a compact integer linear programming formulation of the SDCARP.

Possible extensions of the SDCARP include time windows and U-turns in the middle of an edge. Time windows are often present in real applications like municipal waste collection: for instance, operations must be stopped during rush hours or in case of works in some streets. The U-turns are allowed on the nodes but not on the edges in the SDCARP version studied in this paper. Nevertheless, U-turns on edges are possible in some applications, like treating railroads with herbicides. In such applications, the volume of pesticide sprayed is proportional to the length treated: the distribution of demand along each edge is then uniform and a formulation with U-turns would be possible.

Acknowledgements

This research is part of a larger project on complex vehicle routing problems, funded by the Champagne – Ardenne Regional Council (France) and the European Social Fund. The authors wish to thank J.-M. Belenguer and E. Benavent for fruitful discussions on the SDCARP.

18

A Permutation Coding with Heuristics for the Uncapacitated Facility Location Problem

Bryant A. Julstrom

Department of Computer Science, St. Cloud State University,
St. Cloud, MN 56301 USA
julstrom@stcloudstate.edu

Summary. Given a collection of warehouse locations, each with a fixed cost, and customers to be served from those warehouses, each with a cost associated with both the customer and the one warehouse from which the customer is served, the uncapacitated facility location problem seeks to identify a subset of the warehouse locations that minimizes the total cost. A genetic algorithm for this NP-hard problem encodes candidate subsets of the warehouse locations as permutations of all the available locations; a greedy decoder identifies the subset that such a chromosome represents. Three heuristic extensions reorder chromosomes so that they list included locations before excluded; mutate chromosomes by always swapping an included location with an arbitrary one; and re-scan the included locations to exclude any whose exclusion reduces the total cost. Four versions of the GA implement none, one, two, or all of these extensions. In tests on 235 publicly available problem instances whose optimum solutions are known, all the versions outperform a straightforward binary-coded GA, and the heuristic extensions enable the version that uses them all to find optimum solutions very quickly on almost every trial on the test instances. The heuristic techniques should be effective in permutation-coded evolutionary algorithms for other problems that seek subsets of initially unknown sizes.

Keywords: Permutation Coding, Greedy Decoder, Heuristics, Facility Location, Warehouse Location, Plant Location.

18.1 Introduction

Diversified Products Company (DiPCo) is opening a new service area. The company has contracted with a number of customers in the area and it has identified locations at which it might open warehouses from which to serve those customers. Its managers must now choose the particular locations at which DiPCo will establish its facilities.

The company incurs fixed costs for each warehouse it establishes, and these costs differ; some locations are more expensive than others. Each customer will be served from one warehouse, and the costs of providing that service vary across both the warehouse locations and the customers. The managers seek to open warehouses at a set of locations that minimizes their total cost, which is the sum of the selected locations' fixed costs and the costs of serving customers from

C. Cotta and J. van Hemert (Eds.): Recent Advances in Evol. Comp., SCI 153, pp. 295–307, 2008.
springerlink.com

those locations. DiPCo's managers are facing an instance of the *uncapacitated facility location problem*.

The standard evolutionary coding of candidate solutions to this problem is binary strings of length equal to the number of warehouse locations; 1 indicates that a warehouse will be established at the corresponding location; 0 that one will not. Here, however, a genetic algorithm encodes candidate sets of warehouse locations as permutations of all the available locations. A greedy decoder identifies the subset of the available locations that each such chromosome represents.

The permutation coding supports three heuristic extensions. The first reorders each chromosome so that it lists the locations included in the subset that the chromosome represents before it lists the excluded locations. The second modifies the traditional swapping mutation operator to guarantee that at least one included location will always be swapped. The third re-scans the included locations that a chromosome lists and excludes any whose exclusion reduces the total cost of maintaining the warehouses and serving the customers.

A standard genetic algorithm represents candidate sets of locations with binary strings, and to these chromosomes it applies two-point crossover and position-by-position mutation. Four versions of the permutation-coded GA implement none, one, two, or all of the heuristic extensions. All five algorithms are compared on 235 publicly available instances of the uncapacitated facility location problem whose optimum solutions are known. All the GAs are at least fairly effective on the test instances, though all the versions of the permutation-coded GA return better results on average than does the binary-coded algorithm. In the permutation-coded GA, the reordering step alone makes little difference, but it supports the other two extensions, which together enable the algorithm to find optimum solutions very quickly on almost every trial.

The following sections of this chapter describe the uncapacitated facility location problem formally and in more detail; present the permutation coding, operators appropriate to it, and the three heuristic extensions; summarize the genetic algorithm's structure and its several versions; and report on trials of the binary-coded GA and the versions of the permutation-coded GA on the test problem instances.

18.2   The Problem

Let I be a set of n locations at which warehouses may be established. Associated with each location $i \in I$ is the fixed cost w_i of operating a warehouse at that location. Let J be a set of m customers. Associated with each customer $j \in J$ and each warehouse location $i \in I$ is the cost $c_{i,j}$ of serving customer j from location i. Each customer will be served from exactly one warehouse, and every warehouse, wherever established, has the capacity to serve all the customers that may be assigned to it. We seek a set of warehouse locations that minimizes the total cost of serving the customers. Note that the number of warehouses to be established is not specified and that once their locations are chosen, each customer j will be served by the warehouse i for which $c_{i,j}$ is smallest.

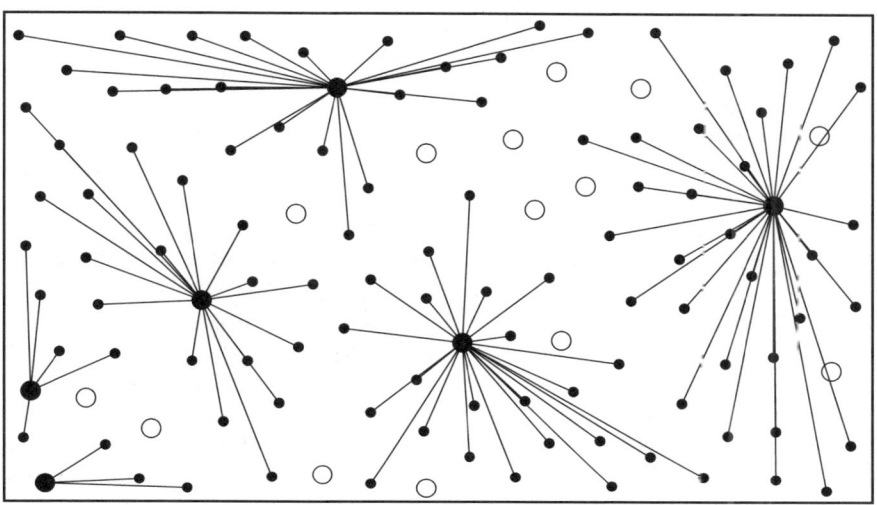

Fig. 18.1. An instance of the uncapacitated facility location problem with $n = 30$ warehouse locations and $m = 100$ customers and its solution. Small dots are customers, solid disks are chosen warehouse locations, open circles are other warehouse locations, and the Euclidean distance between each customer and each location represents the cost of serving that customer from that location.

More formally, given the sets I and J and the costs w_i and $c_{i,j}$, we seek a subset $S \subseteq I$ that minimizes the sum

$$C(S) = \sum_{i \in S} w_i + \sum_{j=1}^{m} \min_{i \in S}\{c_{i,j}\}. \tag{18.1}$$

This is the uncapacitated facility location problem (UFLP), also called the unconstrained or uncapacitated warehouse location problem and the simple plant location problem.

Figure 18.1 shows an instance of the UFLP in which $n = 30$, $m = 100$, the warehouses' fixed costs increase from left to right, and $c_{i,j}$ is the Euclidean distance between warehouse location i and customer j. In the figure, small dots represent customers, and solid larger disks represent chosen warehouse locations. Open disks represent locations at which warehouses are not established; though they are close to many customers, their fixed costs exclude them.

The UFLP is similar to but distinct from the p-median (or k-median) problem. In the latter, there is only one set of locations, and the number p of them that will be chosen as "near" the others is a fixed parameter of the problem. In the *capacitated* location problem, each customer has nonzero demand and each warehouse has a capacity; the sum of a warehouse's customers' demands cannot exceed the warehouse's capacity. Though some simple instances of the UFLP can be solved in polynomial time [327], in general the problem is, like the p-median and capacitated facility location problems, NP-hard [328] [329].

The uncapacitated facility location problem has many applications in operations research and other areas and so has been extensively studied. Researchers have applied to it a variety of exact and heuristic techniques. Dearing [330], Gao, Robinson, and Powell [331], and Krarup and Pruzan [329] have all surveyed algorithms for the problem.

Among the heuristics applied to the UFLP have been evolutionary algorithms by Kratica, Filipovič, and Tošić [332], Horng, Lin, Liu, and Kao [333], Kratica, Tošić, Filipovič, and Ljubić [334], and Jaramillo, Bhadury, and Batta [335]. Researchers have also described EAs for the capacitated problem (e.g. Rao, Hanagandi, and Buescher [336]). Each of these algorithms represented candidate sets of warehouse locations with binary strings of length n, as described above.

Recently, Yigit, Aydin, and Turkbey [337] encoded candidate subsets of warehouse locations directly as sets; that is, as lists of included locations. Three operators on these chromosomes add a location, remove a location, or both, respectively. These operators identify neighboring candidate solutions in short runs of simulated annealing that replace the usual genetic operators in evolutionary algorithms for the UFLP; there is no crossover operator. Yigit, Aydin, and Turkbey report good results on a variety of test instances, including some used here.

18.3   A Permutation Coding

Candidate subsets of warehouse locations in the UFLP can also be represented to good effect by permutations of all the available locations, with a decoding algorithm that identifies the subsets such chromosomes represent. This section describes such a coding, a greedy decoder and operators appropriate to it, and three heuristic extensions of these algorithms.

18.3.1   The Coding and Its Evaluation

A candidate solution to an instance of the uncapacitated facility location problem is a subset of the instance's n warehouse locations. Let a chromosome be a permutation of all the locations, only some of which, in general, are members of the subset that the permutation represents.

A greedy algorithm decodes and evaluates chromosomes. The subset of the warehouse locations that a chromosome represents always includes the first location the chromosome lists, and initially a warehouse at this location serves every customer; that is, the cost is sum of the fixed cost of the first location and the sum of the costs of serving all the customers from there. The decoder then scans the remaining locations, in their order in the chromosome. If a warehouse at a location reduces the total cost of serving the customers, then that location is included in the subset; otherwise, it is not. When all the locations have been considered in chromosome order, the chromosome's fitness is the total cost (18.1) incurred by the chosen warehouse locations.

Every chromosome—every permutation of the warehouse locations—represents a candidate solution to the target UFLP instance; there are no invalid chromosomes. Also, the coding is redundant: A chromosome determines the order in which the decoder examines the warehouse locations; rearranging the chromosome does not always change the subset of locations to which it decodes, so several different chromosomes may represent the same subset.

The following sketch summarizes the decoder. In it, chr[·] is a chromosome; that is, a permutation of the potential warehouse locations. S is the developing subset of warehouse locations, and C is the cost of serving the customers from the locations in S.

$S \leftarrow \{ \, \text{chr[0]} \, \}$;
$C \leftarrow w_{\text{chr[0]}} + \sum_{j=1}^{m} c_{\text{chr[0]},j}$;
for p from 2 to $n-1$
$\quad S' \leftarrow S \cup \{ \, \text{chr[p]} \, \}$;
$\quad C' \leftarrow \sum_{i \in S} w_i + \sum_{j=1}^{m} \min_{i \in S} \{ c_{i,j} \}$;
$\quad$ if $C' < C$
$\quad\quad S \leftarrow S'$;
$\quad\quad C \leftarrow C'$;
fitness $\leftarrow C$;

Note that the computation of C' can be done efficiently by keeping track of the currently-included warehouse location that us closest to each customer. Consideration of each location p then requires, for each customer, only two array look-ups and a comparison.

18.3.2 Operators

A wide variety of crossover operators have been described for chromosomes that are permutations [338, pp.52–56] [339, pp.216–233] [340]. However, researchers developed these operators for problems that sought optimum *orderings* of problem elements, such as job-shop scheduling and the familiar Travelling Salesman Problem. The UFLP seeks an optimum *subset* of warehouse locations.

Here, a chromosome determines the order in which the decoder considers locations, and locations listed early in a chromosome are more likely to be included in the subset the chromosome represents than are locations listed later. An appropriate crossover for these permutations is alternation: build an offspring by alternating the locations listed in two parents, then eliminate duplicates; each location's first appearance survives. For example, if $n = 6$, the alternation of (4 2 1 0 3 5) and (2 5 3 0 1 4) is (4 2 2 5 1 3 0 0 3 1 5 4); removing duplicates yields (4 2 5 1 3 0). This operator tends to keep locations that appear early in parent chromosomes near the front of offspring, so that locations likely to be included in parents' subsets are likely to be included in the subsets their offspring represent.

A mutation operator regularly used with permutations simply swaps the warehouse locations at two random positions in the parent chromosome: (4 2 <u>1</u> 0 <u>3</u> 5) might become (4 2 <u>3</u> 0 <u>1</u> 5). It is possible that an offspring of mutation will represent the same solution as its parent, particularly when both swapped locations fall near the end of the chromosome and are therefore unlikely to be included in the chromosome's subset.

18.3.3 Heuristic Extensions

The permutation coding of candidate subsets, the greedy decoder, and the operators just described admit several heuristic extensions that—we will see—markedly improve the coding's effectiveness.

First, locations listed early in a chromosome are likely to be included in the subset the decoder identifies, and those listed later are not, but in general a chromosome will interleave included and excluded locations. An extension of the decoder reorders the locations in each chromosome so that those in the chromosome's subset precede the excluded locations, though within each group, order is preserved.

For example, if the chromosome (4 5 0 2 1 3) decodes to the subset $\{4, 5, 2\}$, this step reorders the chromosome to become (**4 5 2** 0 1 3); the included warehouse locations precede the excluded ones, without changing the included subset. The chromosome itself is augmented with the number of included locations, thus the length of the chromosome's prefix that they occupy.

Second, with chromosomes rearranged by the extended decoder, mutation can easily be modified to always swap an included warehouse location with an arbitrary one. This mutation will make an included location less likely to be included and an excluded location more likely, or reverse the order in which two included locations are examined. In either case, it increases the likelihood that the subset an offspring chromosome represents will differ from its parent's subset.

The decoder includes a location in a subset when the location lowers the total cost of serving the customers. Locations listed and included later may render earlier inclusions unnecessary; removing some earlier locations might further reduce the total cost. One more step, then, can follow evaluation and reordering. This step re-examines the included locations, which now form a prefix of the chromosome. When the removal of a warehouse from one of these locations reduces the total cost, that location is swapped with the last location in the prefix and the subset size is reduced by one; that is, that location is no longer part of the solution. This step can be thought of as a directed local search around the original chromosome's solution.

The following sketch summarizes the local search step. Again, chr[·] is a chromosome, S is the subset of included warehouse locations, and C is the cost of serving the customers from the locations in S. In addition, ℓ is the number of locations in S; that is, the length of the prefix of chr[·] that they occupy.

> for p from 0 to $\ell - 1$
> $\quad S' \leftarrow S - \{\text{chr}[p]\};$
> $\quad C' \leftarrow \sum_{i \in S'} w_i + \sum_{j=1}^{m} \min_{i \in S'}\{c_{i,j}\};$

if $C' < C$
   $S \leftarrow S'$;
   $C \leftarrow C'$;
   swap(chr$[p]$, chr$[\ell - 1]$);
   $\ell \leftarrow \ell - 1$;

For example, suppose that $n = 10$ and the chromosome (**7 3 8 6 1 2** 0 9 5 4) represents the subset $\{7, 3, 8, 6, 1, 2\}$, so that $\ell = 6$. Rescanning the first six locations that the chromosome lists reveals that excluding location 8 reduces the total cost. The resulting chromosome is then (**7 3 2 6 1** 8 0 9 5 4), representing the subset $\{7, 3, 2, 6, 1\}$ and with $\ell = 5$.

In the early nineteenth century, Lamarck suggested that characteristics organisms acquired during their lifetimes might be passed on to their offspring [341, pp.76–80]. Though biologically impossible—no natural mechanism purposefully modifies an organism's genotype—Lamarckian modification can aid seach in evolutionary algorithms. Here, both the reordering and re-scanning heuristic steps modify the GA's chromosomes outside of its Darwinian mechanisms and so can be considered Lamarckian.

Memetic algorithms [342] are hybrid heuristics that combine evolutionary, population-based search and local, individual search. Here, reordering is not memetic, since it does not investigate or change the represented solution, but re-scanning is. Re-scanning looks at neighboring solutions, seeking improvements and recording in the chromosome those it finds.

The three heuristic extensions were effective in a permutation-coded genetic algorithm for the minimum-label spanning tree problem [343]; here too they significantly improve a GA's performance, as we will see.

18.4   A Genetic Algorithm

The permutation coding of candidate subsets of warehouse locations, the variation operators, and the three heuristic steps described in the previous section were implemented in a straightforward generational genetic algorithm. The GA initializes its population with random permutations of the target UFLP instance's warehouse locations. It selects chromosomes to be parents in k-tournaments without replacement, and it generates each offspring by applying either mutation or crossover to one or two parents, respectively. The algorithm is 1-elitist; it copies the best chromosome of the current generation unchanged into the next. It runs through a fixed number of generations.

Four versions of the GA differ in the number of heuristic steps from Section 18.3.3 that they apply. The first, called GA0, applies none of these steps. The second, GA1, reorders chromosomes so that the locations included in the subset it represents precede the excluded ones. The third, GA2, extends GA1 with heuristic mutation; recall that this operator takes advantage of the reordering of chromosomes to always exchange an included location with an arbitrary one. The fourth version, GA3, augments GA2 by re-scanning each reordered chromosome's included locations to exclude those whose exclusion reduces the

Table 18.1. The heuristic steps implemented by the three versions of the genetic algorithm

Version	Reordering	Heuristic mutation	Re-scanning
GA0	No	No	No
GA1	Yes	No	No
GA2	Yes	Yes	No
GA3	Yes	Yes	Yes

total cost. Table 18.1 summarizes these variations. All four versions apply the alternation crossover operator.

GA0, GA1, GA2, and GA3 were implemented in C++, and in the comparisons that the next section describes, the GA's parameter values were constant across the four versions. On a UFLP instance with n warehouse locations, its population contained $2n$ chromosomes. The size of its selection tournaments was two, and each tournament's winner was selected with probability 0.90, its loser with probability 0.10, so selection pressure was moderate. The probability that crossover generated each offspring chromosome was 0.70, that of mutation 0.30, and the GA ran through $10n$ generations or until it encountered a known optimum solution.

For comparison, a binary-coded genetic algorithm was also implemented. It represented candidate sets of warehouse locations as binary strings whose bits indicated the inclusion or exclusion of the locations, and it applied two-point crossover and position-by-position mutation to these chromosomes. Its structure and parameters were otherwise identical to those of the permutation-coded GA, and in its mutation operator, the probability that any one bit would be inverted was $1/n$. We can call this algorithm BGA.

18.5 Comparisons

The binary-coded GA and the four versions of the permutation-coded GA, with their various levels of heuristic enhancement, were compared on 235 instances of the uncapacitated facility location problem. Fifteen of the instances are found in Beasley's OR-Library[1] [344], as described in [345]. Twelve of these instances are small, with 16, 25, or 50 warehouse locations and 50 customers. They are named `cap71` through `cap74`, `cap101` through `cap104`, and `cap131` through `cap134`, respectively. The last three are larger, with 100 locations and 1000 customers; these are named `capa`, `capb`, and `capc`.

The remaining 220 UFLP instances were described by Bilde and Krarup[2] [346]. 100 of these instances have 30 warehouse locations and 80 customers; they are named `Di.j`, with $i, j \in \{1, 2, \ldots, 10\}$. The other 120 instances have 50

---

[1] http://people.brunel.ac.uk/ mastjjb/jeb/info.html
[2] http://www.mpi-inf.mpg.de/departments/d1/projects/benchmarks/UflLib/

locations and 100 customers; they are named B1.j, C1.j, and Ei.j, with i, j
∈ {1, 2, . . . , 10}. All these instances are non-metric and were randomly generated.
The costs of optimum solutions for both the OR-Library and the Bilde-Krarup
instances are known, so we can easily evaluate the performances of the several
genetic algorithms and the contributions of the heuristic techniques.

BGA, GA0, GA1, GA2, and GA3 were each run 40 independent times on each
UFLP instance, on a Pentium 4 processor with 1Gb of memory, running at 2.56
GHz under Red Hat Linux 9.0. Table 18.2 summarizes the results of these trials.

It is impractical—and not particularly instructive—to present the results for
each of the 235 instances individually. For each *group* of instances, then, the table
lists its numbers of warehouse locations and customers and the total number of
trials on the group's instances. For each GA on each group of instances, the
table lists the number of trials that found optimum solutions ('Hits"), the mean
percentage by which the GA's best solutions exceeded the known optimum costs,
and the mean standard deviation of those percentages. Note that three of the
OR-Library groups contain four instances and one contains three, while each
group of Bilde-Krarup instances contains ten individual problems.

All the genetic algorithms return good results on the test UFLP instances.
The worst performance is that of the binary-coded GA on the E2 instances
of Bilde and Krarup, where the costs of the solutions its returns exceed the
optimum costs by an average of 1.425%. All the versions of the permutation-
coded GA consistently outperform the binary-coded algorithm. Even GA0, with
no heuristic enhancements, finds optimum solutions more often than does BGA.
Moreover, GA0's mean errors are less than those of BGA on every set of instances
except two (Bilde and Krarup's D7 and E7 instances).

The consistent superiority of even GA0 to BGA can, I think, be attributed
to the decoding algorithm that identifies the set of locations that a chromosome
represents. The decoder is greedy, including a location only if it reduces the total
cost computed at each step, while the the binary coding represents all subsets—
including those of high cost—uniformly. With the permutation coding and its
decoder, a GA avoids representing and examining many poor solutions.

Though the binary-coded GA sometimes fails to find optimum solutions to
the smaller ($m = 50$) problem instances from the OR-Library, these instances
offer no challenge to any version of the permutation-coded GA. All four versions
find optimum solutions on every trial when $n = 16$ and 25, and when $n = 50$,
GA0 and GA1 miss on only six trials out of 160 and GA1 and GA2 hit every
time.

More generally, all four versions of the permutation-coded GA perform well
on all the instances, though it is no surprise that GA0, with no heuristic steps,
is the weakest of them. The worst performance is delivered by GA0 on the E1
and E2 instances of Bilde and Krarup. On the former, it finds optimum solutions
on 164 out of 400 trials, with mean error—excess cost above the optima—of
0.736%. On the latter, the number of hits is 168, with mean error of 0.775%.
On all the other instances, the proportion of hits is larger and the mean error is

Table 18.2. Summary of the trials of the binary-coded GA and the four versions of the permutation-coded GA on the OR-Library UFLP instances and the Bilde-Krarup instances. For each set of instances, the table lists its numbers of warehouse locations (n) and customers (m) and the total number of trials on the instances. For each GA on each set of instances, the table lists the total number of runs that found optima (Hits), the average percentage by which the trials' solutions exceeded the optima ($+\%$), and the average standard deviations of those percentages ($s(\%)$).

Inst.	n	m	Trials	BGA			GA0			GA1			GA2			GA3		
				Hits	$+\%$	$s(\%)$	Hits	$+\%$	$s(\%)$	Hits	$+\%$	$s(\%)$	Hits	$+\%$	$s(\%)$	Hits	$+\%$	$s(\%)$
cap7*	16	50	160	154	0.007	0.016	160	0.000	0.000	160	0.000	0.000	160	0.000	0.000	160	0.000	0.000
cap10*	25	50	160	79	0.033	0.034	160	0.000	0.000	160	0.000	0.000	160	0.000	0.000	160	0.000	0.000
cap13*	50	50	160	91	0.041	0.061	158	0.001	0.007	156	0.002	0.006	160	0.000	0.000	160	0.000	0.000
cap[abc]	100	1000	120	53	0.385	0.507	61	0.261	0.291	62	0.311	0.257	100	0.011	0.009	100	0.011	0.008
B1	50	100	400	346	0.069	0.208	374	0.020	0.035	368	0.022	0.023	400	0.000	0.000	400	0.000	0.000
C1	50	100	400	217	0.660	0.972	253	0.371	0.669	265	0.303	0.499	310	0.106	0.109	378	0.019	0.036
D1	30	80	400	152	0.828	0.882	219	0.406	0.622	232	0.320	0.529	273	0.144	0.241	352	0.026	0.037
D2	30	80	400	134	0.950	0.916	204	0.503	0.603	220	0.451	0.571	330	0.169	0.268	379	0.031	0.088
D3	30	80	400	199	0.672	0.937	266	0.313	0.556	264	0.296	0.481	376	0.040	0.096	400	0.000	0.000
D4	30	80	400	250	0.537	0.964	313	0.201	0.396	306	0.216	0.404	391	0.011	0.024	400	0.000	0.000
D5	30	80	400	293	0.348	0.626	357	0.125	0.288	352	0.115	0.276	400	0.000	0.000	400	0.000	0.000
D6	30	80	400	214	0.632	0.700	344	0.164	0.259	338	0.205	0.271	389	0.033	0.066	400	0.000	0.000
D7	30	80	400	328	0.362	0.820	376	0.083	0.180	376	0.083	0.180	400	0.000	0.000	400	0.000	0.000
D8	30	80	400	252	0.779	0.989	330	0.317	0.521	333	0.307	0.515	400	0.000	0.000	400	0.000	0.000
D9	30	80	400	334	0.299	0.764	381	0.048	0.213	383	0.054	0.230	400	0.000	0.000	400	0.000	0.000
D10	30	80	400	242	0.825	1.181	321	0.307	0.471	321	0.306	0.474	400	0.000	0.000	400	0.000	0.000
E1	50	100	400	130	1.213	1.224	164	0.736	0.786	195	0.637	0.767	294	0.200	0.276	374	0.026	0.069
E2	50	100	400	91	1.425	1.238	168	0.775	0.829	163	0.775	0.864	329	0.049	0.402	377	0.031	0.074
E3	50	100	400	135	1.141	1.204	207	0.586	0.747	203	0.607	0.764	354	0.119	0.233	397	0.008	0.037
E4	50	100	400	183	0.755	0.972	250	0.322	0.493	249	0.320	0.481	379	0.026	0.104	396	0.003	0.015
E5	50	100	400	194	0.977	1.166	257	0.479	0.742	256	0.496	0.756	380	0.006	0.020	400	0.000	0.000
E6	50	100	400	179	0.982	1.146	260	0.495	0.730	264	0.489	0.754	383	0.018	0.030	400	0.000	0.000
E7	50	100	400	218	0.632	0.867	283	0.267	0.420	281	0.283	0.435	383	0.010	0.029	391	0.004	0.011
E8	50	100	400	212	0.890	1.121	315	0.282	0.415	318	0.263	0.402	385	0.045	0.059	387	0.039	0.057
E9	50	100	400	244	0.627	0.793	328	0.249	0.465	329	0.234	0.484	400	0.000	0.000	400	0.000	0.000
E10	50	100	400	247	0.592	0.795	310	0.247	0.466	321	0.203	0.453	400	0.000	0.000	400	0.000	0.000

smaller; on no other group of instances is GA0's mean error greater than 0.586% or is its proportion of hits less that 51%.

Reordering chromosomes so that they list included warehouse locations before excluded ones distinguishes GA1 from GA0, but this step alone has little visible effect on the algorithm's performance. Of the 26 sets of instances, GA1 finds optimum solutions more often than does GA0 on 11, less often on 11, and the same number of times on the remaining four. (These last include the two smallest sets from the OR-Library, on which every version always hits.) The mean error percentages are similarly inconsistent. Though the reordering step does not itself make the GA more effective, reordering enables the other heuristic steps, which do consistently improve the GA's performance.

GA2 implements the heuristic mutation, which swaps an included warehouse location with an arbitrary one. With this operator, GA2 identifies optimum solutions more often than GA0 or GA1 on every set of instances except those on which the simpler algorithms' performances were already perfect. Often the number of hits increases by half or more; for example, from 204 out of 400 for GA0 to 330 for GA2 on the D2 instances and from 168 out of 400 for GA0 to 329 for GA2 on the E2 instances. On no set of instances is GA2's mean error greater than 0.2% (E1), and it hits on every trial on eleven instance sets of the 26.

The degree to which the heuristic mutation operator improves the GA's performance might seem surprising. However, the sets of included warehouse locations are usually small with respect to all the available locations, so naive mutation often reorders excluded locations, reproducing the parental solution and adding little to the search. By almost always changing the subset of included locations that a chromosome represents, heuristic mutation implements a wider and more aggressive investigation of the search space. Indeed, the heuristic mutation carries out approximately the same local search that makes the hybrid of simulated annealing and an evolutionary algorithm described by Yigit, Aydin, and Turkbey [337] so effective.

GA3 augments GA2 by re-scanning chromosomes' included locations, and this step yields marked improvement over GA2. On the cap[abc] instances, the largest used here, GA3 identifies optimum solutions slightly less often than does GA2 (106 hits out of 120 vs. 100), and its mean error is the same, but on all the other sets of instances, GA3 identifies optimum solutions more often (if GA2's performance was not already perfect), and its mean error is smaller. In particular, GA3 hits on every trial on 16 instance sets out of 26, and its mean error is never more that 0.039% (the E8 instances of Bilde and Krarup).

In addition to hitting on every trial on the smaller OR-Library instances, all the versions of the GA found optimum solutions on every trial on twelve other individual instances. These allow us to examine the performance improvements occasioned by the three heuristic techniques in another way: by comparing the mean numbers of generations and the mean clock times that the GA versions expended to find those optima. Table 18.3 lists these values for the four GA versions applied to these twelve instances.

Table 18.3. For the individual UFLP instances on which every version of the GA always found optimum solutions, the mean numbers of generations $\overline{\text{Gens}}$ and the mean clock times $\bar{t}$ in seconds required by the four versions to find those optima, and the standard deviations $s(t)$ of the times

Instance	n	m	GA0 Gens	$\bar{t}$	$s(t)$	GA1 Gens	$\bar{t}$	$s(t)$	GA2 Gens	$\bar{t}$	$s(t)$	GA3 Gens	$\bar{t}$	$s(t)$
cap134	50	50	48.6	0.233	0.360	30.6	0.151	0.167	10.6	0.058	0.036	0.8	0.013	0.005
B1.8	50	100	84.9	0.854	0.779	69.6	0.694	0.786	15.1	0.163	0.103	0.6	0.027	0.010
D4.5	30	80	26.7	0.077	0.108	16.8	0.050	0.060	5.6	0.019	0.015	0.8	0.008	0.005
D5.2	30	80	30.9	0.091	0.102	32.6	0.096	0.115	10.0	0.033	0.027	0.7	0.007	0.006
D6.5	30	80	46.8	0.136	0.209	33.4	0.097	0.149	9.6	0.032	0.032	2.2	0.013	0.014
D7.3	30	80	18.2	0.057	0.095	14.9	0.048	0.076	4.4	0.017	0.020	0.3	0.006	0.006
D8.2	30	80	18.2	0.058	0.100	11.0	0.037	0.053	4.5	0.018	0.015	0.2	0.005	0.005
D9.6	30	80	21.6	0.069	0.092	15.3	0.050	0.074	4.9	0.019	0.013	0.2	0.005	0.005
D10.7	30	80	9.0	0.031	0.042	9.2	0.032	0.040	3.2	0.014	0.010	2.6	0.015	0.011
E5.7	50	100	25.3	0.270	0.327	33.4	0.350	0.803	7.3	0.088	0.048	1.0	0.027	0.013
E7.8	50	100	16.2	0.188	0.359	18.0	0.206	0.314	4.8	0.066	0.035	1.8	0.040	0.012
E9.3	50	100	23.9	0.295	0.443	26.6	0.326	0.484	5.6	0.078	0.058	0.6	0.023	0.011
E10.3	50	100	19.9	0.251	0.369	25.1	0.312	0.622	5.1	0.073	0.069	0.2	0.018	0.009

Though the simple reordering of chromosomes that GA1 implements does not always reduce the number of generations or the time required to find an optimum set of warehouse locations, the trend is otherwise monotonic and decisive. The addition of heuristic mutation (GA2) and then re-scanning (GA3) significantly reduces the computational effort and the clock time that the GA requires to find (in these cases) optimum solutions. The results on instance E5.7 are typical. GA0 finds optimum solutions in an average of 25.3 generations and an average of 0.271 seconds. GA1 takes slightly longer, both in generations and in time, but GA2 is faster than GA0 and GA1, and GA3 is faster than GA2. On average, GA3 requires only one generation and less than three one-hundredths of a second to find an optimal solution to UFLP instance E5.7. The differences between GA1 and GA2 and between GA2 and GA3 lie on the edge of significance, but GA3 is decisively and significantly faster than the original permutation-coded GA with no heuristic enhancements.

Average generation counts less than one indicate that some trials of GA3 found optimum solutions in the initial population. This indicates the power of the re-scanning heuristic.

Our conclusions are clear. Even in its naive form (GA0), the permutation-coded GA is fairly effective on the test UFLP instances and superior to a straightforward binary-coded GA. The three heuristic enhancements—reordering chromosomes so that they list included locations before excluded, heuristic mutation, and re-scanning included locations—consistently and decisively improve the GA's performance. The version that implements them all—GA3—often finds optimum solutions very quickly. In all its trials on all the test UFLP instances, it found optimum sets of warehouse locations on 9211 trials out of 9400 (98%), and its mean error was less than 0.008%.

Further, though this project has primarily investigated the benefits that the several heuristics confer on a permutation-coded GA that searches a space of subsets of varying sizes, the GA's final version is competitive with other recent EAs for the problem. For example, on the cap[abc] instances from the OR-Library, GA3 is as effective as and faster than the evolutionary simulated annealing algorithm of Yigit, Aydin, and Turkbey [337].

18.6 Conclusion

A genetic algorithm for the uncapacitated facility location problem encodes candidate subsets of warehouse locations as permutations of all the locations. A greedy decoder identifies the subset of locations that a chromosome represents; it scans the locations in chromosome order and includes those whose inclusion reduces the total cost of serving the customers.

A base version of the GA implements crossover by alternating the elements of two parent chromosomes followed by removing duplicate elements and mutation by swapping two random elements of a parent chromosome. Heuristic enhancements yield other versions. These enhancements include reordering chromosomes so that the locations included in the subsets they represent are listed before the excluded locations, mutating chromosomes by swapping an included location with an arbitrary one, and re-scanning the prefix of a chromosome that lists its included locations to remove any whose removal reduces the total cost. A binary-coded GA, with two-point crossover and position-by-position mutation, was also implemented as a standard of comparison.

All the versions of the permutation-coded GA outperformed the binary-coded algorithm. Within the permutation-coded GA, the reordering heuristic alone has only a limited effect, but it enables the other two heuristics, which improve the algorithm's performance significantly. In tests on 235 UFLP instances, the version that implements all three heuristic techniques almost always finds an optimum solution and almost always very quickly.

More generally, the heuristic techniques presented here could be applied in an evolutionary algorithm for any problem that seeks an optimum subset, of initially unknown size, of problem elements and that encodes candidate solutions as permutations of those elements. The improved performance that these techniques provide in the present GA suggests that they would be effective on other problems as well.

References

1. Booth, R.F., Bormotov, D.Y., Borovik, A.V.: Genetic Algorithms and Equations in Free Groups and Semigroups. Contemp. Math. 349, 63–80 (2004)
2. Borovik, A.V., Esyp, E.S., Kazatchkov, I.V., Remeslennikov, V.N.: Divisibility Theory and Complexity of Algorithms for Free Partially Commutative Groups. Contemp. Math. 378 (2005); (Groups, Languages, Algorithms)
3. Bremermann, H.J.: Optimization Through Evolution and Recombination. In: Yovits, M.C., et al. (eds.) Self-Organizing Systems, Washington, Spartan Books, pp. 93–106 (1962)
4. Craven, M.J.: Genetic Algorithms for Word Problems in Partially Commutative Groups. In: Cotta, C., van Hemert, J.I. (eds.) EvoCOP 2007. LNCS, vol. 4446, pp. 48–59. Springer, Heidelberg (2007)
5. Diekert, V., Matiyasevich, Y., Muscholl, A.: Solving Word Equations Modulo Partial Commutations. Theor. Comp. Sci. 224, 215–235 (1999)
6. Diekert, V., Muscholl, A.: Solvability of Equations in Free Partially Commutative Groups is Decidable. Int. J. Algebra and Computation 16, 1047–1070 (2006)
7. Holland, J.: Adaptation in Natural and Artificial Systems. MIT Press, Cambridge (1998) (5th printing)
8. Knuth, D., Bendix, P.: Simple Word Problems in Universal Algebra. In: Leech, J. (ed.) Computational Problems in Abstract Algebras, pp. 263–297. Pergamon Press, Oxford (1970)
9. Ko, K.-H.: Braid Group and Cryptography. In: 19th SECANTS, Oxford (2002)
10. Miasnikov, A.D.: Genetic Algorithms and the Andrews-Curtis Conjecture. Int. J. Algebra Comput. 9(6), 671–686 (1999)
11. Miasnikov, A.D., Myasnikov, A.G.: Whitehead Method and Genetic Algorithms. Contemp. Math. 349, 89–114 (2004)
12. Michalewicz, Z.: Genetic Algorithms + Data Structures = Evolution Programs. Springer, Berlin (1996) (3rd rev. and extended ed.)
13. VanWyk, L.: Graph Groups are Biautomatic. J. Pure Appl. Algebra 94(3), 341–352 (1994)
14. Vershik, A., Nechaev, S., Bikbov, R.: Statistical Properties of Braid Groups in Locally Free Approximation. Comm. Math. Phys. 212, 59–128 (2000)
15. Wrathall, C.: Free partially commutative groups, Combinatorics, Computing and Complexity. Math. Appl. (Chinese Ser. 1) 195–216 (1989)
16. Wrathall, C.: The Word Problem for Free Partially Commutative Groups. J. Symbolic Comp. 6, 99–104 (1988)

17. Dorigo, M., Stützle, T.: Ant Colony Optimization. MIT Press, Massachusetts (2004)
18. Cheeseman, P., Kanefsky, B., Taylor, W.M.: Where the Really Hard Problems Are. In: Proceedings of the Twelfth International Conference on Artificial Intelligence, vol. 1, pp. 331–337. Morgan Kaufmann Publishers, Inc., USA (1991)
19. Dorigo, M., Gambardella, L.M.: Ant Colony System: A Cooperative Learning Approach to the Traveling Salesman Problem. IEEE Transactions on Evolutionary Computation 1(1), 53–66 (1997)
20. Stützle, T., Hoos, H.H.: Max-Min Ant System. Future Generation Computer Systems 16(8), 889–914 (2000)
21. Johnson, D.S., Papadimitriou, C.H.: Computational Complexity. In: Lawler, E.L., Lenstra, J.K., Kan, A.H.G.R., Shmoys, D.B. (eds.) The Traveling Salesman Problem. Wiley Series in Discrete Mathematics and Optimization, pp. 37–85. John Wiley and Sons, Chichester (1995)
22. Montgomery, D.C.: Design and Analysis of Experiments, 6th edn. John Wiley and Sons Inc., Chichester (2005)
23. Eiben, A., Jelasity, M.: A critical note on experimental research methodology in EC. In: Proceedings of the 2002 IEEE Congress on Evolutionary Computation, pp. 582–587. IEEE, Los Alamitos (2002)
24. Ridge, E., Curry, E.: A Roadmap of Nature-Inspired Systems Research and Development. Multi-Agent and Grid Systems 3(1) (2007)
25. Hooker, J.N.: Testing heuristics: We have it all wrong. Journal of Heuristics 1, 33–42 (1996)
26. Bang-Jensen, J., Chiarandini, M., Goegebeur, Y., Jørgensen, B.: Mixed Models for the Analysis of Local Search Components. In: Stützle, T., Birattari, M., H. Hoos, H. (eds.) SLS 2007. LNCS, vol. 4638, pp. 91–105. Springer, Heidelberg (2007)
27. Lawler, E.L., Lenstra, J.K., Kan, A.H.G.R., Shmoys, D.B. (eds.): The Traveling Salesman Problem – A Guided Tour of Combinatorial Optimization. Wiley Series in Discrete Mathematics and Optimization. John Wiley and Sons, New York (1995)
28. Hoos, H., Stützle, T.: Stochastic Local Search, Foundations and Applications. Morgan Kaufmann, San Francisco (2004)
29. Johnson, D.S.: A Theoretician's Guide to the Experimental Analysis of Algorithms. In: Proceedings of the Fifth and Sixth DIMACS Implementation Challenges, pp. 215–250. American Mathematical Society (2002)
30. Ridge, E.: Design of Experiments for the Tuning of Optimisation Algorithms. PhD thesis, Department of Computer Science, The University of York (2007)
31. Birattari, M.: The Problem of Tuning Metaheuristics. PhD, Faculté des Sciences Appliquées, Université Libre de Bruxelles (2006)
32. Zemel, E.: Measuring the quality of approximate solutions to zero-one programming problems. Mathematics of Operations Research 6, 319–332 (1981)
33. Zlochin, M., Dorigo, M.: Model based search for combinatorial optimization: a comparative study. In: Guervós, J.J.M., Adamidis, P.A., Beyer, H.-G., Fernández-Villacañas, J.-L., Schwefel, H.-P. (eds.) PPSN 2002. LNCS, vol. 2439, pp. 651–661. Springer, Heidelberg (2002)
34. Cohen, P.R.: Empirical Methods for Artificial Intelligence. MIT Press, Cambridge (1995)
35. Applegate, D., Bixby, R., Chvatal, V., Cook, W.: Implementing the Dantzig-Fulkerson-Johnson algorithm for large traveling salesman problems. Mathematical Programming Series B 97(1-2), 91–153 (2003)

36. Goldwasser, M., Johnson, D.S., McGeoch, C.C. (eds.): Proceedings of the Fifth and Sixth DIMACS Implementation Challenges. American Mathematical Society (2002)
37. Reinelt, G.: TSPLIB – A traveling salesman problem library. ORSA Journal of Computing 3, 376–384 (1991)
38. Ridge, E., Kudenko, D.: Analyzing Heuristic Performance with Response Surface Models: Prediction, Optimization and Robustness. In: Proceedings of the Genetic and Evolutionary Computation Conference, vol. 1, pp. 150–157. ACM New York (2007)
39. Ridge, E., Kudenko, D.: Tuning the Performance of the MMAS Heuristic. In: Stützle, T., Birattari, M., Hoos, H. (eds.) SLS 2007. LNCS, vol. 4638, pp. 46–60. Springer, Heidelberg (2007)
40. Kauffman, S.A.: The Origins of Order: Self-Organization and Selection in Evolution, pp. 33–67. Oxford Univ. Press, Oxford (1993)
41. Altenberg, L.: Fitness Landscapes: NK Landscapes. In: Handbook of Evolutionary Computation, pp. B2.7: 5–10. Institute of Physics Publishing & Oxford Univ. Press, Oxford (1997)
42. Heckendorn, R., Rana, S., Whitley, D.: Test Function Generators as Embedded Landscapes. In: Foundations of Genetic Algorithms, vol. 5, pp. 183–198. Morgan Kaufmann, San Francisco (1999)
43. Smith, R.E., Smith, J.E.: An Examination of Tunable, Random Search Landscapes. In: Foundations of Genetic Algorithms, vol. 5, pp. 165–182. Morgan Kaufmann, San Francisco (1999)
44. Smith, R.E., Smith, J.E.: New Methods for Tunable, Random Landscapes. In: Foundations of Genetic Algorithms, vol. 6, pp. 47–67. Morgan Kaufmann, San Francisco (2001)
45. Davidor, Y.: Epistasis Variance: A Viewpoint of GA-Hardness. In: Foundations of Genetic Algorithms, pp. 23–35. Morgan Kaufmann, San Francisco (1991)
46. Manderick, B., de Weger, M., Spiessens, P.: The Genetic Algorithm and the Structure of the Fitness Landscape. In: Proc. 4th Int'l Conf. on Genetic Algorithms, pp. 143–150. Morgan Kaufmann, San Francisco (1991)
47. Altenberg, L.: Evolving Better Representations through Selective Genome Growth. In: Proc. 1st IEEE Conf. on Evolutionary Computation, pp. 182–187. IEEE Press, Los Alamitos (1994)
48. De Jong, K.A., Potter, M.A., Spears, W.M.: Using Problem Generators to Explore the Effects of Epistasis. In: Proc. 7th Int'l Conf. Genetic Algorithms, pp. 338–345. Morgan Kaufmann, San Francisco (1997)
49. Smith, J.E.: Self Adaptation in Evolutionary Algorithms. PhD Thesis University of the West of England, Bristol (1998)
50. Merz, P., Freisleben, B.: On the Effectiveness of Evolutionary Search in High-Dimensional NK-Landscapes. In: Proc. of the 1998 IEEE Int'l Conf. on Evolutionary Computation, pp. 741–745. IEEE Press, Los Alamitos (1998)
51. Mathias, K.E., Eshelman, L.J., Schaffer, D.: Niches in NK-Landscapes. In: Foundations of Genetic Algorithms, vol. 6, pp. 27–46. Morgan Kaufmann, San Francisco (2001)
52. Eshelman, L.J.: The CHC Adaptive Search Algorithm: How to Have a Save Search When Engaging in Nontraditional Genetic Recombination. In: Foundations of Genetic Algorithms, pp. 265–283. Morgan Kaufmann, San Francisco (1991)
53. Davis, L.: Bit-Climbing, Representation Bias, and Test Suite Design. In: Proc. 4th Int'l Conf. on Genetic Algorithms, pp. 18–23. Morgan Kaufmann, San Francisco (1991)

54. Aguirre, H., Tanaka, K., Sugimura, T.: Cooperative Model for Genetic Operators to Improve GAs. In: Proc. IEEE Int'l Conf. on Information, Intelligence, and Systems, pp. 98–106. IEEE Press, Los Alamitos (1999)

55. Aguirre, H., Tanaka, K., Sugimura, T., Oshita, S.: Cooperative-Competitive Model for Genetic Operators: Contributions of Extinctive Selection and Parallel Genetic Operators. In: Proc. Late Breaking Papers Genetic and Evolutionary Computation Conference, pp. 6–14. Morgan Kaufmann, San Francisco (2000)

56. Shinkai, M., Aguirre, H., Tanaka, K.: Mutation Strategy Improves GA's Performance on Epistatic Problems. In: Proc. 2002 IEEE World Congress on Computational Intelligence, pp. 795–800. IEEE Press, Los Alamitos (2002)

57. Bäck, T.: Evolutionary Algorithms in Theory and Practice. Oxford Univ. Press, Oxford (1996)

58. Whitley, D.: The GENITOR Algorithm and Selection Pressure: Why Rank-Based Allocation of Reproductive Trials is Best. In: Proc. 3rd Intl. Conf. on Genetic Algorithms, pp. 116–121. Morgan Kaufmann, San Francisco (1989)

59. Goldberg, D.E.: Genetic Algorithms in Search, Optimization and Machine Learning. Addison-Wesley, Reading (1989)

60. Eshelman, L., Schaffer, D.: Preventing Premature Convergence in Genetic Algorithms by Preventing Incest. In: Proc. 4th Intl. Conf. on Genetic Algorithms, pp. 115–122. Morgan Kaufmann, San Francisco (1991)

61. Schaffer, D., Mani, M., Eshelman, L., Mathias, K.: The Effect of Incest Prevention on Genetic Drift. In: Foundations of Genetic Algorithms, vol. 5, pp. 235–243. Morgan Kaufmann, San Francisco (1999)

62. Liang, K.H., Yao, X., Newton, C., Hoffman, D.: Solving Cutting Stock Problems by Evolutionary Programming. In: Porto, V.W., Waagen, D. (eds.) EP 1998. LNCS, vol. 1447, pp. 291–300. Springer, Heidelberg (1998)

63. He, J., Yao, X.: From an Individual to a Population: An Analysis of the First Hitting Time of Population-Based Evolutionary Algorithms. IEEE Transactions on Evolutionary Computation 6(5), 495–511 (2002)

64. Czarn, A., MacNish, C., Vijayan, K., Turlach, B., Gupta, R.: Statistical Exploratory Analysis of Genetic Algorithms. IEEE Transactions on Evolutionary Computation 8(4), 405–421 (2004)

65. Lenth, R.V.: Java Applets for Power and Sample Size (2006)

66. Fischer, T., Stützle, T., Hoos, H., Merz, P.: An Analysis Of The Hardness Of TSP Instances For Two High Performance Algorithms. In: Proceedings of the Sixth Metaheuristics International Conference, pp. 361–367 (2005)

67. Helsgaun, K.: An effective implementation of the Lin-Kernighan traveling salesman heuristic. European Journal of Operational Research 126(1), 106–130 (2000)

68. van Hemert, J.I.: Property Analysis of Symmetric Travelling Salesman Problem Instances Acquired Through Evolution. In: Raidl, G.R., Gottlieb, J. (eds.) EvoCOP 2005. LNCS, vol. 3448, pp. 122–131. Springer, Heidelberg (2005)

69. Ridge, E., Kudenko, D.: Screening the Parameters Affecting Heuristic Performance. In: Proceedings of the Genetic and Evolutionary Computation Conference, vol. 1, ACM, New York (2007)

70. Jaillet, P.: Probabilistic Traveling Salesman Problems. PhD thesis, Massachusetts Institute of Technology, Cambridge, MA (1985)

71. Bertsimas, D., Jaillet, P., Odoni, A.: A priori optimization. Operations Research 38(6), 1019–1033 (1990)

72. Bertsimas, D.: Probabilistic Combinatorial Optimization Problems. PhD thesis, Massachusetts Institute of Technology, Cambridge, MA (1988)

73. Bianchi, L., Knowles, J., Bowler, N.: Local search for the probabilistic traveling salesman problem: Correction to the 2-p-opt and 1-shift algorithms. European Journal of Operational Research 162(1), 206–219 (2005)
74. Bianchi, L.: Ant Colony Optimization and Local Search for the Probabilistic Traveling Salesman Problem: A Case Study in Stochastic Combinatorial Optimization. PhD thesis, Université Libre de Bruxelles, Brussels, Belgium (2006)
75. Johnson, D.S., McGeoch, L.A.: The travelling salesman problem: A case study in local optimization. In: Aarts, E.H.L., Lenstra, J.K. (eds.) Local Search in Combinatorial Optimization, pp. 215–310. John Wiley & Sons, Chichester (1997)
76. Hoos, H., Stützle, T.: Stochastic Local Search: Foundations and Applications. Morgan Kaufmann, San Francisco (2005)
77. Bentley, J.L.: Fast algorithms for geometric traveling salesman problems. ORSA Journal on Computing 4(4), 387–411 (1992)
78. Johnson, D.S., McGeoch, L.A., Rego, C., Glover, F.: 8th DIMACS implementation challenge (2001)
79. Birattari, M., Balaprakash, P., Stützle, T., Dorigo, M.: Estimation-based local search for stochastic combinatorial optimization. Technical Report TR/IRIDIA/2007–003, IRIDIA, Université Libre de Bruxelles, Brussels, Belgium (2007)
80. Gutjahr, W.: S-ACO: An ant based approach to combinatorial optimization under uncertainity. In: Dorigo, M., Birattari, M., Blum, C., Gambardella, L.M., Mondada, F., Stützle, T. (eds.) ANTS 2004. LNCS, vol. 3172, pp. 238–249. Springer, Heidelberg (2004)
81. Martin, O., Otto, S.W., Felten, E.W.: Large-step Markov chains for the traveling salesman problem. Complex Systems 5(3), 299–326 (1991)
82. Balaprakash, P., Birattari, M., Stützle, T., Dorigo, M.: Adaptive sample size and importance sampling in estimation-based local search for stochastic combinatorial optimization: A complete analysis. Technical Report TR/IRIDIA/2007–014, IRIDIA, Université Libre de Bruxelles, Brussels, Belgium (2007)
83. Balaprakash, P., Birattari, M., Stützle, T.: Improvement strategies for the F-Race algorithm: Sampling design and iterative refinement. In: Bartz-Beielstein, T., Blesa Aguilera, M.J., Blum, C., Naujoks, B., Roli, A., Rudolph, G., Sampels, M. (eds.) HCI/ICCV 2007. LNCS, vol. 4771, pp. 113–127. Springer, Heidelberg (2007)
84. Birattari, M.: The Problem of Tuning Metaheuristics as Seen from a Machine Learning Perspective. PhD thesis, Université Libre de Bruxelles, Brussels, Belgium (2004)
85. Lourenço, H.R., Martin, O., Stützle, T.: Iterated local search. In: Glover, F., Kochenberger, G. (eds.) Handbook of Metaheuristics, pp. 321–353. Kluwar Academic Publishers, Norwell (2002)
86. Bianchi, L., Gambardella, L.: Ant colony optimization and local search based on exact and estimated objective values for the probabilistic traveling salesman problem. Technical Report IDSIA-06-07, IDSIA, Lugano, Switzerland (2007)
87. Dorigo, M., Stützle, T.: Ant Colony Optimization. MIT Press, Cambridge (2004)
88. Yamada, T., Watanabe, K., Katakoa, S.: Algorithms to solve the knapsack constrained maximum spanning tree problem. Int. Journal of Computer Mathematics 82(1), 23–34 (2005)
89. Pirkwieser, S., Raidl, G.R., Puchinger, J.: Combining Lagrangian decomposition with an evolutionary algorithm for the knapsack constrained maximum spanning tree problem. In: Cotta, C., van Hemert, J. (eds.) EvoCOP 2007. LNCS, vol. 4446, pp. 176–187. Springer, Heidelberg (2007)

90. Aggarwal, V., Aneja, Y., Nair, K.: Minimal spanning tree subject to a side constraint. Comput. & Operations Res. 9(4), 287–296 (1982)

91. Jörnsten, K., Migdalas, S.: Designing a minimal spanning tree network subject to a budget constraint. Optimization 19(4), 475–484 (1988)

92. Ravi, R., Goemans, M.X.: The constrained minimum spanning tree problem. In: Karlsson, R., Lingas, A. (eds.) SWAT 1996. LNCS, vol. 1097, pp. 66–75. Springer, Heidelberg (extended abstract, 1996)

93. Xue, G.: Primal-dual algorithms for computing weight-constrained shortest paths and weight-constrained minimum spanning trees. In: IEEE International Performance, Computing & Communications Conference, pp. 271–277 (2000)

94. Jüttner, A.: On resource constrained optimization problems. In: 4th Japanese-Hungarian Symposium on Discrete Mathematics and Its Applications (2005)

95. Pirkwieser, S.: A Lagrangian Decomposition Approach Combined with Metaheuristics for the Knapsack Constrained Maximum Spanning Tree Problem. Master's thesis, Vienna University of Technology, Institute of Computer Graphics and Algorithms (October 2006)

96. Henn, S.T.: Weight-constrained minimal spanning tree problem. Master's thesis, University of Kaiserslautern, Department of Mathematics (May 2007)

97. Beasley, J.E.: An Exact Two-dimensional Non-guillotine Cutting Tree Search Procedure. Operations Research 33(1), 49–64 (1985)

98. Beasley, J.E.: A Population Heuristic for Constrained Two-dimensional Non-guillotine Cutting. European Journal of Operational Research 156(3), 601–627 (2004)

99. Blesa, M.J., Blum, C., Roli, A., Sampels, M.: Hybrid Metaheuristics. In: Blesa, M.J., Blum, C., Roli, A., Sampels, M. (eds.) HM 2005. LNCS, vol. 3636. Springer, Heidelberg (2005)

100. Blum, C., Roli, A.: Metaheuristics in Combinatorial Optimization: Overview and Conceptual Comparison. ACM Computing Surveys 35(3), 268–308 (2003)

101. Christofides, N., Whitlock, C.: An Algorithm for Two-dimensional Cutting Problems. Operations Research 25, 31–44 (1977)

102. Chu, P.C.: A Genetic Algorithm Approach for Combinatorial Optimization Problems. PhD Thesis, University of London (1997)

103. Cook, W., Seymour, P.: Tour Merging via Branch-Decomposition. INFORMS Journal on Computing 15(3), 233–248 (2003)

104. Dowsland, K.A., Dowsland, W.B.: Packing Problems. European Journal of Operational Research 56(1), 2–14 (1992)

105. Dumitrescu, I., Stützle, T.: Combinations of Local Search and Exact Algorithms. In: Raidl, G.R., Cagnoni, S., Cardalda, J.J.R., Corne, D.W., Gottlieb, J., Guillot, A., Hart, E., Johnson, C.G., Marchiori, E., Meyer, J.-A., Middendorf, M. (eds.) EvoIASP 2003, EvoWorkshops 2003, EvoSTIM 2003, EvoROB/EvoRobot 2003, EvoCOP 2003, EvoBIO 2003, and EvoMUSART 2003. LNCS, vol. 2611, pp. 211–224. Springer, Heidelberg (2003)

106. Fekete, S.P., Schepers, J.: A New Exact Algorithm for General Orthogonal D-dimensional Knapsack Problems. In: Burkard, R.E., Woeginger, G.J. (eds.) ESA 1997. LNCS, vol. 1284, pp. 144–156. Springer, Heidelberg (1997)

107. Fischetti, M., Lodi, A.: Local branching. Mathematical Programming 98, 23–47 (2003)

108. Glover, F., Kochenberger, G.A.: Handbook of Metaheuristics. Kluwer, Dordrecht (2003)

109. Hadjiconstantinou, E., Christofides, N.: An Exact Algorithm for General, Orthogonal, Two-dimensional Knapsack Problems. European Journal of Operational Research 83, 39–56 (1995)

110. Mahfoud, S.W., Goldberg, D.E.: Parallel Recombinative Simulated Annealing: A Genetic Algorithm. Parallel Computing 21(1), 1–28 (1995)

111. Martin, O., Otto, S.W., Felten, E.W.: Large-step Markov Chains for the TSP: Incorporating Local Search Heuristics. Operations Research Letters 11(4), 219–224 (1992)

112. Moscato, P.: Memetic algorithms: A short introduction. In: New Ideas in Optimization, pp. 219–234. McGraw Hill, New York (1999)

113. Nepomuceno, N.V., Pinheiro, P.R., Coelho, A.L.V.: Tackling the Container Loading Problem: A Hybrid Approach Based on Integer Linear Programming and Genetic Algorithms. In: Cotta, C., van Hemert, J. (eds.) EvoCOP 2007. LNCS, vol. 4446, pp. 154–165. Springer, Heidelberg (2007)

114. Puchinger, J., Raidl, G.: Combining Metaheuristics and Exact Algorithms in Combinatorial Optimization: A Survey and Classification. In: Mira, J., Álvarez, J.R. (eds.) IWINAC 2005. LNCS, vol. 3562, pp. 41–53. Springer, Heidelberg (2005)

115. Raidl, G.R.: A Unified View on Hybrid Metaheuristics. In: Almeida, F., Blesa Aguilera, M.J., Blum, C., Moreno Vega, J.M., Pérez Pérez, M., Roli, A., Sampels, M. (eds.) HM 2006. LNCS, vol. 4030, pp. 1–12. Springer, Heidelberg (2006)

116. Schrage, L.: Optimization Modeling with LINGO. LINDO Systems, Inc. (2000)

117. Talbi, E.G.: A Taxonomy of Hybrid Metaheuristics. Journal of Heuristics 8(5), 541–564 (2002)

118. Vasquez, M., Hao, J.K.: A Hybrid Approach for the 0-1 Multidimensional Knapsack Problem. In: Proceedings of the 17th International Joint Conference on Artificial Intelligence, pp. 328–333 (2001)

119. Wang, P.Y.: Two Algorithms for Constrained Two-dimensional Cutting Stock Problems. Operations Research 31(3), 573–586 (1983)

120. Fisher, M.L.: The Lagrangian Relaxation Method for Solving Integer Programming Problems. Management Science 27(1), 1–18 (1981)

121. Fisher, M.L.: An application oriented guide to Lagrangean Relaxation. Interfaces 15, 10–21 (1985)

122. Beasley, J.E.: Lagrangian relaxation. In: Reeves, C.R. (ed.) Modern Heuristic Techniques for Combinatorial Problems, pp. 243–303. John Wiley & Sons, Inc, New York (1993)

123. Kruskal, J.B.: On the shortest spanning subtree of a graph and the travelling salesman problem. In: Proc. of the AMS, vol. 7, pp. 48–50 (1956)

124. Prim, R.C.: Shortest connection networks and some generalizations. Bell Systems Technology Journal 36, 1389–1401 (1957)

125. Fredman, M.L., Sedgewick, R., Sleator, D.D., Tarjan, R.E.: The pairing heap: a new form of self-adjusting heap. Algorithmica 1(1), 111–129 (1986)

126. Kellerer, H., Pferschy, U., Pisinger, D.: Knapsack Problems. Springer, Heidelberg (2004)

127. Martello, S., Pisinger, D., Toth, P.: Dynamic programming and strong bounds for the 0–1 knapsack problem. Management Science 45, 414–424 (1999)

128. Barahona, F., Anbil, R.: The volume algorithm: producing primal solutions with a subgradient method. Mathematical Programming 87(3), 385–399 (2000)

129. Bahiense, L., Barahona, F., Porto, O.: Solving steiner tree problems in graphs with lagrangian relaxation. Journal of Combinatorial Optimization 7(3), 259–282 (2003)

130. Haouari, M., Siala, J.C.: A hybrid Lagrangian genetic algorithm for the prize collecting Steiner tree problem. Comput. & Operations Res. 33(5), 1274–1288 (2006)

131. Magnanti, T.L., Wolsey, L.A.: Optimal trees. In: Ball, M.O., et al. (eds.) Handbooks in Operations Research and Management Science, vol. 7, pp. 503–615. Elsevier Science, Amsterdam (1995)

132. Julstrom, B.A., Raidl, G.R.: Edge sets: an effective evolutionary coding of spanning trees. IEEE Transactions on Evolutionary Computation 7(3), 225–239 (2003)

133. Setubal, J., Meidanis, J.: 4 – Fragment Assembly of DNA. In: Introduction to Computational Molecular Biology, University of Campinas, Brazil, pp. 105–139 (1997)

134. Pop, M.: Shotgun sequence assembly. Advances in Computers 60, 194–248 (2004)

135. Sutton, G.G., White, O., Adams, M.D., Kerlavage, A.R.: TIGR Assembler: A new tool for assembling large shotgun sequencing projects. Genome Science & Technology, 9–19 (1995)

136. Chen, T., Skiena, S.S.: Trie-based data structures for sequence assembly. In: The Eighth Symposium on Combinatorial Pattern Matching, pp. 206–223 (1998)

137. Huang, X., Madan, A.: CAP3: A DNA sequence assembly program. Genome Research 9, 868–877 (1999)

138. Myers, E.W.: Towards simplifying and accurately formulating fragment assembly. Journal of Computational Biology 2(2), 275–290 (2000)

139. Pevzner, P.A.: Computational molecular biology: An algorithmic approach. MIT Press, London (2000)

140. Luque, G., Alba, E.: Metaheuristics for the DNA Fragment Assembly Problem. International Journal of Computational Intelligence Research 1(2), 98–108 (2006)

141. Alba, E., Luque, G.: A new local search algorithm for the dna fragment assembly problem. In: Cotta, C., van Hemert, J. (eds.) EvoCOP 2007. LNCS, vol. 4446. Springer, Heidelberg (2007)

142. Green, P.: Phrap (1994),
http://www.mbt.washington.edu/phrap.docs/phrap.html

143. Goldberg, D.E.: Genetic Algorithms in Search, Optimization and Machine Learning. Addison-Wesley, Reading (1989)

144. Parsons, R., Forrest, S., Burks, C.: Genetic algorithms, operators, and DNA fragment assembly. Machine Learning 21, 11–33 (1995)

145. Li, L., Khuri, S.: A comparison of DNA fragment assembly algorithms. In: International Conference on Mathematics and Engineering Techniques in Medicine and Biological Sciences, pp. 329–335 (2004)

146. Lin, S., Kernighan, B.: An effective heuristic algorithm for TSP. Operations Research 21, 498–516 (1973)

147. Cotta, C.: A study of hybridisation techniques and their application to the design of evolutionary algorithms. AI Communications 11(3-4), 223–224 (1998)

148. Engle, M.L., Burks, C.: Artificially generated data sets for testing DNA fragment assembly algorithms. Genomics 16 (1993)

149. Jing, Y., Khuri, S.: Exact and heuristic algorithms for the DNA fragment assembly problem. In: Proceedings of the IEEE Computer Society Bioinformatics Conference, Stanford Univeristy, pp. 581–582. IEEE Press, Los Alamitos (2003)

150. Alba, E. (ed.): Parallel Metaheuristics: A New Class of Algorithms. In: Parallel and Distributed Computing, Wiley, Chichester (2005)

151. Oram, A. (ed.): Peer-to-Peer: Harnessing the Power of Disruptive Technologies. O'Reilly & Associates, Sebastopol (2001)

152. Lv, Q., Cao, P., Cohen, E., Li, K., Shenker, S.: Search and replication in unstructured peer-to-peer networks. In: Proceedings of the 16th International Conference on Supercomputing, pp. 84–95. ACM Press, New York (2002)

153. Yang, B., Garcia-Molina, H.: Improving search in peer-to-peer networks. In: Proceedings of the 22nd International Conference on Distributed Computing Systems (ICDCS 2002), pp. 5–14 (2002)

154. Kalogeraki, V., Gunopulos, D., Zeinalipour-Yazti, D.: A local search mechanism for peer-to-peer networks. In: Proceedings of the 11th International Conference on Information and Knowledge Management, pp. 300–307. ACM Press, New York (2002)

155. Menascé, D.A.: Scalable P2P search. IEEE Internet Computing 7(2), 83–87 (2003)

156. Fisk, A.: Gnutella dynamic query protocol v0.1. Gnutella Developer's Forum (May 2003)

157. Tsoumakos, D., Roussopoulos, N.: Adaptive probabilistic search for peer-to-peer networks. In: Proceedings of the Third IEEE International Conference on P2P Computing (P2P 2003), pp. 102–109. IEEE Press, Los Alamitos (2003)

158. Crespo, A., Garcia-Molina, H.: Routing indices for peer-to-peer systems. In: Proceedings of the 22nd IEEE International Conference on Distributed Computing Systems (ICDCS 2002), pp. 23–33. IEEE Press, Los Alamitos (2002)

159. Sarshar, N., Boykin, P.O., Roychowdhury, V.P.: Percolation search in power law networks: Making unstructured peer-to-peer networks scalable. In: Proceedings of the Fourth International Conference on P2P Computing (P2P 2004), pp. 2–9. IEEE Press, Los Alamitos (2004)

160. Vapa, M., Kotilainen, N., Auvinen, A., Kainulainen, H., Vuori, J.: Resource discovery in P2P networks using evolutionary neural networks. In: International Conference on Advances in Intelligent Systems – Theory and Applications, paper identification number 067-04 (2004)

161. Engelbrecht, A.: Computational Intelligence – An Introduction. J. Wiley and Sons, Chichester (2002)

162. Kotilainen, N., Vapa, M., Keltanen, T., Auvinen, A., Vuori, J.: P2PRealm – peer-to-peer network simulator. In: International Workshop on Computer-Aided Modeling, Analysis and Design of Communication Links and Networks, pp. 93–99. IEEE Communications Society, Los Alamitos (2006)

163. Deb, K.: Multi-objective Optimization using Evolutionary Algorithms, pp. 147–149. John Wiley and Sons LTD., Chichester (2001)

164. Branke, J.: Evolutionary Optimization in Dynamic Environments, pp. 125–172. Kluwer A. P., Netherlands (2001)

165. Derrida, B., Peliti, L.: Evolution in a flat fitness landscape. Bulletin of Mathematical Biology 53, 355–382 (1991)

166. Neri, F., Toivanen, J., Cascella, G.L., Ong, Y.S.: An adaptive multimeme algorithm for designing HIV multidrug therapies. IEEE/ACM Transactions on Computational Biology and Bioinformatics, Special Issue on Computational Intelligence Approaches in Computational Biology and Bioinformatics 4(2), 264–278 (2007)

167. Chellapilla, K., Fogel, D.: Evolving neural networks to play checkers without relying on expert knowledge. IEEE Transactions on Neural Networks 10(6), 1382–1391 (1999)

168. Chellapilla, K., Fogel, D.: Evolving an expert checkers playing program without using human expertise. IEEE Transactions on Evolutionary Computation 5(4), 422–428 (2001)

169. Jin, Y., Branke, J.: Evolutionary optimization in uncertain environments – a survey. IEEE Transactions on Evolutionary Computation 9(3), 303–317 (2005)
170. Neri, F., Cascella, G.L., Salvatore, N., Kononova, A.V., Acciani, G.: Prudent-daring vs tolerant survivor selection schemes in control design of electric drives. In: Rothlauf, F., Branke, J., Cagnoni, S., Costa, E., Cotta, C., Drechsler, R., Lutton, E., Machado, P., Moore, J.H., Romero, J., Smith, G.D., Squillero, G., Takagi, H. (eds.) EvoWorkshops 2006. LNCS, vol. 3907, pp. 805–809. Springer, Heidelberg (2006)
171. Krasnogor, N.: Toward robust memetic algorithms. In: Hart, W.E., Krasnogor, N., Smith, J.E. (eds.) Recent Advances in Memetic Algorithms, pp. 185–207. Springer, Berlin (2004)
172. Ong, Y.S., Keane, A.J.: Meta-lamarckian learning in memetic algorithms. IEEE Transactions on Evolutionary Computation 8, 99–110 (2004)
173. Kirkpatrick, S., Gelatt, C.D.J., Vecchi, M.P.: Optimization by simulated annealing. Science (220), 671–680 (1983)
174. Cerny, V.: A thermodynamical approach to the traveling salesman problem. Journal of Optimization, Theory and Applications 45(1), 41–51 (1985)
175. Szu, H., Hartley, R.: Fast simulated annealing. Physics Letters A 122, 157–162 (1987)
176. Hooke, R., Jeeves, T.A.: Direct search solution of numerical and statistical problems. Journal of the ACM 8, 212–229 (1961)
177. Kaupe, F.: Algorithm 178: direct search. Communications of the ACM 6(6), 313–314 (1963)
178. Kelley, C.T.: Iterative Methods of Optimization, pp. 212–229. SIAM, Philadelphia (1999)
179. Krasnogor, N., Blackburne, B., Burke, E., Hirst, J.: Multimeme algorithms for proteine structure prediction. In: Guervós, J.J.M., Adamidis, P.A., Beyer, H.-G., Fernández-Villacañas, J.-L., Schwefel, H.-P. (eds.) PPSN 2002. LNCS, vol. 2439, pp. 769–778. Springer, Heidelberg (2002)
180. Caponio, A., Cascella, G.L., Neri, F., Salvatore, N., Sumner, M.: A fast adaptive memetic algorithm for on-line and off-line control design of pmsm drives. IEEE Transactions on System Man and Cybernetics-part B, special issue on Memetic Algorithms 37(1), 28–41 (2007)
181. Neri, F., Mäkinen, R.A.E.: Hierarchical evolutionary algorithms and noise compensation via adaptation. In: Yang, S., Ong, Y.S., Jin, Y. (eds.) Evolutionary Computation in Dynamic and Uncertain Environments. Studies in Computational Intelligence, pp. 345–369. Springer, Heidelberg (2007)
182. Eiben, A.E., Smith, J.E.: Introduction to Evolutionary Computation. Springer, Berlin (2003)
183. Bäck, T.: The interaction rate of mutation rate, selection, and self-adaptation within a genetic algorithm. In: Proceedings of Parallel Problem Solving from Nature (PPSN-II), pp. 85–94. Elsevier Science, Amsterdam (1992)
184. Eiben, A.E., Hinterding, R., Michaelwicz, Z.: Parameter control. In: Baeck, T., Fogel, D.B., Michaelwicz, Z. (eds.) Evolutionary Computation 2, Advanced Algorithms and Operators, pp. 170–187. Institute of Physics Publishing (2000)
185. Fitzpatrick, J.M., Grefenstette, J.J.: Genetic algorithms in noisy enviroments. Machine Learning 3, 101–120 (1988)
186. Miller, B.L., Goldberg, D.E.: Genetic algorithms, selection schemes, and the varying effects of noise. Evolutionary Computation 4(2), 113–131 (1996)

187. Schmidt, C., Branke, J., Chick, S.E.: Integrating techniques from statistical ranking into evolutionary algorithms. In: Rothlauf, F., Branke, J., Cagnoni S., Costa, E., Cotta, C., Drechsler, R., Lutton, E., Machado, P., Mcore, J.H , Romero, J., Smith, G.D., Squillero, G., Takagi, H. (eds.) EvoWorkshops 2006. LNCS, vol. 3907, pp. 752–763. Springer, Heidelberg (2006)
188. Lauritzen, S., Spiegelhalter, D.: Local computations with probabilities on graphical structures and their application to expert systems. Journal of the Royal Statistical Society, Series B 50, 157–224 (1988)
189. Koster, A.M., van Hoesel, S.P., Kolen, A.W.: Optimal solutions for frequency assignment problems via tree decomposition. In: Widmayer, P., Neyer, G., Eidenbenz, S. (eds.) WG 1999. LNCS, vol. 1665, pp. 338–350. Springer, Heidelberg (1999)
190. Alber, J., Dorn, F., Niedermeier, R.: Experimental evaluation of a tree decomposition based algorithm for vertex cover on planar graphs. Discrete Applied Mathematics 145, 210–219 (2004)
191. Koster, A., van Hoesel, S., Kolen, A.: Solving partial constraint satisfaction problems with tree-decomposition. Networks 40(3), 170–180 (2002)
192. Xu, J., Jiao, F., Berger, B.: A tree-decomposition approach to protein structure prediction. In: Proceedings of the IEEE Computational Systems Bioinformatics Conference, pp. 247–256 (2005)
193. Robertson, N., Seymour, P.D.: Graph minors II: Algorithmic aspects of treewidth. Journal Algorithms 7, 309–322 (1986)
194. Koster, A., Bodlaender, H., van Hoesel, S.: Treewidth: Computational experiments. In: Electronic Notes in Discrete Mathematics, vol. 8, Elsevier Science Publishers, Amsterdam (2001)
195. Clautiaux, F., Moukrim, A., Négre, S., Carlier, J.: Heuristic and meta-heurisistic methods for computing graph treewidth. RAIRO Oper. Res. 38, 13–26 (2004)
196. Fulkerson, D.R., Gross, O.: Incidence matrices and interval graphs. Pacific Journal of Mathematics 15, 835–855 (1965)
197. Gavril, F.: Algorithms for minimum coloring, maximum clique, minimum coloring cliques and maximum independent set of a chordal graph. SIAM J. Comput. 1, 180–187 (1972)
198. Arnborg, S., Corneil, D.G., Proskurowski, A.: Complexity of finding embeddings in a k-tree. SIAM J. Alg. Disc. Meth. 8, 277–284 (1987)
199. Shoikhet, K., Geiger, D.: A practical algorithm for finding optimal triangulations. In: Proc. of National Conference on Artificial Intelligence (AAAI 1997), pp. 185–190 (1997)
200. Gogate, V., Dechter, R.: A complete anytime algorithm for treewidth. In: Proceedings of the 20th Annual Conference on Uncertainty in Artificial Intelligence, UAI 2004, pp. 201–208 (2004)
201. Bachoore, E., Bodlaender, H.: A branch and bound algorithm for exact, upper, and lower bounds on treewidth. In: Cheng, S.-W., Poon, C.K. (eds.) AAIM 2006. LNCS, vol. 4041, pp. 255–266. Springer, Heidelberg (2006)
202. Tarjan, R., Yannakakis, M.: Simple linear-time algorithm to test chordality of graphs, test acyclicity of hypergraphs, and selectively reduce acyclic hypergraphs. SIAM J. Comput. 13, 566–579 (1984)
203. Kjaerulff, U.: Optimal decomposition of probabilistic networks by simulated annealing. Statistics and Computing 2(1), 2–17 (1992)
204. Larranaga, P., Kujipers, C., Poza, M., Murga, R.: Decomposing bayesian networks: triangulation of the moral graph with genetic algorithms. Statistics and Computing 7(1), 19–34 (1997)

205. Musliu, N., Schafhauser, W.: Genetic algorithms for generalized hypertree decompositions. European Journal of Industrial Engineering 1(3), 317–340 (2007)
206. Johnson, D.S., Trick, M.A.: The second dimacs implementation challenge: NP-hard problems: Maximum clique, graph coloring, and satisfiability. Series in Discrete Mathematics and Theoretical Computer Science. American Mathematical Society (1993)
207. Bodlaender, H.L.: Discovering treewidth. In: Vojtáš, P., Bieliková, M., Charron-Bost, B., Sýkora, O. (eds.) SOFSEM 2005. LNCS, vol. 3381, pp. 1–16. Springer, Heidelberg (2005)
208. Mouly, M., Paulet, M.B.: The GSM System for Mobile Communications. Mouly et Paulet, Palaiseau (1992)
209. Aardal, K.I., van Hoesel, S.P.M., Koster, A.M.C.A., Mannino, C., Sassano, A.: Models and solution techniques for frequency assignment problems. Annals of Operations Research 153(1), 79–129 (2007)
210. FAP Web: http://fap.zib.de/
211. Eisenblätter, A.: Frequency Assignment in GSM Networks: Models, Heuristics, and Lower Bounds. PhD thesis, Technische Universität Berlin (2001)
212. Hale, W.K.: Frequency assignment: Theory and applications. Proceedings of the IEEE 68(12), 1497–1514 (1980)
213. Blum, C., Roli, A.: Metaheuristics in Combinatorial Optimization: Overview and Conceptual Comparison. ACM Computing Surveys 35(3), 268–308 (2003)
214. Glover, F.W., Kochenberger, G.A.: Handbook of Metaheuristics. Kluwer, Dordrecht (2003)
215. Luna, F., Alba, E., Nebro, A.J., Pedraza, S.: Evolutionary algorithms for real-world instances of the automatic frequency planning problem in GSM networks. In: Cotta, C., van Hemert, J. (eds.) EvoCOP 2007. LNCS, vol. 4446, pp. 108–120. Springer, Heidelberg (2007)
216. Bäck, T.: Evolutionary Algorithms: Theory and Practice. Oxford University Press, New York (1996)
217. Kirkpatrick, S., Gelatt, C.D., Vecchi, M.P.: Optimization by simulated annealing. Science 220, 671–680 (1983)
218. Mishra, A.R.: Radio Network Planning and Optimisation. In: Fundamentals of Cellular Network Planning and Optimisation: 2G/2.5G/3G. Evolution to 4G, pp. 21–54. Wiley, Chichester (2004)
219. Dorne, R., Hao, J.K.: An evolutionary approach for frequency assignment in cellular radio networks. In: Proc. of the IEEE Int. Conf. on Evolutionary Computation, pp. 539–544 (1995)
220. Björklund, P., Värbrand, P., Yuan, D.: Optimal frequency planning in mobile networks with frequency hopping. Computers and Operations Research 32, 169–186
221. Vidyarthi, G., Ngom, A., Stojmenović, I.: A hybrid channel assignment approach using an efficient evolutionary strategy in wireless mobile networks. IEEE Transactions on Vehicular Technology 54(5), 1887–1895 (2005)
222. Demšar, J.: Statistical comparison of classifiers over multiple data sets. Journal of Machine Learning Research 7, 1–30 (2006)
223. Sheskin, D.J.: Handbook of Parametric and Nonparametric Statistical Procedures. CRC Press, Boca Raton (2003)
224. Hochberg, Y., Tamhane, A.C.: Multiple Comparison Procedures. Wiley, Chichester (1987)

225. Cheeseman, P., Kanefsky, B., Taylor, W.M.: Where the really hard problems are. In: Proceedings of the Twelfth International Joint Conference on Artificial Intelligence (IJCAI 1991), Sidney, Australia, pp. 331–337 (1991)
226. Vegdahl, S.R.: Using node merging to enhance graph coloring. In: PLDI 1999: Proceedings of the ACM SIGPLAN 1999 conference on Programming language design and implementation, NY, USA, pp. 150–154. ACM Press, New York (1999)
227. Culberson, J., Gent, I.: Frozen development in graph coloring. Theor. Comput. Sci. 265(1-2), 227–264 (2001)
228. Vassilakis, C.: An optimisation scheme for coalesce/valid time selection operator sequences. SIGMOD Record 29(1), 38–43 (2000)
229. Briggs, P., Cooper, K., Torczon, L.: Improvements to graph coloring register allocation. ACM Trans. Program. Lang. Syst. 16(3), 428–455 (1994)
230. George, L., Appel, A.: Iterated register coalescing. ACM Trans. Program. Lang. Syst. 18(3), 300–324 (1996)
231. Lueh, G.Y., Gross, T., Adl-Tabatabai, A.R.: Global register allocation based on graph fusion. In: Languages and Compilers for Parallel Computing, pp. 246–265 (1996)
232. Park, J., Moon, S.M.: Optimistic register coalescing. ACM Trans. Program. Lang. Syst. 26(4), 735–765 (2004)
233. Juhos, I., Tóth, A., van Hemert, J.: Binary merge model representation of the graph colouring problem. In: Gottlieb, J., Raidl, G.R. (eds.) EvoCOP 2004. LNCS, vol. 3004, pp. 124–134. Springer, Heidelberg (2004)
234. Juhos, I., Tóth, A., van Hemert, J.: Heuristic colour assignment strategies for merge models in graph colouring. In: Raidl, G.R., Gottlieb, J. (eds.) EvoCOP 2005. LNCS, vol. 3448, pp. 132–143. Springer, Heidelberg (2005)
235. Juhos, I., van Hemert, J.: Improving graph colouring algorithms and heuristics by a novel representation. In: Gottlieb, J., Raidl, G.R. (eds.) EvoCOP 2006. LNCS, vol. 3906, pp. 123–134. Springer, Heidelberg (2006)
236. Monasson, R., Zecchina, R., Kirkpatrick, S., Selman, B., Troyansky, L.: Determining computational complexity from characteristic phase transitions. Nature 400, 133–137 (1999)
237. Bollobás, B., Borgs, C., Chayes, J., Kim, J., Wilson, D.: The scaling window of the 2-SAT transition. Random Structures and Algorithms 18(3), 201–256 (2001)
238. Craenen, B., Eiben, A., van Hemert, J.: Comparing evolutionary algorithms on binary constraint satisfaction problems. IEEE Transactions on Evolutionary Computation 7(5), 424–444 (2003)
239. Juhos, I., van Hemert, J.: Increasing the efficiency of graph colouring algorithms with a representation based on vector operations. Journal of Software 1(2), 24–33 (2006)
240. Johnson, D.J., Trick, M.A. (eds.): Cliques, Coloring, and Satisfiability: Second DIMACS Implementation Challenge, Workshop, October 11–13, 1993. American Mathematical Society, Boston (1996)
241. Wigderson, A.: Improving the performance for approximate graph coloring. Journal of the ACM 30(4), 729–735 (1983)
242. Graham, R.L., Grötschel, M., Lovász, L. (eds.): Handbook of combinatorics, vol. 1. MIT Press, Cambridge (1995)
243. Halldórsson, M.M.: A still better performance guarantee for approximate graph coloring. Inf. Process. Lett. 45(1), 19–23 (1993)
244. Blum, A.: New approximation algorithms for graph coloring. J. ACM 41(3), 470–516 (1994)

245. Dutton, R.D., Brigham, R.C.: A new graph coloring algorithm. Computer Journal 24, 85–86 (1981)
246. Bäck, T., Fogel, D., Michalewicz, Z. (eds.): Handbook of Evolutionary Computation. Institute of Physics Publishing Ltd, Bristol and Oxford University Press, Oxford (1997)
247. Culberson, J.: Iterated greedy graph coloring and the difficulty landscape. Technical Report TR 92-07, University of Alberta, Dept. of Computing Science (1992)
248. Garey, M.R., Johnson, D.S.: Computers and Intractability: A Guide to the Theory of NP-Completeness. W. H. Freeman, San Francisco (1979)
249. Padberg, M., Rinaldi, G.: A Branch-and-Cut Algorithm for the Resolution of Large-Scale Symmetric Traveling Salesman Problems. SIAM Review 33(1), 60–100 (1991)
250. Applegate, D., Bixby, R., Chvátal, V., Cook, W.J.: On the Solution of Traveling Salesman Problems. Documenta Mathematica Extra Volume ICM III, 645–656 (1998)
251. Applegate, D., Bixby, R., Chvátal, V., Cook, W.J.: Implementing the Dantzig-Fulkerson-Johnson Algorithm for large Traveling Salesman Problems. Mathematical Programming 97, 91–153 (2003)
252. Lin, S., Kernighan, B.W.: An Effective Heuristic Algorithm for the Traveling-Salesman Problem. Operations Research 21(2), 498–516 (1973)
253. Johnson, D.S.: Local optimization and the traveling salesman problem. In: Paterson, M. (ed.) ICALP 1990. LNCS, vol. 443, pp. 446–461. Springer, Heidelberg (1990)
254. Moscato, P., Norman, M.G.: A Memetic Approach for the Traveling Salesman Problem Implementation of a Computational Ecology for Combinatorial Optimization on Message-Passing Systems. In: Valero, M., Onate, E., Jane, M., Larriba, J.L., Suarez, B. (eds.) Parallel Computing and Transputer Applications, pp. 177–186. IOS Press, Amsterdam (1992)
255. Merz, P., Freisleben, B.: Memetic Algorithms for the Traveling Salesman Problem. Complex Systems 13(4), 297–345 (2001)
256. Applegate, D., Bixby, R., Chvátal, V., Cook, W.J.: Concorde TSP Solver (2005), http://www.tsp.gatech.edu/concorde/
257. Cook, W.J., Seymour, P.: Tour Merging via Branch-Decomposition. INFORMS Journal on Computing 15(3), 233–248 (2003)
258. Applegate, D., Bixby, R., Chvátal, V., Cook, W.J.: Finding Cuts in the TSP (a Preliminary Report). Technical Report 95-05, Rutgers University, Piscataway, NJ, USA (1995)
259. Dorigo, M., Stützle, T.: Ant Colony Optimization. MIT Press, Cambridge (2004)
260. Mühlenbein, H., Gorges-Schleuter, M., Krämer, O.: Evolution Algorithms in Combinatorial Optimization. Parallel Computing 7(1), 65–85 (1988)
261. Nagata, Y.: New eax crossover for large tsp instances. In: Runarsson, T.P., Beyer, H.-G., Burke, E.K., Merelo-Guervós, J.J., Whitley, L.D., Yao, X. (eds.) PPSN 2006. LNCS, vol. 4193, pp. 372–381. Springer, Heidelberg (2006)
262. Cook, W.J.: Log of SW24978 Computation (2004), http://www.tsp.gatech.edu/world/swlog.html
263. Helsgaun, K.: An Effective Implementation of the Lin-Kernighan Traveling Salesman Heuristic. European Journal of Operational Research 126(1), 106–130 (2000)
264. Arenas, M.G., Collet, P., Eiben, A.E., et al.: A Framework for Distributed Evolutionary Algorithms. In: Guervós, J.J.M., Adamidis, P.A., Beyer, H.-G., Fernández-Villacañas, J.-L., Schwefel, H.-P. (eds.) PPSN 2002. LNCS, vol. 2439, pp. 665–675. Springer, Heidelberg (2002)

265. Fischer, T., Merz, P.: Embedding a Chained Lin-Kernighan Algorithm into a Distributed Algorithm. In: MIC 2005: 6th Metaheuristics International Conference, Vienna, Austria (2005)
266. Cook, W.J.: World Traveling Salesman Problem, http://www.tsp.gatech.edu/world/ (2005)
267. Walshaw, C.: Multilevel Refinement for Combinatorial Optimisation Problems. Annals of Operations Research 131, 325–372 (2004)
268. Walshaw, C.: A Multilevel Approach to the Travelling Salesman Problem. Operations Research 50(5), 862–877 (2002)
269. Applegate, D., Cook, W.J., Rohe, A.: Chained Lin-Kernighan for Large Traveling Salesman Problems. INFORMS Journal on Computing 15(1), 82–92 (2003)
270. Kruskal, J.B.: On the Shortest Spanning Subtree of a Graph and the Traveling Salesman Problem. Proceedings of the American Mathematical Society 7, 48–50 (1956)
271. Friedman, J.H., Baskett, F., Shustek, L.H.: An Algorithm for Finding Nearest Neighbors. IEEE Transactions on Computers (TOC) C-24, 1000–1006 (1975)
272. Sproull, R.F.: Refinements to Nearest-Neighbor Searching in k-Dimensional Trees. Algorithmica 6(4), 579–589 (1991)
273. Reinelt, G.: TSPLIB – a traveling salesman problem library. ORSA Journal on Computing 3(4), 376–384 (1991), http://www.iwr.uni-heidelberg.de/groups/comopt/software/TSPLIB95/
274. Cook, W.J.: National Traveling Salesman Problems (2005), http://www.tsp.gatech.edu/world/countries.html
275. Barbarosoglu, G., Ozgur, D.: Tabu search algorithm for the vehicle routing problem. Computers and Operations Research 26, 225–270 (1999)
276. Golden, B., Wong, R.: Capacitated arc routing problems. Networks 11, 305–315 (1981)
277. Dror, M.: ARC ROUTING: Theory, Solutions and Applications. Kluwer Academic Publishers, Dordrecht (2000)
278. Assad, A., Golden, B.: Arc routing methods and applications. In: Ball, M.O., et al. (eds.) Network Routing of Handbooks in Operations Research and Management Science, vol. 8, pp. 375–483. Elsevier, Amsterdam (1995)
279. Hirabayashi, R., Saruwatari, Y., Nishida, N.: Tour construction algorithm for the capacitated arc routing problems. Asia Pacific Journal of Operations Reasearch 9(2), 155–175 (1992)
280. Baldacci, R., Maniezzo, V.: Exact methods based on node-routing formulations for undirected arc-routing problems. Networks 47(1), 52–60 (2006)
281. Belenguer, J.M., Benavent, E.: The capacitated arc routing problem: Valid inequalities and facets. Computational Optimization & Applications 10(2), 165–187 (1998)
282. hlk, S.W.: New lower bound for the capacitated arc routing problem. Comput. Oper. Res. 33(12), 3458–3472 (2006)
283. Belenguer, J.M., Benavent, E., Cognata, F.: A metaheuristic for the capacitated arc routing problem. 11, 305–315 (Unpublished manuscript, 1997)
284. Hertz, A., Laporte, G., Mittaz, M.: A tabu search heuristic for the capacitated arc routing problem. Operations Research 48, 129–135 (2000)
285. Hertz, A., Laporte, G., Mittaz, M.: A variable neighborhood descent algorithm for the undirected capacitated arc routing problem. Transportation Science 35, 425–434 (2001)

286. Lacomme, P., Prins, C., Ramdane-Cherif, W.: A genetic algorithm for the capacitated routing problem and its extensions. In: Boers, E.J.W.et al. (eds.): EvoWorkshop 2001, pp. 473–483 (2001)

287. Hertz, A., Mittaz, M.: Heuristic algorithms. In: Dror, M. (ed.) Arc Routing: Theory, Solutions, and Applications, pp. 327–386. Kluwer Academic Publishers, Dordrecht (2000)

288. Amberg, A., Domschke, W., Voß, S.: Multiple center capacitated arc routing problems: A tabu search algorithm using capacitated trees. European Journal of Operational Research 124, 360–376 (2000)

289. Xin, X.: Hanblen swroute: A gis based spatial decision support system for designing solid waste collection routes in rural countries. Technical report, The University of Tennessee, Knoxville, U.S.A (2000)

290. Snizek, J., Bodin, L., Levy, L., Ball, M.: Capacitated arc routing problem with vehicle-site dependencies: the philadelphia experience. In: Toth, P., Vigo, D. (eds.) The Vehicle Routing Problem, pp. 287–308. SIAM, Philadelphia (2002)

291. Bodin, L., Levy, L.: Scheduling of local delivery carrier routes for the united states postal service. In: Dror, M. (ed.) Arc Routing: theory, solutions and applications, pp. 419–442. Kluwer Acad. Publ., Dordrecht (2000)

292. Codenotti, B., Manzini, G., Margara, L., Resta, G.: Perturbation: An efficient technique for the solution of very large instance of the euclidean tsp. INFORMS Journal on Computing 8, 125–133 (1996)

293. Pearn, W.L., Assad, A., Golden, B.: Transforming arc routing into node routing problems. Computers and Operations Research 14, 285–288 (1987)

294. Mladenović, N., Hansen, P.: Variable neighborhood search. Computers Oper. Res. 24, 1097–1100 (1997)

295. Christofides, N., Mingozzi, A., Toth, P.: The vehicle routing problem. In: Christofides, N., Mingozzi, A., Toth, P., Sandi, C. (eds.) Combinatorial Optimization, pp. 315–338. Wiley, Chichester (1979)

296. Goldberg, D.: Genetic Algorithms in Search, Optimization and Machine Learning. Addison-Wesley Longman Publishing Co., Inc., Amsterdam (1989)

297. Maniezzo, V.: Genetic evolution of the topology and weight distribution of neural networks. IEEE Transactions on Neural Networks 5(1), 39–53 (1994)

298. Martin, O., Otto, S., Felten, E.: Large-step markov chains for the tsp incorporating local search heuristics. Operations Research Letters 11(4), 219–224 (1992)

299. Canuto, S., Resende, M., Ribeiro, C.: Local search with perturbations for the prize-collecting steiner tree problem in graphs. Networks 38, 50–58 (2001)

300. Renaud, J., Boctor, F., Laporte, G.: Perturbation heuristics for the pickup and delivery traveling salesman problem. Computers and Operations Research 29, 1129–1141 (2002)

301. Ribeiro, C., Uchoa, E., Werneck, R.: A hybrid grasp with perturbations for the steiner problem in graphs. INFORMS Journal on Computing 14, 228–246 (2002)

302. DeArmon, J.: A comparison of heuristics for the capacitated chinese postman problem. MSc Thesis, University of Maryland (1981)

303. Belenguer, J., Benavent, E.: A cutting plane algorithm for the capacitated arc routing problem. Comput. Oper. Res. 30(5), 705–728 (2003)

304. Dror, M., Trudeau, P.: Savings by split delivery routing. Transportation Science 23, 141–145 (1989)

305. Dror, M., Trudeau, P.: Split delivery routing. Naval Research Logistics 37, 383–402 (1990)

306. Dror, M., Laporte, G., Trudeau, P.: Vehicle routing with split deliveries. Discrete Applied Mathematics 50, 239–254 (1994)

307. Archetti, C., Mansini, R., Speranza, M.: Complexity and reducibility of the skip delivery problem. Transportation Science 39(2), 182–187 (2005)
308. Belenguer, J., Martinez, M., Mota, E.: A lower bound for the split delivery vehicle routing problem. Operations Research 48(5), 801–810 (2000)
309. Mingzhou, J., Kai, L., Royce, O.: A two-stage algorithm with valid inequalities for the split delivery vehicle routing problem. International Journal of Production Economics 105, 228–242 (2007)
310. Mota, E., Campos, V., Corberan, A.: A new metaheuristic for the vehicle routing problem with split demands. In: Cotta, C., van Hemert, J. (eds.) EvoCOP 2007. LNCS, vol. 4446, pp. 121–129. Springer, Heidelberg (2007)
311. Archetti, C., Hertz, A., Speranza, M.: A tabu search algorithm for the split delivery vehicle routing problem. Transportation Science 40(1), 64–73 (2006)
312. Frizzell, P., Giffin, J.: The bounded split delivery vehicle routing problem with grid networks distances. Asia Pacific Journal of Operations Research 9, 101–116 (1992)
313. Frizzell, P., Giffin, J.: The split delivery vehicle scheduling problem with time windows and grid network distances. Computers and Operations Research 22(6), 655–667 (1995)
314. Mullaseril, P., Dror, M., Leung, J.: Split-delivery routing heuristics in livestock feed distribution. Journal of the Operational Research Society 48, 107–116 (1997)
315. Guéguen, C.: Exact solution methods for vehicle routing problems. PhD thesis, Central School of Paris, France (1999); (in French)
316. Labadi, N., Prins, C., Reghioui, M.: GRASP with path relinking for the capacitated arc routing problem with time windows. In: Giacobini, M. (ed.) EvoWorkshops 2007. LNCS, vol. 4448, pp. 722–731. Springer, Heidelberg (2007)
317. Wøhlk, S.: Contributions to arc routing. PhD thesis, Faculty of Social Sciences, University of Southern Denmark (2005)
318. Golden, B., Wong, R.: Capacitated arc routing problems. Networks 11, 305–315 (1981)
319. Clarke, G., Wright, J.: Scheduling of vehicles from a central depot to a number of delivery points. Operations Research 12, 568–581 (1964)
320. Golden, B., DeArmon, J., Baker, E.: Computational experiments with algorithms for a class of routing problems. Computers and Operations Research 10(1), 47–59 (1983)
321. Moscato, P.: Memetic algorithms: a short introduction. In: Corne, D., Dorigo, M., Glover, F. (eds.) New ideas in optimization, pp. 219–234. McGraw-Hill, New York (1999)
322. Sörensen, K., Sevaux, M.: MA|PM: memetic algorithms with population management. Computers and Operations Research 33, 1214–1225 (2006)
323. Lacomme, P., Prins, C., Ramdane-Chérif, W.: Competitive memetic algorithms for arc routing problems. Annals of Operations Research 131, 159–185 (2004)
324. Martí, R., Laguna, M., Campos, V.: Scatter search vs genetic algorithms: an experimental evaluation with permutation problems. In: Rego, C., Alidaee, B. (eds.) Metaheuristic optimization via memory and evolution, tabu search and scatter search, pp. 263–283. Springer, Heidelberg (2005)
325. Belenguer, J., Benavent, E.: A cutting plane algorithm for the capacitated arc routing problem. Computers and Operations Research 30, 705–728 (2003)
326. Longo, H., de Aragaõ, M.P., Uchoa, E.: Solving capacitated arc routing problems using a transformation to the CVRP. Computers and Operations Research 33, 1823–1837 (2006)

327. Grishukhin, V.P.: On polynomial solvability conditions for the simplest plant location problem. In: Kelmans, A.K., Ivanov, S. (eds.) Selected Topics in Discrete Mathematics. American Mathematical Society, Providence (1994)
328. Cornuejols, G., Nemhauser, G.L., Wolsey, L.A.: The uncapacitated facility location problem. In: Francis, R.L., Mirchandani, P.B. (eds.) Discrete Location Theory, pp. 119–171. Wiley-Interscience, New York (1990)
329. Krarup, J., Pruzan, P.M.: The simple plant location problem: Survey and synthesis. European Journal of Operational Research 12, 36–57 (1983)
330. Dearing, P.M.: Location problems. Operations Research Letters 4, 95–98 (1985)
331. Gao, L.L., Robinson, E.P.: Uncapacitated facility location: General solution procedure and computational experience. European Journal of Operational Research 76, 410–427 (1994)
332. Kratica, J., Filipović, V., Tošić, D.: Solving the uncapacitated warehouse location problem by SGA with add-heuristic. In: Proceedings of the XV ECPD International Conference on Material Handling and Warehousing. 2.28–2.32 (1998)
333. Horng, J.T., Lin, L.Y., Liu, B.J., Kao, C.Y.: Resolution of simple plant location problems using an adapted genetic algorithm. In: Proceedings of the 1999 Congress on Evolutionary Computation, Piscataway, NJ, vol. 2, pp. 1186–1193. IEEE Press, Los Alamitos (1999)
334. Kratica, J., Tosic, D., Filipovic, V., Ljubic, I.: Solving the simple plant location problem by genetic algorithm. RAIRO Operations Research 35, 127–142 (2001)
335. Jaramillo, J.H., Bhadury, J., Batta, R.: On the use of genetic algorithms to solve location problems. Computers and Operations Research 29, 761–779 (2002)
336. Rao, R., Hanagandi, V., Buescher, K.: Solution of the optimal plant location and sizing problem using simulated annealing and genetic algorithms. In: Proceedings: International Federation of Operations Research Societies (October 1995)
337. Yigit, V., Aydin, M.E., Turkbey, O.: Solving large-scale uncapacitated facility location problems with evolutionary simulated annealing. International Journal of Production Research 44(22), 4773–4791 (2006)
338. Eiben, A.E., Smith, J.E.: Introduction to Evolutionary Computing. Springer, Berlin (2003)
339. Michalewicz, Z.: Genetic Algorithms + Data Structures = Evolution Programs, 3rd edn. Springer, Berlin (1996)
340. Whitley, D.: Permutations. In: Bäck, T., Fogel, D.B., Michalewicz, T. (eds.) Evolutionary Computation 1, pp. 274–284. Institute of Physics Publishing, Bristol (2000)
341. Gould, S.J.: The Panda's Thumb. W. W. Norton and Co., New York (1980)
342. Moscato, P.: On evolution, search, optimization, genetic algorithms and martial arts: Towards memetic algorithms. Technical Report Caltech Concurrent Computation Program Report 826, California Institute of Technology (1989)
343. Nummela, J., Julstrom, B.A.: An effective genetic algorithm for the minimum-label spanning tree problem. In: M.K. et al. (eds.): Proceedings of the Genetic and Evolutionary Computation Conference: GECCO 2006, New York, vol. 1, pp. 553–557. The ACM Press, New York (2006)
344. Beasley, J.E.: Obtaining test problems via Internet. Journal of Global Optimization 8, 429–433 (1996)
345. Beasley, J.E.: Langrangean heuristics for location problems. European Journal of Operational Research 65, 383–399 (1993)
346. Bilde, O., Krarup, J.: Sharp lower bounds and efficient algorithms for the simple plant location problem. Annals of Discrete Mathematics 1, 79–97 (1977)

347. Ahuja, H.N.: Construction performance control by networks. John Wiley, New York (1976)
348. Ballestín, F.: When it is worthwhile to work with the stochastic RCPSP? Journal of Scheduling 10(3), 153–166 (2007)
349. Ballestín, F.: A Genetic Algorithm for the Resource Renting Problem with Minimum and Maximum Time Lags. In: Cotta, C., van Hemert, J. (eds.) EvoCOP 2007. LNCS, vol. 4446, pp. 25–35. Springer, Heidelberg (2007)
350. Ballestín, F., Trautmann, N.: An iterated-local-search heuristic for the resource-constrained weighted earliness-tardiness project.International Journal of Production Research (to appear, 2007)
351. Ballestín, F., Valls, V., Quintanilla, S.: Due Dates and RCPSP. In: Józefowska, J., Weglarz, J. (eds.) Perspectives in Modern Project Scheduling. ch. 4, pp. 79–104. Springer, Berlin (2006)
352. Ballestín, F., Schwindt, C., Zimmermann, J.: Resource leveling in make-to-order production: modeling and heuristic solution method. International Journal of Operations Research 4, 50–62 (2007)
353. Ballestín, F., Valls, V., Quintanilla, M.: Preemption in resource-constrained project scheduling. European Journal of Operational Research (to appear, 2007)
354. Brucker, P., Drexl, A., Möhring, R., Neumann, K., Pesch, E.: Resource-Constrained Project Scheduling: Notation, Classification, Models, and Methods. European Journal of Operational Research 112, 3–41 (1999)
355. Drexl, A.: Scheduling of project networks by job assignment. Management Science 37, 1590–1602 (1991)
356. Goldberg, D.E.: Genetic algorithms in search, optimization, and machine learning. Addison-Wesley, Reading (1989)
357. Hartmann, S.: A competitive genetic algorithm for resource-constrained project scheduling. Naval Research Logistics 45, 733–750 (1998)
358. Hartmann, S.: Project scheduling with multiple modes: a genetic algorithm. Annals of Operations Research 102, 111–135 (2001)
359. Herroelen, W., De Reyck, B., Demeulemeester, E.: Resource-Constrained Project Scheduling: A Survey of Recent Developments. Computers and Operations Research 25(4), 279–302 (1998)
360. Kolisch, R., Hartmann, S.: Heuristic algorithms for the resource-constrained project scheduling problem: Classification and computational analysis. In: Weglarz, J. (ed.) Project Scheduling – Recent Models, Algorithms and Applications, pp. 147–178. Kluwer Academic Publishers, Boston (1999)
361. Kolisch, R., Padman, R.: An Integrated Survey of Deterministic Project Scheduling. Omega 29, 249–272 (2001)
362. Kolisch, R., Schwindt, C., Sprecher, A.: Benchmark instances for project scheduling problems. In: Weglarz, J. (ed.) Project scheduling – recent models, algorithms and applications, pp. 197–212. Kluwer, Boston (1999)
363. Lourenço, H.R., Martin, O., Stützle, T.: Iterated Local Search. In: Glover, F., Kochenberger, G. (eds.) Handbook of Metaheuristics, pp. 321–353. Kluwer Academic Publishers, Norwell (2002)
364. Marchiori, E., Steenbeek, A.: An evolutionary algorithm for large set covering problems with applications to airline crew scheduling. In: Oates, M.J., Lanzi, P.L., Li, Y., Cagnoni, S., Corne, D.W., Fogarty, T.C., Poli, R., Smith, G.D. (eds.) EvoIASP 2000, EvoWorkshops 2000, EvoFlight 2000, EvoSCONDI 2000, EvoS-TIM 2000, EvoTEL 2000, and EvoROB/EvoRobot 2000. LNCS, vol. 1803, pp. 367–381. Springer, Heidelberg (2000)

365. Moscato, P.: On Evolution, Search, Optimization, Genetic Algorithms and Martial Arts: Toward Memetic Algorithms. Caltech Concurrent Computation Program, C3P Report 826 (1989)

366. Moscato, P.: Memetic Algorithms. In: Pardalos, P.M., Resende, M.G.C. (eds.) Handbook of Applied Optimization, pp. 157–167. Oxford University Press, Oxford (2002)

367. Neumann, K., Zimmermann, J.: Resource levelling for projects with schedule-dependent time windows. European Journal of Operational Research 117, 591–605 (1999)

368. Neumann, K., Schwindt, C., Zimmermann, J.: Project Scheduling with Time Windows and Scarce Resources. Springer, Berlin (2003)

369. Nübel, H.: A branch-and-bound procedure for the resource investment problem subject to temporal constraints. Technical Report WIOR-516. University of Karlsruhe (1998)

370. Nübel, H.: Minimierung der Ressourcenkosten für Projekte mit planungsabhängigen Zeitfenstern. Gabler, Wiesbaden (1999)

371. Nübel, H.: The resource renting problem subject to temporal constraints. OR Spektrum 23, 359–381 (2001)

372. Ruiz, R., Stützle, T.: A simple and effective iterated greedy algorithm for the permutation flowshop scheduling problem. European Journal of Operational Research (2006)

373. Valls, V., Quintanilla, S., Ballestín, F.: Resource-constrained project scheduling: a critical activity reordering heuristic. European Journal of Operational Research 149(2), 282–301 (2003)

374. Vanhoucke, M., Demeulemeester, E., Herroelen, W.: An exact procedure for the resource-constrained weighted earliness-tardiness project scheduling problem. Annals of Operations Research 102, 179–196 (2001)

375. Vanhoucke, M., Demeulemeester, E., Herroelen, W.: On maximizing the net present value of a project under renewable resource constraints. Management Science 47(8), 1113–1121 (2001)

376. Wilcoxon, F.: Individual Comparisons by Ranking Methods. Biometrics 1, 80–83 (1945)

377. Adams, J., Balas, E., Zawack, D.: The Shifting Bottleneck Procedure for Job Shop Scheduling. Management Science 34, 391–401 (1988)

378. Aiex, R.M., Binato, S., Resende, M.G.C.: Parallel GRASP with Path-Relinking for Job Shop Scheduling. Parallel Computing 29, 393–430 (2002)

379. Applegate, D., Cook, W.: A Computational Study of the Job-Shop Scheduling Problem. ORSA Journal on Computing 3, 149–156 (1991)

380. Balas, E., Vazacopoulos, A.: Guided Local Search with Shifting Bottleneck for Job Shop Scheduling. Management Science 44, 262–275 (1998)

381. Binato, S., Hery, W.J., Loewenstern, D.M., Resende, M.G.C.: A GRASP for Job Shop Scheduling. In: Ribeiro, C.C., Hansen, P. (eds.) Essays and surveys on metaheuristics, pp. 59–79. Kluwer Academic Publishers, Dordrecht (2001)

382. Carlier, J.: The one-machine sequencing problem. European Journal of Operational Research 11, 42–47 (1982)

383. Caseau, Y., Laburthe, F.: Disjunctive scheduling with task intervals. Technical Report LIENS 95-25, Ecole Normale Superieure Paris (1995)

384. Chen, S., Talukdar, S., Sadeh, N.: Job-shop-scheduling by a team of asynchronous agentes. In: Proceedings of the IJCAI 1993 Workshop on Knowledge-Based Production, Scheduling and Control, Chambery France (1993)

385. Danna, E., Rothberg, E., Pape, C.L.: Exploring relaxation induced neighborhoods to improve MIP solutions. Mathematical Programming Ser. A 102, 71–90 (2005)
386. Dell'Amico, M., Trubian, M.: Applying Tabu-Search to the Job-Shop Scheduling Problem. Annals of Operations Research 41, 231–252 (1993)
387. Denzinger, J., Offermann, T.: On Cooperation between Evolutionary Algorithms and other Search Paradigms. In: Proceedings of the CEC 1999 – Congress on Evolutionary Computational 1999, Washington, pp. 2317–2324. IEEE Press, Los Alamitos (1999)
388. Feo, T., Resende, M.G.C.: Greedy Randomized Adaptive Search Procedures. Journal of Global Optimization 6, 109–133 (1995)
389. Fernandes, S., Lourenço. H.R. Optimized Search Heuristics. Universitat Pompeu Fabra, Barcelona, Spain (2007), http://www.econ.upf.edu/~ramalhin/OSHwebpage/
390. Fisher, H., Thompson, G.L.: Probabilistic learning combinations of local job-shop scheduling rules. In: Muth, J.F., Thompson, G.L. (eds.) Industrial Scheduling, pp. 225–251. Prentice Hall, Englewood Cliffs (1963)
391. Garey, M.R., Johnson, D.S.: Computers and Intractability: A Guide to the Theory of NP-Completeness. Freeman, San Francisco (1979)
392. Jain, A.S., Meeran, S.: Deterministic job shop scheduling: Past, present and future. European Journal of Operational Research 133, 390–434 (1999)
393. Lawrence, S.: Resource Constrained Project Scheduling: an Experimental Investigation of Heuristic Scheduling techniques. Graduate School of Industrial Administration, Carnegie-Mellon University (1984)
394. Lourenço, H.R.: Job-shop scheduling: Computational study of local search and large-step optimization methods. European Journal of Operational Research 83, 347–367 (1995)
395. Lourenço, H.R., Zwijnenburg, M.: Combining large-step optimization with tabu-search: Application to the job-shop scheduling problem. In: Osman, I.H., Kelly, J.P. (eds.) Meta-heuristics: Theory & Applications, pp. 219–236. Kluwer Academic Publishers, Dordrecht (1996)
396. Nowicki, E., Smutnicki, C.: Some new tools to solve the job shop problem. Technical Report, 60/2002, Institute of Engineering Cybernetics, Wroclaw University of Technology (2002)
397. Nowicki, E., Smutnicki, C.: An Advanced Tabu Search Algorithm for the Job Shop Problem. Journal of Scheduling 8, 145–159 (2005)
398. Nowicki, E., Smutnicki, C.: A Fast Taboo Search Algorithm for the Job Shop Problem. Management Science 42, 797–813 (1996)
399. Roy, B., Sussman, B.: Les probems d'ordonnancement avec constraintes disjonctives. Note DS 9 bis, SEMA, Paris (1964)
400. Schaal, A., Fadil, A., Silti, H.M., Tolla, P.: Meta heuristics diversification of generalized job shop scheduling based upon mathematical programming techniques. In: Proceedings of the CP-AI-OR 1999, Ferrara, Italy (1999)
401. Schrage, L.: Solving resource-constrained network problems by implicit enumeration: Non pre-emptive case. Operations Research 18, 263–278 (1970)
402. Storer, R.H., Wu, S.D., Vaccari, R.: New search spaces for sequencing problems with application to job shop scheduling. Management Science 38, 1495–1509 (1992)
403. Taillard, E.D.: Benchmarks for Basic Scheduling Problems. European Journal of Operational Research 64, 278–285 (1993)
404. Taillard, E.D.: Parallel Taboo Search Techniques for the Job Shop Scheduling Problem. ORSA Journal on Computing 6, 108–117 (1994)

405. Tamura, H., Hirahara, A., Hatono, I., Umano, M.: An approximate solution method for combinatorial optimisation. Transactions of the Society of Instrument and Control Engineers 130, 329–336 (1994)
406. Yamada, T., Nakano, R.: A genetic algorithm applicable to large-scale job-shop problems. In: Manner, R., Manderick, B. (eds.) Parallel Problem Solving from Nature, vol. 2, pp. 281–290. Elsevier Science, Amsterdam (1992)
407. Abraham, A., Buyya, R., Nath, B.: Nature's heuristics for scheduling jobs on computational grids. In: Proceedings of the 8th IEEE International Conference on Advanced Computing and Communications, pp. 45–52. Tata McGraw-Hill, New York (2000)
408. Alba, E., Almeida, F., Blesa, M., Cotta, C., Diaz, M., Dorta, I., Gabarro, J., Leon, C., Luque, G., Petit, J., Rodriguez, C., Rojas, A., Xhafa, F.: Efficient parallel LAN/WAN algorithms for optimization. the mallba project. Parallel Computing 32(5-6), 415–440 (2006)
409. Blesa, M.J., Moscato, P., Xhafa, F.: A memetic algorithm for the minimum weighted k-cardinality tree subgraph problem. In: de Sousa, J.P. (ed.) Metaheuristic International Conference, vol. 1, pp. 85–91 (2001)
410. Braun, T.D., Siegel, H.J., Beck, N., Boloni, L.L., Maheswaran, M., Reuther, A.I., Robertson, J.P., Theys, M.D., Yao, B.: A comparison of eleven static heuristics for mapping a class of independent tasks onto heterogeneous distributed computing systems. J. of Parallel and Distributed Comp. 61(6), 810–837 (2001)
411. Buyya, R., Abramson, D., Giddy, J.: Nimrod/G: An architecture for a resource management and scheduling system in a global computational grid. In: The 4th Int. Conf. on High Performance Comp., pp. 283–289. IEEE Press, Los Alamitos (2000)
412. Cant-Paz, E.: A Survey of Parallel Genetic Algorithms. Calculateurs Paralleles, Reseaux et Systems Repartis 10(2), 141–171 (1998)
413. Carretero, J., Xhafa, F.: Using genetic algorithms for scheduling jobs in large scale grid applications. Journal of Technological and Economic Development 12(1), 11–17 (2006)
414. Crainic, T., Toulouse, M.: Parallel Metaheuristics. In: Crainic, T.G., Laporte, G. (eds.) Fleet Management and Logistics, pp. 205–251. Kluwer, Dordrecht (1998)
415. Foster, I., Kesselman, C.: The Grid – Blueprint for a New Computing Infrastructure. Morgan Kaufmann Publishers, San Francisco (1998)
416. Foster, I., Kesselman, C., Tuecke, S.: The anatomy of the grid: Enabling scalable virtual organization. International Journal of Supercomputer Applications 15(3), 200–222 (2001)
417. Gruau, F.: Neural networks synthesis using cellular encoding and the genetic algorithm, Ph.D. Thesis, Universit Claude Bernard-Lyon I, France (1994)
418. Holland, J.: Adaptation in Natural and Artificial Systems. University of Michigan Press, Ann Arbor (1975)
419. Hoos, H.H., Stutzle, T.: Stochastic Local Search: Foundations and Applications. Elsevier/Morgan Kaufmann, San Francisco (2005)
420. Di Martino, V., Mililotti, M.: Sub optimal scheduling in a grid using genetic algorithms. Parallel Computing 30, 553–565 (2004)
421. Michalewicz, Z., Fogel, D.B.: How to solve it: modern heuristics. Springer, Heidelberg (2000)
422. Moscato, P.: On evolution, search, optimization, genetic algorithms and martial arts: Towards memetic algorithms. Techrep N. 826, CalTech, USA (1989)

423. Zomaya, A.Y., Teh, Y.H.: Observations on using genetic algorithms for dynamic load-balancing. IEEE Trans. on Parallel and Distributed Sys. 12(9), 899–911 (2001)
424. Xhafa, F.: A Hybrid Evolutionary Heuristic for Job Scheduling on Computational Grids. In: Grosan, Abraham, and Ishibuchi (eds.), Hybrid Evolutionary Algorithms, Studies in Computational Intelligence, ch. 13. Springer (2007)
425. Xhafa, F., Carretero, J., Barolli, L., Durresi, A.: Requirements for a an event-based simulation package for grid systems. Journal of Interconnection Networs 8(2), 63–178 (2007)

Index

Author Index

Printing: Krips bv, Meppel, The Netherlands
Binding: Stürtz, Würzburg, Germany